CAMBRIDGE LIBRARY (

Books of enduring scholarly

T0250432

Philosophy

This series contains both philosophical texts and critical essays about philosophy, concentrating especially on works originally published in the eighteenth and nineteenth centuries. It covers a broad range of topics including ethics, logic, metaphysics, aesthetics, utilitarianism, positivism, scientific method and political thought. It also includes biographies and accounts of the history of philosophy, as well as collections of papers by leading figures. In addition to this series, primary texts by ancient philosophers, and works with particular relevance to philosophy of science, politics or theology, may be found elsewhere in the Cambridge Library Collection.

The Philosophy of the Inductive Sciences

First published in 1840, this two-volume treatise by Cambridge polymath William Whewell (1794–1886) remains significant in the philosophy of science. The work was intended as the 'moral' to his three-volume *History of the Inductive Sciences* (1837), which is also reissued in this series. Building on philosophical foundations laid by Immanuel Kant and Francis Bacon, Whewell opens with the aphorism 'Man is the Interpreter of Nature, Science the right interpretation'. Volume 1 contains the majority of Whewell's section on 'ideas', in which he investigates the philosophy underlying a range of different disciplines, including pure, classificatory and mechanical sciences. Whewell's work upholds throughout his belief that the mind was active and not merely a passive receiver of knowledge from the world. A key text in Victorian epistemological debates, notably challenged by John Stuart Mill and his *System of Logic*, Whewell's treatise merits continued study and discussion in the present day.

Cambridge University Press has long been a pioneer in the reissuing of out-of-print titles from its own backlist, producing digital reprints of books that are still sought after by scholars and students but could not be reprinted economically using traditional technology. The Cambridge Library Collection extends this activity to a wider range of books which are still of importance to researchers and professionals, either for the source material they contain, or as landmarks in the history of their academic discipline.

Drawing from the world-renowned collections in the Cambridge University Library and other partner libraries, and guided by the advice of experts in each subject area, Cambridge University Press is using state-of-the-art scanning machines in its own Printing House to capture the content of each book selected for inclusion. The files are processed to give a consistently clear, crisp image, and the books finished to the high quality standard for which the Press is recognised around the world. The latest print-on-demand technology ensures that the books will remain available indefinitely, and that orders for single or multiple copies can quickly be supplied.

The Cambridge Library Collection brings back to life books of enduring scholarly value (including out-of-copyright works originally issued by other publishers) across a wide range of disciplines in the humanities and social sciences and in science and technology.

The Philosophy of the Inductive Sciences

Founded Upon their History

VOLUME 1

WILLIAM WHEWELL

CAMBRIDGE
UNIVERSITY PRESS

CAMBRIDGE
UNIVERSITY PRESS

University Printing House, Cambridge, CB2 8BS, United Kingdom

Published in the United States of America by Cambridge University Press, New York

Cambridge University Press is part of the University of Cambridge.
It furthers the University's mission by disseminating knowledge in the pursuit of
education, learning and research at the highest international levels of excellence.

www.cambridge.org
Information on this title: www.cambridge.org/9781108064026

© in this compilation Cambridge University Press 2014

This edition first published 1840
This digitally printed version 2014

ISBN 978-1-108-06402-6 Paperback

This book reproduces the text of the original edition. The content and language reflect
the beliefs, practices and terminology of their time, and have not been updated.

Cambridge University Press wishes to make clear that the book, unless originally published
by Cambridge, is not being republished by, in association or collaboration with, or
with the endorsement or approval of, the original publisher or its successors in title.

THE

PHILOSOPHY

OF THE

INDUCTIVE SCIENCES,

FOUNDED UPON THEIR HISTORY.

BY THE

REV. WILLIAM WHEWELL, B.D.,

FELLOW OF TRINITY COLLEGE, AND PROFESSOR OF MORAL PHILOSOPHY IN THE UNIVERSITY
OF CAMBRIDGE, VICE-PRESIDENT OF THE GEOLOGICAL SOCIETY
OF LONDON.

IN TWO VOLUMES.

Λαμπάδια ἔχοντες διαδώσουσιν ἀλλήλοις.

VOLUME THE FIRST.

LONDON:

JOHN W. PARKER, WEST STRAND.

CAMBRIDGE: J. AND J. J. DEIGHTON.

M.DCCC.XL.

TO THE

Rev. ADAM SEDGWICK, M.A.,

SENIOR FELLOW OF TRINITY COLLEGE,

WOODWARDIAN PROFESSOR OF GEOLOGY IN THE UNIVERSITY OF
CAMBRIDGE, AND PREBENDARY OF NORWICH.

My dear Sedgwick,

When I showed you the last sheet of my *History of the Inductive Sciences* in its transit through the press, you told me that I ought to add a paragraph or two at the end, by way of Moral to the story; and I replied that the moral would be as long as the story itself. The present work, the *Moral* which you then desired, I have, with some effort, reduced within a somewhat smaller compass than I then spoke of; and I cannot dedicate it to any one with so much pleasure as to you.

It has always been my wish that, as far and as long as men might know anything of me by my writings, they should hear of me along with the friends with whom I have lived, whom I have loved, and by whose conversation I have been animated to hope that I too might add something to the literature of our country. There is no one whose name has, on such grounds, a better claim than yours to stand in the front of a work, which has been the subject of my labours for no small portion of our long period of friendship. But there is another reason which gives a peculiar propriety to this dedication of my Philosophy to you. I have little doubt that if your life had not been absorbed in struggling with many of the most difficult problems of a difficult science, you would have been my fellow-labourer or master in the work which I have here undertaken. The same spirit which dictated your vigorous protest against some of the errors which I also attempt to expose, would have led you, if your thoughts had been more free, to take a leading share in that Reform of Philosophy,

a 2

which all who are alive to such errors, must see to be now indispensable. To you I may most justly inscribe a work which contains a criticism of the fallacies of the ultra-Lockian school.

I will mention one other reason which enters into the satisfaction with which I place your name at the head of my Philosophy. By doing so, I may consider myself as dedicating it to the College to which we both belong, to which we both owe so much of all that we are, and in which we have lived together so long and so happily; and *that*, be it remembered, the College of Bacon and of Newton. That College, I know, holds a strong place in your affections, as in mine; and among many reasons, not least on this account;—we believe that sound and enduring philosophy ever finds there a congenial soil and a fostering shelter. If the doctrines which the present work contains be really true and valuable, my unhesitating trust is, that they will spread gradually from these precincts to every part of the land.

That this office of being the fosterer and diffuser of truth may ever belong to our common Nursing Mother, and that you, my dear Sedgwick, may long witness and contribute to these beneficial influences, is the hearty wish of

Yours affectionately,

W. WHEWELL.

Trinity College, May 1, 1840.

CONTENTS

OF

THE FIRST VOLUME.

THE PHILOSOPHY OF THE INDUCTIVE SCIENCES.

PART I.—OF IDEAS.

Book I.—Of Ideas in General.

Book II.—The Philosophy of the Pure Sciences.

Book III.—The Philosophy of the Mechanical Sciences.

* The number repeated by mistake.

PREFACE.

THE Work now before the Reader is intended as an application of the Plan of Bacon's *Novum Organum* to the present condition of Physical Science. The progress of such Science during the last three centuries has given us the means of inquiring, with advantages which former generations did not possess, what that *Organ*, or intellectual method, is, by which solid truth is to be extracted from the observation of Nature; and though the attempt to discover this cannot but be an arduous undertaking, it is so plainly required of the present generation, that any one engaging in it with sobriety and industry, may claim to have his labours soberly and tolerantly estimated. I shall, therefore, make no apology for what might otherwise appear the presumption of such a design. My scheme is, however, narrower than Bacon's in this respect, that I have in the present work confined myself to those branches of human knowledge which have external nature for their object, and are often exclusively termed *Sciences*. The reason given for this limitation in the following pages, (namely, that it seemed proper to collect our philosophy of knowledge from the most certain and distinct portions of knowledge alone,) will, I trust, be considered as an adequate justification of the course pursued.

Many writers, both before and since Bacon, have employed themselves upon the subjects with which we are here concerned;—the philosophy of knowledge, and the methods of arriving at science: and I have availed myself of their conclusions, where they appeared to be real additions to sound philosophy. I must add, that I have not scrupled to subject their speculations to a critical examination, and to reject all that was thus found to be erroneous or worthless. A system which professes to give a view of the nature of knowledge, supplies canons of criticism by which all other philosophical doctrines must be tried, in order to determine their import and reality. It is an office essentially connected with the exposition of such a system, to pronounce on the value of previous essays of the same kind. Hence I have not hesitated, on some occasions, to dissent from the great masters of the philosophy of science, from Bacon, from Cuvier, and even from Newton himself; believing that they, upon maturer consideration, would have been led to those doctrines and precepts which I have preferred to theirs. In like manner, although I have adopted Kant's reasoning respecting the nature of Space and Time, it will be found by any one acquainted with the system of that acute metaphysician, that my views differ widely from his. I have also ventured to condemn some of the opinions respecting physical philosophy, published by another eminent German writer (Schelling) to whose works I have in other subjects great obligations.

The present work was announced in the outset of a

History of the Inductive Sciences which was published three years ago. That History was, indeed, the result of labours undertaken with a view to the formation of a *Philosophy of Science*, and was intended from the first as an introduction to a work on that subject. I may therefore take the liberty of saying that I have as yet seen no reason to wish to make any material change in the History, as it is now before the world. I will not omit this opportunity of expressing my obligations to the German translator of the History*. It is a testimony which may well give an author some confidence in the value of his work, that one of the most eminent men of science in Europe should spontaneously postpone other tasks in order to give it, with the most flattering expressions, to the public of his own country. I may add that a Review of my History which has appeared in our own language has tended in no small degree to convince me that the work contains few material errors, and none which are of any importance with regard to its general scope. The Reviewer, obviously an enemy eager to find faults, was able to detect but very few passages which are really mistakes†.

Other critics have made objections of various kinds to

---

* *The History of the Inductive Sciences* has been translated into German by M. Von Littrow, Director of the Imperial Observatory at Vienna, and author of many well-known mathematical works.

† See *Edinburgh Review*, No. cxxxiii. p. 129 ; also No. cxxxvi., p. 274. But I am compelled by justice to acknowledge that the value of this testimony is materially weakened by the Reviewer's extreme laxity and obscurity of view with regard to the nature of science ;—defects which make his judgment on such subjects nearly worthless.

that part of the History which relates to Physiology; but none which it is necessary to notice here. I regret that my plan should again lead me, in the present work, to trespass upon the domain of the physiologists. Those who have well studied that subject, feel a persuasion, a very natural and just one, that nothing less than a life professionally devoted to the science, can entitle a person to decide the still controverted questions which it involves; and hence they look, with a reasonable jealousy, upon attempts to discuss such questions, made by a *lay* speculator. I trust it will be found that I have not, in the present work, asserted any opinions on such subjects, without alleging sufficient reasons. Such discussions as I have introduced, appeared to me to be requisite to complete the philosophy of science : the value of the opinions thus delivered, it must be left to physiologists, present and future, to decide.

In writing the History, I was led, on several occasions, to pass on from the facts to the lessons of philosophy which they suggest. I have now, in three or four instances, taken the liberty of restoring reflections so made to their proper place, by incorporating a few phrases, or here and there a sentence, from the History, in the present work. I think it right to mention this, that those who read both works may not deem me guilty of careless repetition.

Perhaps I shall be charged with having employed the term *Idea* in an unusual manner in these pages. Almost every writer who has introduced that term into his speculations, has been accused, by succeeding critics, of some degree of vagueness and vacillation in its use. The mode in which I have applied it appears to me very definite. The grounds of the universal and necessary truths which we are able to assert in various departments of knowledge, reside in certain general forms of apprehension, or relations of our conceptions; as *Space, Time, Cause;* and these I term *Ideas;* or, when ambiguity is to be guarded against, *Fundamental Ideas.* If I could have found any other word or phrase which, in common usage, came nearer to this meaning, I should have been glad to adopt it. I have employed the word *Conception* to express that which is, I think, its common signification;— our Conceptions are *that,* in the mind, which we denote by our General Terms, as a *Triangle,* a *Square Number,* a *Force.* But still, this term, in the present work, implies principles which have not been employed, at least not commonly, by previous writers. For in the course of my speculations, I am further led to speak of such Conceptions as Modifications of our Fundamental Ideas; and as deriving from the Ideas their power of leading to universal and necessary truths. In instances in which no obscurity appeared likely to arise, I may, perhaps, occasionally have employed the terms *Idea, Conception, Notion,* and others, with less discrimination.

Bacon's purpose and promise was, that his New Organ should produce material as well as intellectual profit;— works, as well as knowledge. That the study of the order of nature does add to man's power, the history of the sciences since Bacon has abundantly shown. But though this hope of derivative advantages may stimulate our exertions, it cannot govern our methods of seeking knowledge, without leading us away from the most general and genuine forms of knowledge. The nature of knowledge must be studied in itself and for its own sake, before we attempt to learn what external rewards it will bring us. I have, therefore, not aimed at imitating Bacon in those parts of his work, in which he contemplates the increase of man's dominion over nature, as the main object of natural philosophy; being fully persuaded that if Bacon himself had had unfolded before him the great theories which have been established since his time, he would have acquiesced in the contemplation of them; and would have readily proclaimed the real reason for aiming at the knowledge of such truths to be, that they are *true*. Thus I have ventured to separate his first Aphorism* into two; to consider the *Interpretation* as our primary object, not the *ministration;* the *knowing*, not the *doing;* the *Intelligence*, not the *Power*.

The mode of delivering the philosophy of science in Aphorisms which Bacon has adopted, would not well suffice for the treatment of the subject at present, since many questions must be discussed, many perplexities explained. No writer upon such subjects can expect to be either understood or assented to, beyond the limits of a

* Homo naturæ minister et interpres tantum facit et intelligit quantum de naturæ ordine re vel mente observaverit; nec amplius scit aut potest.

narrow school, who is not prepared with good arguments, as well as magisterial decisions, upon the many obscure and controverted points which the subject brings before him. But though an Aphoristic Philosophy, unsupported by reasoning, is thus unsuited to the time, it may be satisfactory to many readers to see the Philosophy to which in the present work we are led, presented in the Aphoristic form. I have, therefore, expressed in Aphorisms a large part of the doctrines resulting from the discussions which occupy the following pages. These Aphorisms are given at the end of this Preface.

Along with these, I shall add some other Aphorisms on the subject of the Language of Science; a subject in which it appears to be time to collect, from the usage of the most judicious writers, some rules which may tend to preserve the purity and analogies of scientific language from wanton and needless violation. As this subject is not discussed in the work itself, I have given, along with these Aphorisms, such examples as may tend to confirm and illustrate them, and have applied them to some cases at present unsettled.

APHORISMS CONCERNING IDEAS.

I.

MAN is the Interpreter of Nature, Science the right interpretation.

II.

The *Senses* place before us the *Characters* of the Book of Nature; but these convey no knowledge to us, till we have discovered the Alphabet by which they are to be read.

III.

The *Alphabet*, by means of which we interpret Phenomena, consists of the *Ideas* existing in our own minds; for these give to the phenomena that coherence and significance which is not an object of sense.

IV.

The antithesis of *Sense* and *Ideas* is the foundation of the Philosophy of Science. No knowledge can exist without the union, no philosophy without the separation, of these two elements.

V.

Fact and *Theory* correspond to Sense on the one hand, and to Ideas on the other, so far as we are *conscious* of our Ideas: but all facts involve ideas *unconsciously*; and thus the distinction of Facts and Theories is not tenable, as that of Sense and Ideas is.

VI.

Sensations and Ideas in our knowledge are like Matter and Form in bodies. Matter cannot exist without Form, nor Form without Matter : yet the two are altogether distinct and opposite. There is no possibility either of separating, or of confounding them. The same is the case with Sensations and Ideas.

VII.

Ideas are not *trans*formed, but *in*formed Sensations ; for without ideas, sensations have no foim.

VIII.

The Sensations are the *Objective*, the Ideas the *Subjective* part of every act of perception or knowledge.

IX.

General terms denote *Ideal Conceptions*, as a *circle*, an *orbit*, a *rose*. These are not *images* of real things, as was held by the Realists, but conceptions: yet they are conceptions, not bound together by mere *name*, as the Nominalists held, but by an idea.

X.

It has been said by some, that all Conceptions are merely *states* or *feelings of the mind*, but this assertion only tends to confound what it is our business to distinguish.

XI.

Observed Facts are connected so as to produce new truths, by superinducing upon them an Idea: and such truths are obtained *by Induction*.

XII.

Truths once obtained by legitimate Induction are Facts: these Facts may be again connected, so as to produce higher truths: and thus we advance to *Successive Generalizations*.

XIII.

Truths obtained by Induction are made compact and permanent by being expressed in *Technical Terms*.

XIV.

Experience cannot conduct us to universal and necessary truths:—Not to universal, because she has not tried all cases:— Not to necessary, because necessity is not a matter to which experience can testify.

XV.

Necessary truths derive their necessity from the *Ideas* which they involve; and the existence of necessary truths proves the existence of Ideas not generated by experience.

XVI.

In Deductive Reasoning, we cannot have any truth in the conclusion which is not virtually contained in the premises.

XVII.

In order to acquire any exact and solid knowledge, the student must possess with perfect precision the ideas appropriate to that part of knowledge: and this precision is tested by the student's *perceiving* the axiomatic evidence of the *axioms* belonging to each *Fundamental Idea*.

XVIII.

The Fundamental Ideas which it is most important to consider, as being the Bases of the Material Sciences, are the Ideas of *Space*, *Time* (including Number), *Cause* (including Force and Matter), *Outness* of Objects, and *Media* of Perception of Secondary Qualities, *Polarity* (Contrariety), Chemical *Composition* and *Affinity*, *Substance*, *Likeness* and Natural *Affinity*, *Means and Ends* (whence the notion of Organization), *Symmetry*, and the Ideas of *Vital Powers*.

XIX.

The Sciences which depend upon the Ideas of Space and Number are *Pure* Sciences, not *Inductive* Sciences: they do not infer special Theories from Facts, but deduce the conditions of all theory from Ideas. The Elementary Pure Sciences, or Elementary Mathematics, are Geometry, Theoretical Arithmetic and Algebra.

XX.

The Ideas on which the Pure Sciences depend, are those of *Space* and *Number*; but Number is a modification of the conception of Repetition, which belongs to the Idea of *Time*.

XXI.

The *Idea of Space* is not derived from experience, for experience of external objects *pre*supposes bodies to exist in Space. Space is a condition under which the mind receives the impressions of sense, and therefore the relations of space are necessarily and universally true of all perceived objects. Space is a *form* of our perceptions, and regulates them, whatever the *matter* of them may be.

XXII.

Space is not a general notion collected by abstraction from particular cases; for we do not speak of *Spaces* in general, but of universal or absolute *Space*. Absolute space is infinite. All special spaces are *in* absolute space, and are parts of it.

XXIII.

Space is not a real object or thing, distinct from the objects which exist in it; but it is a real condition of the existence of external objects.

XXIV.

We have an *Intuition* of objects in space; that is, we contemplate objects as *made up* of spatial parts, and apprehend their spatial relations by the same act by which we apprehend the objects themselves.

XXV.

Form or figure is space limited by boundaries. Space has necessarily *three* dimensions, length, breadth, depth; and no others which cannot be resolved into these.

XXVI.

The Idea of Space is exhibited for scientific purposes, by the *Definitions* and *Axioms* of Geometry; such, for instance, as these:—the *Definition of a Right Angle*, and *of a Circle;*—the *Definition of Parallel Lines*, and the *Axiom* concerning them;—the Axiom that *two straight lines cannot inclose a space*. These Definitions are necessary, not arbitrary; and the Axioms are needed as well as the Definitions, in order to express the necessary conditions which the Idea of Space imposes.

XXVII.

The Definitions and Axioms of Elementary Geometry do not *completely* exhibit the Idea of Space. In proceeding to the Higher Geometry, we may introduce other additional and independent Axioms; such as that of Archimedes, that *a curve line which joins two points is less than any broken line joining the same points and including the curve line*.

XXVIII.

The perception of a *solid object* by sight requires that *act* of mind by which, from figure and shade, we infer distance and position in space. The perception of *figure* by sight requires that act of mind by which we give an outline to each object.

XXIX.

The perception of form by touch is not an impression on the passive sense, but requires an *act* of our muscular frame by which we become aware of the position of our own limbs. The perceptive faculty involved in this act has been called *the muscular sense*.

XXX.

The *Idea of Time* is not derived from experience, for experience of changes *presupposes* occurrences to take place in Time. Time is a condition under which the mind receives the impressions of sense, and therefore the relations of time are necessarily and universally true of all perceived occurrences. Time is a *form* of our perceptions, and regulates them, whatever the *matter* of them may be.

XXXI.

Time is not a general notion collected by abstraction from particular cases. For we do not speak of particular *Times* as examples of time in general, but as parts of a single and infinite *Time*.

XXXII.

Time, like Space, is a form, not only of perception, but of *Intuition*. We consider the whole of any time as *equal* to the *sum* of the parts; and an occurrence as *coinciding* with the portion of time which it occupies.

XXXIII.

Time is analogous to Space of *one dimension*: portions of both have a beginning and an end, are long or short. There is nothing in Time which is analogous to Space of two, or of three, dimensions, and thus nothing which corresponds to Figure.

XXXIV.

The Repetition of a set of occurrences, as, for example, strong and weak, or long and short sounds, according to a steadfast order, produces *Rhythm*, which is a conception peculiar to Time, as Figure is to Space.

XXXV.

The simplest form of Repetition is that in which there is no variety, and this gives rise to the conception of *Number*.

XXXVI.

The simplest numerical truths are seen by Intuition; when we endeavour to deduce the more complex from these simplest, we employ such maxims as these:—*If equals be added to equals the wholes are equal:*—*If equals be subtracted from equals the remainders are equal:*—*The whole is equal to the sum of all its parts.*

XXXVII.

The Perception of Time involves a constant and latent kind of memory, which may be termed a *Sense of Succession*. The Perception of Number also involves this Sense of Succession, although in small numbers we appear to apprehend the units simultaneously and not successively.

XXXVIII.

The Perception of Rhythm is not an impression on the passive sense, but requires an *act* of thought by which we connect and group the strokes which form the Rhythm.

XXXIX.

Intuitive is opposed to *discursive* reason. In intuition, we obtain our conclusions by dwelling upon *one* aspect of the fundamental Idea; in discursive reasoning, we combine *several* aspects of the Idea, (that is, several axioms,) and reason from the combination.

XL.

Geometrical deduction (and deduction in general) is called *synthesis*, because we introduce, at successive steps, the results of new principles. But in reasoning on the relations of space, we sometimes go on *separating* truths into their component truths, and these into other component truths; and so on; and this is geometrical *analysis*.

XLI.

Among the foundations of the Higher Mathematics, is the *Idea of Symbols* considered as general *Signs* of Quantity. This idea of a Sign is distinct from, and independent of other ideas. The axiom to which we refer in reasoning by means of Symbols of quantity is this :—*The interpretation of such symbols must be perfectly general.* This Idea and Axiom are the bases of Algebra in its most general form.

XLII.

Among the foundations of the Higher Mathematics is also the *Idea of a Limit.* The Idea of a Limit cannot be superseded by any other definitions or Hypotheses. The Axiom which we employ in introducing this Idea into our reasoning is this:— *What is true up to the Limit is true at the Limit.* This Idea and Axiom are the bases of all Methods of Limits, Fluxions, Differentials, Variations, and the like.

XLIII.

There is a *pure* Science of Motion, which does not depend upon observed facts, but upon the Idea of motion. It may also be termed *Pure Mechanism,* in opposition to Mechanics Proper, or *Machinery,* which involves the mechanical conceptions of force and matter. It has been proposed to name this Pure Science of Motion, *Kinematics.*

XLIV.

The pure mathematical sciences must be successfully cultivated, in order that the progress of the principal inductive sciences may take place. This appears in the case of Astronomy, in which Science, both in ancient and in modern times, each advance of the theory has depended upon the previous solution of problems in pure mathematics. It appears also inversely in the Science of the Tides, in which, at present, we cannot advance in the theory, because we cannot solve the requisite problems in the Integral Calculus.

XLV.

The *Idea of Cause,* modified into the conceptions of mechanical cause, or Force, and resistance to force, or Matter, is the foundation of the Mechanical Sciences; that is, Mechanics, (including Statics and Dynamics,) Hydrostatics, and Physical Astronomy.

XLVI.

The Idea of Cause is not derived from experience; for in judging of occurrences which we contemplate, we consider them as being, universally and necessarily, Causes and Effects, which a finite experience could not authorize us to do. The Axiom, that every event must have a cause, is true independently of experience, and beyond the limits of experience.

XLVII.

The Idea of Cause is expressed for purposes of science by these three Axioms:—*Every Event must have a Cause:—Causes are measured by their Effects:—Reaction is equal and opposite to Action.*

XLVIII.

The Conception of Force involves the Idea of Cause, as applied to the motion and rest of bodies. The conception of *force* is suggested by muscular action exerted: the conception of *matter* arises from muscular action resisted. We necessarily ascribe to all bodies solidity and inertia, since we conceive Matter as that which cannot be compressed or moved without resistance.

XLIX.

Mechanical Science depends on the Conception of Force; and is divided into *Statics*, the doctrine of Force preventing motion, and *Dynamics*, the doctrine of Force producing motion.

L.

The Science of Statics depends upon the Axiom, that Action and Reaction are equal, which in Statics assumes this form:—*When two equal weights are supported on the middle point between them, the pressure on the fulcrum is equal to the sum of the weights.*

LI.

The Science of Hydrostatics depends upon the Fundamental Principle that *fluids press equally in all directions.* This Principle necessarily results from the conception of a Fluid, as a body of which the parts are perfectly moveable in all directions. For since the Fluid is a body, it can transmit pressure; and the transmitted pressure is equal to the original pressure, in virtue of the

Axiom that Reaction is equal to Action.   That the Fundamental
Principle is not derived from experience, is plain both from its
evidence and from its history.

LII.

The Science of Dynamics depends upon the three Axioms
above stated respecting Cause.   The First Axiom,—that every
change must have a Cause,—gives rise to the First Law of
Motion,—that *a body not acted upon by a force will move with a
uniform velocity in a straight line.*   The Second Axiom,—that
Causes are measured by their Effects,—gives rise to the Second
Law of Motion,—that *when a force acts upon a body in motion,
the effect of the force is compounded with the previously existing
motion.*   The Third Axiom,—that *Reaction is equal and opposite
to Action,*—gives rise to the Third Law of Motion, which is
expressed in the same terms as the Axiom ; Action and Reaction
being understood to signify momentum gained and lost.

LIII.

The above Laws of Motion, historically speaking, were established by means of experiment : but since they have been discovered and reduced to their simplest form, they have been considered by many philosophers as self-evident.   This result is
principally due to the introduction and establishment of terms
and definitions, which enable us to express the Laws in a very
simple manner.

LIV.

In the establishment of the Laws of Motion, it happened, in
several instances, that Principles were assumed as self-evident
which do not now appear evident, but which have since been demonstrated from the simplest and most evident principles.   Thus
it was assumed that *a perpetual motion is impossible ;*—that *the
velocities of bodies acquired by falling down planes or curves of the
same vertical height are equal ;*—that *the actual descent of the
centre of gravity is equal to its potential ascent.*   But we are not
hence to suppose that these assumptions were made without
ground : for since they really follow from the laws of motion,

they were probably, in the minds of the discoverers, the results of undeveloped demonstrations which their sagacity led them to divine.

LV.

It is a *Paradox* that Experience should lead us to truths confessedly universal, and apparently necessary, such as the Laws of Motion are. The *Solution* of this paradox is, that these laws are interpretations of the Axioms of Causation. The Axioms are universally and necessarily true, but the right interpretation of the terms which they involve, is learnt by experience. Our Idea of Cause supplies the *Form*, Experience, the *Matter*, of these Laws.

LVI.

Primary Qualities of Bodies are those which we can conceive as directly perceived ; *Secondary* Qualities are those which we conceive as perceived by means of a·Medium.

LVII.

We necessarily perceive bodies as *without* us : the Idea of *Externality* is one of the conditions of perception.

LVIII.

We necessarily assume a *Medium* for the perceptions of Light, Colour, Sound, Heat, Odours, Tastes ; and this Medium *must* convey impressions by means of its mechanical attributes.

LIX.

Secondary Qualities are not *extended* but *intensive ;* their effects are not augmented by addition of parts, but by increased operation of the medium. Hence they are not measured directly, but by *scales;* not by *units*, but by *degrees*.

LX.

In the Scales of Secondary Qualities, it is a condition (in order that the scale may be complete,) that every example of the quality must either *agree* with one of the degrees of the Scale, or lie between two *contiguous* degrees.

c 2

LXI.

We perceive *by means of* a medium and *by means of* impressions on the nerves: but we do not (by our senses,) perceive either the medium or the impressions on the nerves.

LXII.

The *Prerogatives of the Sight* are, that by this sense we necessarily and immediately apprehend the *position* of its objects: and that from visible circumstances, we *infer* the *distance* of objects from us, so readily that we seem to perceive and not to infer.

LXIII.

The *Prerogatives of the Hearing* are, that by this sense we perceive relations perfectly precise and definite between two notes, namely, *Musical Intervals* (as an *Octave*, a *Fifth*); and that when two notes are perceived together, they are apprehended as distinct, (a *Chord*,) and as having a certain relation, (*Concord* or *Discord*.)

LXIV.

The Sight cannot decompose a compound colour into simple colours, or distinguish a compound from a simple colour. The Hearing cannot directly perceive the place, still less the distance, of its objects. We infer these obscurely and vaguely from audible circumstances.

LXV.

The *First Paradox of Vision* is, that we see objects *upright*, though the images on the retina are *inverted*. The solution is, that we do not see the image on the retina at all, we only see by means of it.

LXVI.

The *Second Paradox of Vision* is, that we see objects *single*, though there are two images on the retinas, one in each eye. The explanation is, that it is a Law of Vision that we see (small or distant) objects single, when their images fall on *corresponding points* of the two retinas.

LXVII.

The law of single vision for *near* objects is this :—When the two images in the two eyes are situated, part for part, nearly but not exactly, upon corresponding points, the object is apprehended as single and solid if the two images are such as would be produced by a single solid object seen by the eyes separately.

LXVIII.

The ultimate object of each of the Secondary Mechanical Sciences is, to determine the nature and laws of the processes by which the impression of the Secondary Quality treated of is conveyed : but before we discover the *cause*, it may be necessary to determine the *laws* of the phenomena ; and for this purpose a *Measure* or *Scale* of each quality is necessary.

LXIX.

Secondary qualities are measured by means of such effects as can be estimated in number or space.

LXX.

The Measure of Sounds, as high or low, is the *Musical Scale*, or *Harmonic Canon*.

LXXI.

The Measures of Pure Colours are the *Prismatic Scale ;* the same, including *Fraunhofer's Lines ;* and *Newton's Scale* of Colours. The principal Scales of Impure Colours are *Werner's Nomenclature* of Colours, and *Merimée's Nomenclature* of Colours.

LXXII.

The Idea of *Polarity* involves the conception of contrary properties in contrary directions :—the properties being, for example, attraction and repulsion, darkness and light, synthesis and analysis ; and the contrary directions being those which are directly opposite, or, in some cases, those which are at right angles.

LXXIII.   (*Doubtful.*)

Coexistent polarities are fundamentally identical.

LXXIV.

The Idea of Chemical *Affinity*, as implied in Elementary Composition, involves peculiar conceptions. It is not properly expressed by assuming the qualities of bodies to *resemble* those of the elements, or to depend on the *figure* of the elements, or on their *attractions*.

LXXV.

Attractions take place between bodies, affinities between the particles of a body. The former may be compared to the alliances of states, the latter to the ties of family.

LXXVI.

The governing principles of chemical affinity are, that it is *elective*; that it is *definite*; that it *determines the properties* of the compound; and that *analysis is possible*.

LXXVII.

We have an Idea of *Substance*: and an axiom involved in this Idea is, that *the weight of a body is the sum of the weights of all its elements*.

LXXVIII.

Hence Imponderable Fluids are not to be admitted as chemical elements.

LXXIX.

The Doctrine of Atoms is admissible as a mode of expressing and calculating laws of nature; but is not proved by any fact, chemical or physical, as a philosophical truth.

LXXX.

We have an Idea of *Symmetry*; and an axiom involved in this Idea is, that in a symmetrical natural body, if there be a tendency to modify any member in any manner, there is a tendency to modify all the corresponding members in the same manner.

LXXXI.

All hypotheses respecting the manner in which the elements of inorganic bodies are arranged in space, must be constructed with regard to the general facts of crystallization.

LXXXII.

When we consider any object as *one*, we give unity to it by an act of thought. The condition which determines what this unity shall include, and what it shall exclude, is this;—that assertions concerning the one thing shall be possible.

LXXXIII.

We collect individuals into *kinds* by applying to them the Idea of Likeness. Kinds of things are not determined by definitions, but by this condition;—that general assertions concerning such kinds of things shall be possible.

LXXXIV.

The *names* of kinds of things are governed by their use; and that may be a right name in one use which is not so in another. A whale is not a *fish* in natural history, but it is a *fish* in commerce and law.

LXXXV.

We take for granted that each kind of things has a special *character* which may be expressed by a Definition. The ground of our assumption is this;—that reasoning must be possible.

LXXXVI.

The "Five Words," *genus, species, difference, property, accident,* were used by the Aristotelians, in order to express the subordination of kinds, and to describe the nature of definitions and propositions. In modern times, these technical expressions have been more referred to by Natural Historians than by Metaphysicians.

LXXXVII.

The construction of a Classificatory Science includes *Terminology*, the formation of a descriptive language;—*Diataxis*, the Plan of the System of Classification, called also the *Systematick;*—*Diagnosis*, the Scheme of the Characters by which the different Classes are known, called also the *Characteristick*. *Physiography* is the knowledge which the System is employed to convey. Diataxis includes *Nomenclature*.

LXXXVIII.

Terminology must be conventional, precise, constant; copious in words, and minute in distinctions, according to the needs of the science. The student must understand the terms, *directly* according to the convention, not through the medium of explanation or comparison.

LXXXIX.

The *Diataxis*, or Plan of the System, may aim at a Natural or an Artificial System. But no classes can be absolutely artificial, for if they were, no assertions could be made concerning them.

XC.

An *Artificial System* is one in which the *smaller* groups (the Genera) are *natural;* and in which the *wider* divisions (Classes, Orders) are constructed by the *peremptory* application of selected Characters; (selected, however, so as not to break up the smaller groups.)

XCI.

A *Natural System* is one which attempts to make *all* the divisions *natural*, the widest as well as the narrowest; and therefore applies *no* characters *peremptorily*.

XCII.

Natural Groups are best described, not by any definition which marks their boundaries, but by a *Type* which marks their centre. The Type of any natural group is an example which possesses in a marked degree all the leading characters of the class.

XCIII.

A Natural Group is steadily fixed, though not precisely limited; it is given in position, though not circumscribed; it is determined, not by a boundary without, but by a central point within;—not by what it strictly excludes, but by what it eminently includes;—by a Type, not by a Definition.

XCIV.

The prevalence of Mathematics as an element of education has made us think Definition the philosophical mode of fixing the meaning of a word: if (Scientific) Natural History were introduced into education, men might become familiar with the fixation of the signification of words by Types; and this agrees more nearly with the common processes by which words acquire their significations.

XCV.

The attempts at Natural Classification are of three sorts; according as they are made by the process of *blind trial*, of *general comparison*, or of *subordination of characters*. The process of Blind Trial professes to make its classes by attention to *all* the characters, but without proceeding methodically. The process of General Comparison professes to enumerate all the characters, and forms its classes by the *majority*. Neither of these methods can really be carried into effect. The method of Subordination of Characters considers some characters as *more important* than others; and this method gives more consistent results than the others. This method, however, does not depend upon the Idea of Likeness only, but introduces the Idea of Organization or Function.

XCVI.

A *Species* is a collection of individuals which are descended from a common stock, or which resemble such a collection as much as these resemble each other: the resemblance being opposed to a *definite* difference.

XCVII.

A *Genus* is a collection of species which resemble each other more than they resemble other species: the resemblance being opposed to a *definite* difference.

XCVIII.

The *Nomenclature* of a Classificatory Science is the collection of the names of the Species, Genera, and other divisions. The *binary* nomenclature, which denotes a species by the *generic* and *specific* name, is now commonly adopted in Natural History.

XCIX.

The *Diagnosis*, or Scheme of the Characters, comes, in the order of philosophy, after the Classification. The characters do not *make* the classes, they only enable us to *recognize* them. The Diagnosis is an Artificial Key to a Natural System.

C.

The basis of all Natural Systems of Classification is the Idea of Natural Affinity. The Principle which this Idea involves is this:—Natural arrangements, obtained from *different* sets of characters, must *coincide* with each other.

CI.

In order to obtain a Science of Biology, we must analyse the Idea of Life. It has been proved by the biological speculations of past time, that organic Life cannot rightly be resolved into mechanical or chemical forces, or the operation of a vital fluid, or of a soul.

CII.

Life is a System of Vital Forces; and the conception of such Forces involves a peculiar Fundamental Idea.

CIII.

Mechanical, chemical, and vital Forces form an ascending progression, each including the preceding. Chemical Affinity includes in its nature Mechanical Force, and may often be practically resolved into Mechanical Force. (Thus the ingredients of gunpowder, liberated from their chemical union, exert great mechanical Force: a galvanic battery acting by chemical process does the like.) Vital Forces include in their nature both chemical Affinities and mechanical Forces: for Vital Powers produce both chemical changes, (as digestion,) and motions which imply considerable mechanical force, (as the motion of the sap and of the blood.)

CIV.

In *voluntary* motions, Sensations produce Actions, and the connexion is made by means of Ideas: in *reflected* motions, the connexion neither seems to be nor is made by means of Ideas: in *instinctive* motions, the connexion is such as requires Ideas, but we cannot believe the Ideas to exist.

CV.

The assumption of a Final Cause in the structure of each part of animals and plants is as inevitable as the assumption of an Efficient Cause for every event. The maxim that in organized bodies nothing is *in vain*, is as necessarily true as the maxim that nothing happens *by chance*.

CVI.

The idea of living beings as subject to *disease* includes a recognition of a Final Cause in organization; for disease is a state in which the vital forces do not attain their *proper ends*.

CVII.

The Palætiological Sciences depend upon the Idea of Cause; but the leading conception which they involve is that of *historical cause*, not mechanical cause.

CVIII.

Each Palætiological Science, when complete, must possess three members: the *Phenomenology*, the *Ætiology*, and the *Theory*.

CIX.

There are, in the Palætiological Sciences, two antagonist doctrines: *Catastrophes* and *Uniformity*. The doctrine of a *uniform course of nature* is tenable only when we extend the notion of uniformity so far that it shall include catastrophes.

CX.

The Catastrophist constructs Theories, the Uniformitarian demolishes them. The former adduces evidence of an Origin, the latter explains the evidence away. The Catastrophist's dogmatism is undermined by the Uniformitarian's skeptical hypotheses. But when these hypotheses are asserted dogmatically, they cease to be consistent with the doctrine of uniformity.

CXI.

In each of the Palætiological Sciences, we can ascend to remote periods by a chain of causes, but in none can we ascend to a *beginning* of the chain.

CXII.

In contemplating the series of causes and effects which constitutes the world, we necessarily assume a *First Cause* of the whole series.

CXIII.

The Palætiological Sciences point backwards with lines which are broken, but which all converge to the *same* invisible point: and this point is the Origin of the Moral and Spiritual, as well as of the natural world.

APHORISMS CONCERNING SCIENCE.

I.

The two processes by which Science is constructed are the *Explication of Conceptions* and the *Colligation of Facts.*

II.

The Explication of Conceptions, as requisite for the progress of science, has been effected by means of discussions and controversies among scientists; often by debates concerning definitions; these controversies have frequently led to the establishment of a Definition; but along with the Definition, a corresponding Proposition has always been expressed or implied. The essential requisite for the advance of science is the clearness of the Conception, not the establishment of a Definition. The construction of an exact Definition is often very difficult. The requisite conditions of clear Conceptions may often be expressed by Axioms as well as by Definitions.

III.

Conceptions, for purposes of science, must be *appropriate* as well as clear: that is, they must be modifications of that Fundamental Idea, by which the phenomena can really be interpreted. This maxim may warn us from error, though it may not lead to discovery. Discovery depends upon the previous cultivation or natural clearness of the appropriate Idea, and therefore *no discovery is the work of accident.*

IV.

Facts are the materials of science, but all Facts involve Ideas. Since, in observing Facts, we cannot exclude Ideas, we must, for the purposes of science, take care that the Ideas are clear and rigorously applied.

V.

The last Aphorism leads to such Rules as the following:—
That Facts, for the purposes of material science, must involve
Conceptions of the Intellect only, and not Emotions:—That Facts
must be observed with reference to our most exact conceptions,
Number, Place, Figure, Motion:—That they must also be observed with reference to any other exact conceptions which the
phenomena suggest, as Force, in mechanical phenomena, Concord,
in musical.

VI.

The resolution of complex Facts into precise and measured
partial Facts, we call the *Decomposition of Facts.* This process
is requisite for the progress of science, but does not necessarily
lead to progress.

VII.

Science begins with *common* observation of facts ; but even at
this stage, requires that the observations be precise. Hence the
sciences which depend upon space and number were the earliest
formed. After common Observation, come scientific *Observation*
and *Experiment.*

VIII.

The Conceptions by which Facts are bound together, are suggested by the sagacity of discoverers. This sagacity cannot be
taught. It commonly succeeds by guessing ; and this success
seems to consist in framing several *tentative hypotheses* and selecting the right one. But a supply of appropriate hypotheses cannot
be constructed by rule, nor without inventive talent.

IX.

The truth of tentative hypotheses must be tested by their
application to facts. The discoverer must be ready, carefully to
try his hypotheses in this manner, and to reject them if they will
not bear the test, in spite of indolence and vanity.

X.

The process of scientific discovery is cautious and rigorous, not by abstaining from hypotheses, but by rigorously comparing hypotheses with facts, and by resolutely rejecting all which the comparison does not confirm.

XI.

Hypotheses may be useful, though involving much that is superfluous, and even erroneous: for they may supply the true bond of connexion of the facts ; and the superfluity and error may afterwards be pared away.

XII.

It is a test of true theories not only to account for, but to predict phenomena.

XIII.

Induction is a term applied to describe the *process* of a true Colligation of Facts by means of an exact and appropriate Conception. *An Induction* is also employed to denote the *proposition* which results from this process.

XIV.

The Consilience of Inductions takes place when an Induction, obtained from one class of facts, coincides with an Induction, obtained from another different class. This Consilience is a test of the truth of the Theory in which it occurs.

XV.

An Induction is not the mere *sum* of the Facts which are colligated. The Facts are not only brought together, but seen in a new point of view. (A new mental Element is *superinduced ;* and a peculiar constitution and discipline of mind are requisite in order to make this Induction.)

XVI.

Although in Every Induction a new conception is superinduced upon the Facts ; yet this once effectually done, the novelty of the conception is overlooked, and the conception is considered as a part of the fact.

XVII.

The *Logic of Induction* consists in stating the Facts and the Inference in such a manner, that the evidence of the Inference is manifest; just as the Logic of Deduction consists in stating the Premises and the Conclusion in such a manner that the Evidence of the Conclusion is manifest.

XVIII.

The Logic of Deduction is exhibited by means of a certain Formula; namely, a Syllogism; and every train of deductive reasoning, to be demonstrative, must be capable of resolution into a series of such Formulæ legitimately constructed. In like manner, the Logic of Induction may be exhibited by means of certain *Formulæ*; and every train of inductive inference, to be sound, must be capable of resolution into a scheme of such Formulæ, legitimately constructed.

XIX.

The *inductive act of thought* by which several Facts are colligated into one Proposition, may be expressed by saying : *The several Facts are exactly expressed as one Fact, if, and only if, we adopt the Conceptions and the Assertion* of the Proposition.

XX.

The One Fact, thus inductively obtained from several Facts, may be combined with other Facts, and colligated with them by a new act of Induction. This process may be indefinitely repeated : and these successive processes are the *Steps* of Induction, or of *Generalization*, from the lowest to the highest.

XXI.

The relation of the successive Steps of Induction may be exhibited by means of an *Inductive Table*, in which the several Facts are indicated, and tied together by a Bracket, and the Inductive Inference placed on the other side of the Bracket; and this arrangement repeated, so as to form a genealogical Table of each Induction, from the lowest to the highest.

XXII.

The Logic of Induction is the *Criterion of Truth* inferred from Facts, as the Logic of Deduction is the Criterion of Truth deduced from necessary Principles. The Inductive Table enables us to apply such a Criterion; for we can determine whether each Induction is verified and justified by the Facts which its Bracket includes; and if each induction in particular be sound, the highest, which merely combines them all, must necessarily be sound also.

XXIII.

The distinction of *Fact* and *Theory* is only relative. Events and phenomena, considered as particulars which may be colligated by Induction, are *Facts*; considered as generalities already obtained by colligation of other Facts, they are *Theories*. The same event or phenomenon is a Fact or a Theory, according as it is considered as standing on one side or the other of the Inductive Bracket.

XXIV.

Inductive truths are of two kinds, *Laws of Phenomena*, and *Theories of Causes*. It is necessary to begin in every science with the Laws of Phenomena; but it is impossible that we should be satisfied to stop short of a Theory of Causes. In Physical Astronomy, Physical Optics, Geology, and other sciences, we have instances showing that we can make a great advance in inquiries after true Theories of Causes.

XXV.

Art and Science differ. The object of Science is Knowledge; the objects of Art, are Works. In Art, truth is a means to an end; in Science, it is the only end. Hence the Practical Arts are not to be classed among the Sciences.

XXVI.

Practical Knowledge, such as Art implies, is not Knowledge such as Science includes. Brute animals have a practical knowledge of relations of space and force; but they have no knowledge of Geometry or Mechanics.

XXVII.

The Methods by which the construction of Science is promoted are, *Methods of Observation, Methods of obtaining clear Ideas,* and *Methods of Induction.*

XXVIII.

The Methods of Observation of Quantity in general, are *Numeration,* which is precise by the nature of Number; the *Measurement of Space* and *of Time,* which are easily made precise; the *Conversion of Space and Time,* by which each aids the measurement of the other; the *Method of Repetition;* the *Method of Coincidences* or *Interferences.* The measurement of Weight is made precise by the *Method of Double-weighing.* Secondary Qualities are measured by means of *Scales of Degrees;* but in order to apply these Scales, the student requires the *Education of the Senses.* The Education of the Senses is forwarded by the practical study of *Descriptive Natural History, Chemical Manipulation,* and *Astronomical Observation.*

XXIX.

The Methods by which the acquisition of clear Scientific Ideas is promoted, are mainly two; *Intellectual Education* and *Discussion of Ideas.*

XXX.

The Idea of Space becomes more clear by studying *Geometry;* the Idea of Force, by studying *Mechanics;* the Ideas of Likeness, of Kind, of subordination of Classes, by studying *Natural History.*

XXXI.

Elementary Mechanics should now form a part of intellectual education, in order that the student may understand the Theory of Universal Gravitation: for an intellectual education should cultivate such ideas as enable the student to understand the most complete and admirable portions of the knowledge which the human race has attained to.

XXXII.

Natural History ought to form a part of intellectual education, in order to correct certain prejudices which arise from cultivating the intellect by means of mathematics alone; and in order to lead the student to see that the division of things into kinds, and the attribution and use of names, are processes susceptible of great precision.

XXXIII.

The conceptions involved in scientific truths have attained the requisite degree of clearness by means of the *Discussions* respecting ideas which have taken place among discoverers and their followers. Such discussions are very far from being unprofitable to science. They are metaphysical, and must be so: the difference between discoverers and barren reasoners is, that the former employ good, and the latter bad metaphysics.

XXXIV.

The Process of Induction may be resolved into three steps; the *Selection of the Idea*, the *Construction of the Conception*, and the *Determination of the Magnitudes*.

XXXV.

These three steps correspond to the determination of the *Independent variable*, the *Formula*, and the *Coefficients*, in mathematical investigations; or to the *Argument*, the *Law*, and the *Numerical Data*, in a Table of an Inequality.

XXXVI.

The Selection of the Idea depends mainly upon inventive sagacity: which operates by suggesting and trying various hypotheses. Some inquirers try erroneous hypotheses; and thus, exhausting the forms of error, form the Prelude to Discovery.

XXXVII.

The following Rules may be given, in order to the selection of the Idea for purposes of Induction:—the Idea and the Facts must be *homogeneous;* and the Rule must be *tested by the Facts.*

XXXVIII.

The Construction of the Conception very often includes, in a great measure, the Determination of the Magnitudes.

XXXIX.

When a series of *progressive* numbers is given as the result of observation, it may generally be reduced to law by combinations of arithmetical and geometrical progressions.

XL.

A true formula for a progressive series of numbers cannot commonly be obtained from a *narrow range* of observations.

XLI.

Recurrent series of numbers must, in most cases, be expressed by circular formulæ.

XLII.

The true construction of the conception is frequently suggested by some hypothesis; and in these cases, the hypothesis may be useful, though containing superfluous parts.

XLIII.

There are special Methods of Induction applicable to Quantity; of which the principal are, the *Method of Curves*, the *Method of Means*, the *Method of Least Squares*, and the *Method of Residues*.

XLIV.

The Method of Curves consists in drawing a curve, of which the observed quantities are the ordinates, the quantity on which the change of these quantities depends being the abscissa. Its efficacy depends upon the faculty which the eye possesses, of readily detecting regularity and irregularity in forms. It may be used to detect the laws which the observed quantities follow; and also, when the observations are inexact, it may be used to correct these observations, so as to obtain data more true than the observed facts themselves.

XLV.

The Method of Means gets rid of irregularities by taking the arithmetical mean of a great number of observed quantities. Its efficacy depends upon this ; that in cases in which observed quantities are affected by other inequalities, besides that of which we wish to determine the law, the excesses *above* and defects *below* the quantities which the law in question would produce, will, in a collection of many observations, *balance* each other.

XLVI.

The Method of Least Squares is a Method of Means, in which the mean is taken according to the condition, that the sum of the squares of the errors of observation shall be the least possible which the law of the facts allows. It appears, by the doctrine of chances, that this is the *most probable* mean.

XLVII.

The Method of Residues consists in subtracting, from the quantities given by observation, the quantity given by any law already discovered ; and then examining the remainder, or *Residue*, in order to discover the leading law which it follows. When this second law has been discovered, the quantity given by it may be subtracted from the first Residue ; thus giving a *Second Residue*, which may be examined in the same manner ; and so on. The efficacy of this method depends principally upon the circumstance of the laws of variation being successively smaller and smaller in amount (or at least in their mean effect) ; so that the ulterior undiscovered laws do not prevent the law in question from being *prominent* in the observations.

XLVIII.

The Method of Means and the Method of Least Squares cannot be applied without our *knowing the Arguments* of the Inequalities which we seek. The Method of Curves and the Method of Residues, when the Arguments of the principal Inequalities are known, often make it easy to find the others.

XLIX.

The Law of Continuity is this:—that a quantity cannot pass from one amount to another by any change of conditions, without passing through all intermediate magnitudes according to the intermediate conditions. It may often be employed to disprove distinctions which have no real foundation.

L.

The Method of Gradation consists in taking a number of stages of a property in question, intermediate between two extreme cases which appear to be different. It is employed to determine whether the extreme cases are really distinct or not.

LI.

The Method of Gradation, applied to decide the question, whether the existing geological phenomena arise from existing causes, leads to this result:—That the phenomena do appear to arise from existing causes, but that the action of existing causes may, in past times, have transgressed, to any extent, their recorded limits of intensity.

LII.

The Method of Natural Classification consists in classing cases, not according to any assumed definition, but according to the connexion of the facts themselves, so as to make them the means of asserting general truths.

LIII.

In the *Induction of Causes* the principal maxim is, that we must be careful to possess, and to apply, with perfect clearness, the Fundamental Idea on which the Induction depends.

LIV.

The Induction of Substance, of Force, of Polarity, go beyond mere laws of phenomena, and may be considered as the Induction cf Causes.

LV.

The Cause of certain phenomena being inferred, we are led to inquire into the Cause of this Cause, which inquiry must be conducted in the same manner as the previous one; and thus we have the Induction of Ulterior Causes.

LVI.

In contemplating the series of Causes which are themselves the effects of other causes, we are necessarily led to assume a Supreme Cause in the Order of Causation, as we assume a First Cause in Order of Succession.

APHORISMS

CONCERNING THE LANGUAGE OF SCIENCE.

INTRODUCTION.

IT has been shown in the History of Science, and will further appear in the course of the present work, that almost every step in the progress of science is marked by the formation or appropriation of a technical term. Common language has, in most cases, a certain degree of looseness and ambiguity; as common knowledge has usually something of vagueness and indistinctness. In common cases too, knowledge usually does not occupy the intellect alone, but more or less interests some affection, or puts in action the fancy; and common language, accommodating itself to the office of expressing such knowledge, contains, in every sentence, a tinge of emotion or of imagination. But when our knowledge becomes perfectly exact and purely intellectual, we require a language which shall also be exact and intellectual;—which shall exclude alike vagueness and fancy, imperfection and superfluity;—in which each term shall convey a meaning steadily fixed and rigorously limited. Such a language that of science becomes through the use of technical terms. And we must now endeavour to lay down some maxims and suggestions, by attention to which technical terms may be better fitted to answer their purpose. In order to do this, we shall in the first place take a rapid survey of the manner in which technical terms have been employed from the earliest periods of scientific history.

The progress of the use of technical scientific language offers to our notice two different and successive periods; in the first of which, technical terms were formed casually, as convenience in each case prompted; while in the second period, technical language was constructed intentionally, with set purpose, with a regard to its connexion, and with a view of constructing a system. Though the casual and systematic formation of technical terms cannot be separated by any precise date of time, (for at all periods some terms in some sciences have been framed unsystematically,) we may, as a general description, call the former the *ancient* and the latter the *modern* period. In illustrating the two following Aphorisms, I will give examples of the course followed in each of these periods.

Aphorism I.

In the Ancient Period of Science, Technical Terms were formed in three different ways:—by appropriating common words and fixing their meaning;—by constructing terms containing a description;—by constructing terms containing reference to a theory.

THE earliest sciences offer the earliest examples of technical terms. These are Geometry, Arithmetic, and Astronomy; to which we have soon after to add Harmonics, Mechanics, and Optics. In these sciences, we may notice the above-mentioned three different modes in which technical terms were formed.

I. The simplest and first mode of acquiring technical terms, is to take words current in common usage, and by rigorously defining or otherwise fixing their meaning, to fit them for the expression of scientific truths. In this manner almost all the fundamental technical terms of Geometry were formed. A *sphere*, a *cone*, a *cylinder*, had among the Greeks, at first, meanings less precise than those which geometers gave to these words, and besides the mere designation of form, implied some use or application. A sphere (σφαῖρα) was a hand-ball used in games; a *cone* (κῶνος) was a boy's spinning-top, or the crest of a helmet; a *cylinder* (κύλινδρος) was a roller; a *cube* (κύβος) was a die: till these words were adopted by the geometers, and made to signify among them pure modifications of space. So an *angle* (γωνία) was only a corner ; a *point* (σημεῖον) was a signal ; a *line* (γραμμὴ) was a mark ; a *straight* line (εὐθεῖα) was marked by an adjective which at first meant only *direct*. A *plane* (ἐπίπεδον) is the neuter form of an adjective, which by its derivation means *on the ground*, and hence *flat*. In all these cases, the word adopted as a term of science has its sense rigorously fixed ; and where the common use of the term is in any degree vague, its meaning may be modified at the same time that it is thus limited. Thus a *rhombus* (ῥόμβος) by its derivation, might mean any figure which is *twisted* out of a regular form ; but it is confined by geometers to that figure which has four equal sides, its angles being oblique. In like manner, a *trapezium* (τραπέζιον) originally signifies a *table*, and thus might denote any form ; but as the tables of the Greeks had one side shorter than the opposite one, such a figure was at first called a trapezium. Afterwards the term was made to signify any figure with four unequal sides ; a name being more needful in geometry for this kind of figure than for the original form.

This class of technical terms, namely, words adopted from common language, but rendered precise and determinate for purposes of science, may also be exemplified in other sciences. Thus, as was observed in the early portion of the history of astronomy*, a *day*, a *month*, a *year*, described at first portions of time marked by familiar changes, but afterwards portions determined by rigorous mathematical definitions. The conception of the heavens as a revolving sphere, is so obvious, that we may consider the terms which involve this conception as parts of common language ; as the *pole* (πόλος) of the *arctic circle*, which includes the stars that never set†; the *horizon* (ὁρίζων) a boundary, applied technically to the circle bounding the visible earth and sky. The *turnings of the sun* (τροπαί ἠελίοιο), which are mentioned by Hesiod, gave occasion to the term *tropics*, the circles at which the sun in his annual motion turns back from his northward or southward advance. The *zones* of the earth, (the torrid, temperate, and frigid ;) the *gnomon* of a dial ; the *limb* (or border) of the moon, or of a circular

---

* *Hist. Ind. Sci.*, i. 112.   † *Hist. Ast.*, i. 144.

instrument, are terms of the same class. An *eclipse* (ἔκλειψις) is originally a deficiency or disappearance, and joined with the name of the luminary, an *eclipse of the sun* or *of the moon*, described the phenomenon; but when the term became technical, it sufficed, without addition, to designate the phenomenon.

In Mechanics, the Greeks gave a scientific precision to very few words : we may mention *weights* (βαρέα), the *arms of a lever* (μήχεα), its *fulcrum* (ὑπομοχλίον), and the verb *to balance* (ἰσορροπεῖν). Other terms which they used, as *momentum* (ῥοπή) and *force* (δύναμις), did not acquire a distinct and definite meaning till the time of Galileo, or later. We may observe that all abstract terms, though in their scientific application expressing mere conceptions, were probably at first derived from some word describing external objects. Thus the Latin word for force, *vis*, seems to be connected with a Greek word, *ἲς*, or *Fὶς*, which often has nearly the same meaning; but originally, as it would seem, signified a sinew or muscle, the obvious seat of animal strength.

In later times, the limitation imposed upon a word by its appropriation to scientific purposes, is often more marked than in the cases above described. Thus the *variation* is made to mean, in astronomy, the second inequality of the moon's motion ; in magnetism, the *variation* signifies the angular deviation of the compass-needle from the north ; in pure mathematics, the *variation* of a quantity is the formula which expresses the result of any small change of the most general kind. In like manner, *parallax* (παράλλαξις) denotes a *change* in general, but is used by astronomers to signify the change produced by the spectator's being removed from the centre of the earth, his theoretical place, to the surface. *Alkali* at first denoted the ashes of a particular plant, but afterwards, all bodies having a certain class of chemical properties ; and, in like manner, *acid*, the class opposed to alkali, was modified in signification by chemists, so as to refer no longer to the taste.

Words thus borrowed from common language, and converted by scientific writers into technical terms, have some advantages and some disadvantages. They possess this great convenience, that they are understood after a very short explanation, and retained in the memory without effort. On the other hand, they lead to some inconvenience; for since they have a meaning in common language, a careless reader is prone to disregard the technical limitation of this meaning, and to attempt to collect their import in scientific books, in the same vague and conjectural manner in which he collects the purpose of words in common cases. Hence the language of science, when thus resembling common language, is liable to be employed with an absence of that scientific precision which alone gives it value. Popular writers and talkers, when they speak of *force, momentum, action and reaction*, and the like, often afford examples of the inaccuracy thus arising from the scientific appropriation of common terms.

II. Another class of technical terms, which we find occurring as soon as speculative science assumes a distinct shape, consists of those which are intentionally constructed by speculators, and which contain some description or indication distinctive of the conception to which they are applied. Such are a *parallelogram* (παραλληλόγραμμον), which denotes a plane figure bounded by two pairs of parallel lines; a *parallelopiped* (παραλληλοπίπεδον), which signifies a solid figure bounded by three pairs of parallel planes. A triangle (τρίγωνος) and a quadrangle (τετράγωνος) were perhaps words invented independently of the mathematicians: but such words extended to other cases, *pentagon, decagon, heccædecagon, polygon*, are inventions of scientific men. Such also are *tetrahedron, hexahedron, dodecahedron, tesseracontaoctohedron, polyhedron*, and the like. These words being constructed by speculative writers, explain themselves, or at least require only some conventional limitation, easily adopted. Thus *parallelogram* might mean a figure bounded by any number of sets of parallel lines, but it is conventionally restricted to a figure of *four* sides. So a *great circle* in a sphere means one which passes through the centre of the sphere; and a *small circle* is any other. So in trigonometry, we have the *hypotenuse* (ὑποτείνουσα), or *subtending* line, to designate the line subtending an angle. In this branch of mathematics we have many invented technical terms; as *complement, supplement, cosine, cotangent*, a *spherical angle*, the *pole of a circle*, or of a sphere. The word *sine* itself appears to belong to the class of terms already described as scientific appropriations of common terms, although its origin is somewhat obscure.

Mathematicians were naturally led to construct these and many other terms by the progress of their speculations. In like manner, when astronomy took the form of a speculative science, words were invented to denote distinctly the conceptions thus introduced. Thus the sun's annual path among the stars, in which not only solar, but also all lunar eclipses occur, was termed the *ecliptic*. The circle which the sun describes in his diurnal motion, when the days and nights are equal, the Greeks called the *equidiurnal* (ἰσημερινὸς,) the Latin astronomers the *equinoctial*, and the corresponding circle on the earth was the *equator*. The ecliptic intersected the equinoctial in the *equinoctial points*. The *solstices* (in Greek τροπαί) were the times when the sun arrested his motion northwards or southwards; and the *solstitial points* (τὰ τροπίκα σημεῖα) were the places in the ecliptic where he then was. The name of *meridians* was given to circles passing through the poles of the equator; the *solstitial colure* (κόλουρος, curtailed), was one of these circles which passes through the solstitial points, and is intercepted by the horizon.

We have borrowed from the Arabians various astronomical terms, as *Zenith, Nadir, Azimuth, Almacantar*. And these words, which among the Arabians probably belonged to the first class, of appropriated scientific terms, are for us examples of the second class, invented scientific terms; although they differ from most that we have mentioned, in not containing an etymology corresponding to their meaning in any language with which European cultivators of science are generally familiar. Indeed, the distinction of our two classes, though convenient, is in a great measure, casual. Thus most of the words we formerly mentioned, as *parallax, horizon, eclipse*, though appropriated technical terms among the Greeks, are to us invented technical terms.

In the construction of such terms as we are now considering, those languages have a great advantage which possess a power of forming words by composition. This was eminently the case with the Greek language; and hence most of the ancient terms of science in that language, when their origin is once explained, are clearly understood and easily retained. Of modern European languages, the German possesses the greatest facility of composition; and hence scientific authors in that language are able to invent terms which it is impossible to imitate in the other languages of Europe. Thus Weiss distinguishes his various systems of crystals as *zwei-und-zwei-gliedrig, ein-und-zwei-gliedrig, drey-und-drey-gliedrig,* &c., (two-and-two-membered, one-and-two-membered, three-and-three-membered.) And Hessel, also a writer on crystallography, speaks of *doubly-one-membered edges, four-and-three spaced rays,* and the like.

How far the composition of words, in such cases, may be practised in the English language, and the general question, what are the best rules and artifices in such cases, I shall afterwards consider. In the mean time, I may observe that this list of invented technical terms might easily be much enlarged. Thus in harmonics we have the various intervals, as a *Fourth,* a *Fifth,* an *Octave, (Diatessaron, Diapente, Diapason,)* a *Comma,* which is the difference of a *major* and *minor Tone;* we have the various *Moods* or *Keys,* and the notes of various lengths, as *Minims, Breves, Semibreves, Quavers.* In chemistry, *gas* was at first a technical term invented by Van Helmont, though it has now been almost adopted into common language. I omit many words which will perhaps suggest themselves to the reader, because they belong rather to the next class, which I now proceed to notice.

III. The third class of technical terms consists of such as are constructed by men of science, and involve some theoretical idea in the meaning which their derivation implies. They do not merely describe, like the class last spoken of, but describe with reference to some doctrine or hypothesis which is accepted as a portion of science. Thus *latitude* and *longitude,* according to their origin, signify breadth and length; they are used, however, to denote measures of the distance of a place on the earth's surface from the equator, and from the first meridian, of which distances, one cannot be called *length* more properly than the other. But this appropriation of these words may be explained by recollecting that the earth, as known to the ancient geographers, was much further extended from east to west than from north to south. The *Precession* of the equinoxes is a term which implies that the stars are fixed, while the point which is the origin of the measure of celestial longitude moves backward. The *Right Ascension* of a star is a measure of its position corresponding to terrestrial longitude; this quantity is identical with the angular ascent of the equinoctial point, when the star is in the horizon in a *right* sphere; that is, a sphere which supposes the spectator to be at the equator. The *Oblique Ascension* (a term now little used), is derived in like manner from an oblique sphere. The motion of a planet is *direct* or *retrograde, in consequentia (signa)*, or *in antecedentia*, in reference to a certain assumed standard direction for celestial motions, namely, the direction opposite to that of the sun's daily motion, and agreeing with his annual motion among the stars; or with what is much more evident, the moon's monthly motion. The *equation of time* is the quantity which must be added to or subtracted from the time marked by the sun, in order to reduce it to a theoretical condition of equable progress. In like manner the *equation of the centre* of the sun or of the moon is the angle which must be added to, or subtracted from, the actual advance of the luminary in the heavens, in order to make its motion equable. Besides the equation of the centre of the moon, which represents the first and greatest of her deviations from equable motion, there are many other *equations*, by the application of which her motion is brought nearer and nearer to perfect uniformity. The second of these equations is called the *erection*, the third the *variation*, the fourth the *annual equation*. The motion of the sun as affected by its inequalities is called his *anomaly*, which term denotes inequality. In the History of Astronomy, we find that the inequable motions of the sun, moon, and planets were, in a great measure, reduced to rule and system by the Greeks, by the aid of an hypothesis of circles, revolving, and carrying in their motion other circles which also revolved. This hypothesis introduced many technical terms, as *deferent, epicycle, eccentric*. In like manner, the theories which have more recently taken the place of the theory of epicycles have introduced other technical terms, as the *elliptical orbit*, the *radius vector*, and the *equable description of areas* by this radius, which phrases express the true laws of the planetary motions.

There is no subject on which theoretical views have been so long and so extensively prevalent as astronomy, and therefore no other science in which there are so many technical terms of the kind we are now considering. In other subjects, so far as theories have been established, they have been accompanied by the introduction or fixation of technical terms. Thus, as we have seen in the examination of the foundations of mechanics, the terms *force* and *inertia* derive their precise meaning from a recognition of the first law of motion; *accelerating force* and *composition of motion* involve the second law; *moving force, momentum, action* and *reaction*, are expressions which imply the third law. The term *vis viva* was introduced to express a general property of moving bodies; and other terms have been introduced for like purposes, as *impetus* by Smeaton, and *work done*, by other engineers. The proposition which was termed the *hydrostatic paradox* had this name in reference to its violating a supposed law of the action of forces. The verb to *gravitate*, and the abstract term *gravitation*, sealed the establishment of Newton's theory of the solar system.

In some of the sciences, opinions, either false or disguised in very fantastical imagery, have prevailed; and the terms which have been introduced during the reign of such opinions, bear the impress of the time. Thus in the days of alchemy, the substances with which the operator dealt were personified; and a metal when exhibited pure and free from all admixture was considered as a little king, and was hence called a *regulus*, a term not yet quite obsolete. In like manner, a substance from which nothing more of any value could be extracted, was dead, and was called a *caput mortuum*. Quick silver, that is, live silver (*argentum vivum*), was killed by certain admixtures, and was *revived* when restored to its pure state.

We find a great number of medical terms which bear the mark of opinions formerly prevalent among physicians; and though these opinions hardly form a part of the progress of science, and were not presented in our History, we may notice some of these terms as examples of the mode in which words involve in their derivation obsolete opinions. Such words as *hysterics, hypochondriac, melancholy, cholera, colic, quinsey* (*squinantia,* συνάγχη, a suffocation), *megrim, migraine* (*hemicranium,* the middle of the skull), *rickets,* (*rachitis,* from ῥαχὶς, the backbone), *palsy,* (*paralysis,* παραλύσις,) *apoplexy* (ἀποπληξία, a stroke), *emrods* (ἁιμορροίδες, *hemorrhoids,* a flux of blood), *imposthume,* (corrupted from *aposteme,* ἀπόϲημα, an abscess), *phthisic* (φθίσις, consumption), *tympany* (τυμπανία, swelling), *dropsy* (*hydropsy,* ὕδρωψ), *sciatica,* isciatica (ἰσχιαδικὴ, from ἰσχίον, the hip), *catarrh* (κατάρρους, a flowing down), *diarrhœa* (διαρροία, a flowing through), *diabetes* (διαβήτης, a passing through), *dysentery* (δυσεντερία, a disorder of the entrails), *arthritic* pains (from ἄρθρα, the joints), are names derived from the supposed or real seat and circumstances of the diseases. The word from which the first of the above names is derived (ὑστερα, the last place,) signifies the womb, according to its order in a certain systematic enumeration of parts. The second word, *hypochondriac,* means something affecting the viscera below the cartilage of the breastbone, which cartilage is called χόνδρος; *melancholy* and *cholera* derive their names from supposed affections of χολὴ, the bile. *Colic* is that which affects the *colon* (κῶλον), the largest member of the bowels. A disorder of the eye is called *gutta serena* (the "drop serene" of Milton), in contradistinction to *gutta turbida,* in which the impediment to vision is perceptibly opake. Other terms also record the opinions of the ancient anatomists, as *duodenum,* a certain portion of the intestines, which they estimated as twelve inches long. We might add other allusions, as the *tendon of Achilles.*

Astrology also supplied a number of words founded upon fanciful opinions; but this study having been expelled from the list of sciences, such words now survive only so far as they have found a place in common language. Thus men were termed *mercurial, martial, jovial,* or *saturnine,* accordingly as their characters were supposed to be determined by the influence of the planets, Mercury, Mars, Jupiter, or Saturn. Other expressions, such as *disastrous, ill-starred, exorbitant, lord of the ascendant,* and hence *ascendancy, influence,* a *sphere of action,* and the like, may serve to show how extensively astrological opinions have affected language, though the doctrine is no longer a recognized science.

The preceding examples will make it manifest that opinions, even of a recondite and complex kind, are often implied in the derivation of words; and thus will show how scientific terms, framed by the cultivators of science, may involve received hypotheses and theories. When terms are thus constructed, they serve not only to convey with ease, but to preserve steadily and to diffuse widely, the opinions which they thus assume. Moreover, they enable the speculator to employ these complex conceptions, the creations of science, and the results of much labour and thought, as readily and familiarly as if they were convictions borrowed at once from the senses. They are thus powerful instruments in enabling philosophers to ascend from one step of induction and generalization to another; and hereby contribute powerfully to the advance of knowledge and truth.

It should be noticed, before we proceed, that the names of natural objects, when they come to be considered as the objects of a science, are selected according to the processes already enumerated. For the most part, the natural historian adopts the common names of animals, plants, minerals, gems, and the like, and only endeavours to secure their steady and consistent application. But many of these names imply some peculiar, often fanciful, belief respecting the object.

Various plants derive their names from their supposed virtues, as *herniaria, rupture-wort;* or from legends, as *herba Sancti Johannis, St. John's wort.* The same is the case with minerals: thus the *topaz* was asserted to come from an island so shrouded in mists that navigators could only *conjecture* (τοπάζειν) where it was. In these latter cases, however, the legend appears not to be the true origin of the name, but to be suggested by it.

The privilege of constructing names where they are wanted, belongs to natural historians no less than to the cultivators of physical science; yet in the ancient world, writers of the former class appear rarely to have exercised this privilege, even when they felt the imperfections of the current language. Thus Aristotle repeatedly mentions classes of animals which have no name, as co-ordinate with classes that have names; but he hardly ventures to propose names which may supply these defects*. The vast importance of nomenclature in natural history was not recognized till the modern period.

We have, however, hitherto considered only the formation or appropriation of single terms in science; except so far as several terms may in some instances be connected by reference to a common theory. But when the value of technical terms began to be fully appreciated, philosophers proceeded to introduce them into their sciences more copiously and in a more systematic manner. In this way, the modern history of technical language has some features of a different aspect from the ancient; and must give rise to a separate Aphorism.

Aphorism II.

In the Modern Period of Science, besides the three processes anciently employed in the formation of technical terms, there have been introduced Systematic Nomenclature, Systematic Terminology, and the Systematic Modification of Terms to express theoretical relations†.

Writers upon science have gone on up to modern times forming such technical terms as they had occasion for, by the three processes above described; — namely, appropriating and limiting words in common use;—constructing for themselves words descriptive of the conception which they wished to convey;—or framing terms which by their signification imply the

* In his *History of Animals*, (book i. chap. 6), he says that the great classes of animals are Quadrupeds, Birds, Fishes, Whales (*Cetaceans*), Oysters (*Testaceans*), animals like crabs which have no general name (*Crustaceans*), soft animals (*Mollusks* and *Insects*). He does, however, call the Crustaces by a name (*Malacostraca*, soft-shelled) which has since been adopted by Naturalists.

† On the subject of Terminology and Nomenclature, see also Aphorisms lxxxviii and xcviii concerning Ideas, and book viii. chap. 2 of the Philosophy.

adoption of a theory. Thus among the terms introduced by
the study of the connexion between magnetism and electricity,
the word *pole* is an example of the first kind; the name of the
subject, *electro-magnetism*, of the second; and the term *current*,
involving an hypothesis of the motion of a fluiu, is an instance
of the third class. In chemistry, the term *salt* was adopted
from common language, and its meaning extended to denote
any compound of a certain kind; the term *neutral* salt implied
the notion of a balanced opposition in the two elements of the
compound; and such words as *subacid* and *superacid*, invented
on purpose, were introduced to indicate the cases in which this
balance was not attained. Again, when the phlogistic theory of
chemistry was established, the term *phlogiston* was introduced to
express the theory, and from this such terms as *phlogisticated* and
dephlogisticated were derived, exclusively words of science. But
in such instances as have just been given, we approach towards a
systematic modification of terms, which is a peculiar process of
modern times. Of this, modern chemistry forms a prominent
example, which we shall soon consider, but we shall first notice
the other processes mentioned in the Aphorism.

I. In ancient times, no attempt was made to invent or select
a Nomenclature of the objects of Natural History which should
be precise and permanent. The omission of this step by the
ancient naturalists gave rise to enormous difficulty and loss of
time when the sciences resumed their activity. We have seen
in the history of the sciences of classification, and of botany in
especial*, that the early cultivators of that study in modern times
endeavoured to identify all the plants described by Greek and
Roman writers with those which grow in the north of Europe;
and were involved in endless confusion†, by the multiplication
of names of plants, at the same time superfluous and ambiguous.
The *Synonymies* which botanists (Bauhin and others) found it
necessary to publish, were the evidences of these inconveniences.
In consequence of the defectiveness of the ancient botanical
nomenclature, we are even yet uncertain with respect to the iden-

---

* *Hist. Ind. Sci.*, iii. 272.   † *Ib.*, 293.

tification of some of the most common trees mentioned by classical writers *. The ignorance of botanists respecting the importance of nomenclature operated in another manner to impede the progress of science. As a good nomenclature presupposes a good system of classification, so, on the other hand, a system of classification cannot become permanent without a corresponding nomenclature. Cæsalpinus, in the sixteenth century †, published an excellent system of arrangement for plants; but this, not being connected with any system of names, was never extensively accepted, and soon fell into oblivion. The business of framing a scientific botanical classification was in this way delayed for about a century. In the same manner, Willoughby's classification of fishes, though, as Cuvier says, far better than any which preceded it, was never extensively adopted, in consequence of having no nomenclature connected with it.

II. Probably one main cause which so long retarded the work of fixing at the same time the arrangement and the names of plants, was the great number of minute and diversified particulars in the structure of each plant which such a process implied. The stalks, leaves, flowers, and fruits of vegetables, with their appendages, may vary in so many ways, that common language is quite insufficient to express clearly and precisely their resemblances and differences. Hence botany required not only a fixed system of *names* of plants, but also an artificial system of phrases fitted to *describe* their parts: not only a *Nomenclature*, but also a *Terminology*. The Terminology was, in fact, an instrument indispensably requisite in giving fixity to the Nomenclature. The recognition of the kinds of plants must depend upon the exact comparison of their resemblances and differences; and to become a part of permanent science, this comparison must be recorded in words.

The formation of an exact descriptive language for botany was thus the first step in that systematic construction of the technical language of science, which is one of the main features

* For instance whether the *fagus* of the Latins be the beech or the chesnut.

† *Hist. Ind. Sci.*, iii. 281.

in the intellectual history of modern times. The ancient botanists, as Decandolle* says, did not make any attempt to select
terms of which the sense was rigorously determined ; and each
of them employed in his descriptions the words, metaphors, or
periphrases which his own genius suggested. In the History of
Botany†, I have noticed some of the persons who contributed
to this improvement. " Clusius," it is there stated, " first taught
botanists to describe well. He introduced exactitude, precision,
neatness, elegance, method: he says nothing superfluous ; he
omits nothing necessary." This task was further carried on by
Jung and Ray‡. In these authors we see the importance which
began to be attached to the exact definition of descriptive terms ;
for example, Ray quotes Jung's definition of *Caulis*, a stalk.

The improvement of descriptive language, and the formation
of schemes of classification of plants, went on gradually for some
time, and was much advanced by Tournefort. But at last
Linnæus embodied and followed out the convictions which had
gradually been accumulating in the breasts of botanists ; and by
remodelling throughout both the terminology and the nomenclature of botany, produced one of the greatest reforms which ever
took place in any science. He thus supplied a conspicuous
example of such a reform, and a most admirable model of a language, from which other sciences may gather great instruction.
I shall not here give any account of the terms and words introduced by Linnæus. They have been exemplified in the *History
of Science*§ ; and the principles which they involve I shall consider separately hereafter. I will only remind the reader that
the great simplification in *nomenclature* which was the result of
his labours, consisted in designating each kind of plant by a *binary*
term consisting of the name of the *genus* combined with that of
the *species :* an artifice seemingly obvious, but more convenient
in its results than could possibly have been anticipated.

Since Linnæus, the progress of Botanical Anatomy and of

---

* *Theor. Elem. de la Bot.*, p. 327.
† *Hist. Ind. Sci.*, iii. 289.   ‡ *Ib.*, 297 (about A.D. 1660).
§ *Ib.*, 307—311.

Descriptive Botany have led to the rejection of several inexact expressions, and to the adoption of several new terms, especially in describing the structure of the fruit and the parts of cryptogamous plants. Hedwig, Medikus, Necker, Desvaux, Mirbel, and especially Gærtner, Link, and Richard, have proposed several useful innovations, in these as in other parts of the subject; but the general mass of the words now current consists still, and will probably continue to consist, of the terms established by the Swedish Botanist*.

When it was seen that botany derived so great advantages from a systematic improvement of its language, it was natural that other sciences, and especially classificatory sciences, should endeavour to follow its example. This attempt was made in Mineralogy by Werner, and afterwards further pursued by Mohs. Werner's innovations in the descriptive language of Mineralogy were the result of great acuteness, an intimate acquaintance with minerals, and a most methodical spirit: and were in most respects great improvements upon previous practices. Yet the introduction of them into Mineralogy was far from regenerating that science, as Botany had been regenerated by the Linnæan reform. It would seem that the perpetual scrupulous attention to most minute differences, (as of lustre, colour, fracture,) the greater part of which are not really important, fetters the mind, rather than disciplines or arms it for generalization. Cuvier has remarked† that Werner, after his first *Essay on the Characters of Minerals*, wrote little; as if he had been afraid of using the system which he had created, and desirous of escaping from the chains which he had imposed upon others. And he justly adds, that Werner dwelt least, in his descriptions, upon that which is really the most important feature of all, the crystalline structure. This, which is truly a definite character, like those of Botany, does, when it can be clearly discerned, determine the place of the mineral in a system. This, therefore, is the character which, of all others, ought to be most carefully expressed by an appropriate language. This task, hardly begun by Werner, has since been fully executed by others, especially by Romé de l'Isle,

* DECANDOLLE, *Th. Elem.*, p. 307.   † *Eloges*, ii. 314.

Haüy, and Mohs. All the forms of crystals can be described
in the most precise manner by the aid of the labours of these
writers and their successors. But there is one circumstance
well worthy our notice in these descriptions. It is found that
the language in which they can best be conveyed is not that of
words, but of *symbols*. The relations of space which are involved
in the forms of crystalline bodies, though perfectly definite, are so
complex and numerous, that they cannot be expressed, except in
the language of mathematics: and thus we have an extensive
and recondite branch of mathematical science, which is, in fact,
only a part of the terminology of the mineralogist.

The terminology of Mineralogy being thus reformed, an attempt was made to improve its nomenclature also, by following the
example of Botany. Professor Mohs was the proposer of this
innovation. The names framed by him were, however, not composed of two but of three elements, designating respectively the
Species, the Genus, and the Order*: thus he has such species as
*Rhombohedral Lime Haloide, Octahedral Fluor Haloide, Prismatic
Hal Baryte*. These names have not been generally adopted ; nor
is it likely that any names constructed on such a scheme will find
acceptance among mineralogists, till the higher divisions of the
system are found to have some definite character. We see no
real mineralogical significance in Mohs's Genera and Orders, and
hence we do not expect them to retain a permanent place in
the science.

The only systematic names which have hitherto been generally
admitted in Mineralogy, are those expressing the chemical constitution of the substance ; and these belong to a system of technical
terms different from any we have yet spoken of, namely to terms
formed by systematic modification.

III. The language of Chemistry was already, as we have seen,
tending to assume a systematic character, even under the reign of
the phlogiston theory. But when the oxygen theory succeeded to
the throne, it very fortunately happened that its supporters had the
courage and the foresight to undertake a completely new and systematic recoinage of the terms belonging to the science. The new

* *Hist. Ind. Sci.*, iii. 240.

nomenclature was constructed upon a principle hitherto hardly applied in science, but eminently commodious and fertile ; namely, the principle of indicating a modification of relations of elements, by a change in the termination of the word. Thus the new chemical school spoke of sulph*uric* and sulph*urous* acids ; of sulph*ates* and sulph*ites* of bases ; and of sulph*urets* of metals ; and in like manner, of phos*phoric* and phos*phorous* acids, of phos*phates*, phos*phites*, phos*phurets*. In this manner a nomenclature was produced, in which the very name of a substance indicated at once its constitution and place in the system.

The introduction of this chemical language can never cease to be considered one of the most important steps ever made in the improvement of technical terms ; and as a signal instance of the advantages which may result from artifices apparently trivial, if employed in a manner conformable to the laws of phenomena, and systematically pursued. It was, however, proved that this language, with all its merits, had some defects. The relations of elements in composition were discovered to be more numerous than the modes of expression which the terminations supplied. Besides the sulphurous and sulphuric acids, it appeared there were others ; these were called the *hyposulphurous* and *hyposulphuric :* but these names, though convenient, no longer implied, by their form, any definite relation. The compounds of Nitrogen and Oxygen are, in order, the *Protoxide*, the *Deutoxide* or *Binoxide*; *Hyponitrous* Acid, *Nitrous* Acid, and *Nitric* Acid. The nomenclature here ceases to be systematic. We have three oxides of Iron, of which we may call the first the *Protoxide*, but we cannot call the others the *Deutoxide* and *Tritoxide*, for by doing so we should convey a perfectly erroneous notion of the proportions of the elements. They are called the *Protoxide*, the *Black* Oxide, and the *Peroxide*. We are here thrown back upon terms quite unconnected with the system.

Other defects in the nomenclature arose from errors in the theory ; as for example the names of the muriatic, oxymuriatic, and hyperoxymuriatic acids ; which, after the establishment of the new theory of chlorine, were changed to *hydrochloric* acid, *chlorine*, and *chloric* acid.

Thus the chemical system of nomenclature, founded upon the oxygen theory, while it shows how much may be effected by a good and consistent scheme of terms, framed according to the real relations of objects, proves also that such a scheme can hardly be permanent in its original form, bnt will almost inevitably become imperfect and anomalous, in consequence of the accumulation of new facts, and the introduction of new generalizations. Still, we may venture to say that such a scheme does not, on this account, become worthless ; for it not only answers its purpose in the stage of scientific progress to which it belongs :—so far as it is not erroneous, or merely conventional, but really systematic and significant of truth, its terms can be translated at once into the language of any higher generalization which is afterwards arrived at. If terms express relations really ascertained to be true, they can never lose their value by any change of the received theory. They are like coins of pure metal, which, even when carried into a country which does not recognize the sovereign whose impress they bear, are still gladly received, and may, by the addition of an explanatory mark, continue part of the common currency of the country.

These two great instances of the reform of scientific language, in Botany and in Chemistry, are much the most important and instructive events of this kind which the history of science offers. It is not necessary to pursue our historical survey further. Our remaining Aphorisms respecting the Language of Science will be collected and illustrated indiscriminately, from the precepts and the examples of preceding philosophers of all periods.

We may, however, remark that Aphorisms III., IV., V., VI., VII., respect peculiarly the Formation of Technical Terms by the Appropriation of Common Words, while the remaining ones apply to the Formation of New Terms.

It does not appear possible to lay down a system of rules which may determine and regulate the construction of all technical terms, on all the occasions on which the progress of science makes them necessary or convenient. But if we can collect a few maxims such as have already offered themselves to the minds of philosophers, or such as may be justified by the instances by which we shall illustrate them, these maxims may avail to guide us in doubtful cases, and to prevent our aiming at advantages which are unattainable, or being disturbed by seeming imperfections which are really no evils. I shall therefore state such maxims of this kind as seem most sound and useful.

Aphorism III.

In framing scientific terms, the appropriation of old words is preferable to the invention of new ones.

This maxim is stated by Bacon in his usual striking manner. After mentioning *Metaphysic*, as one of the divisions of Natural Philosophy, he adds*: "Wherein I desire it may be conceived that I use the word metaphysic in a differing sense from that that is received: and in like manner I doubt not but it will easily appear to men of judgment that in this and other particulars, wheresoever my conception and notion may differ from the ancient, yet I am studious to keep the ancient terms. For, hoping well to deliver myself from mistaking by the order and perspicuous expressing of that I do propound ; I am otherwise zealous and affectionate to recede as little from antiquity, either in terms or opinions, as may stand with truth, and the proficience of knowledge. . . . To me, that do desire, as much as lieth in my pen, to ground a sociable intercourse between antiquity and proficience, it seemeth best to keep a way with antiquity *usque ad aras*; and therefore to retain the ancient terms, though I sometimes alter the uses and definitions ; according to the moderate proceeding in civil governments, when, although there be some alteration, yet that holdeth which Tacitus wisely noteth, *eadem magistratuum vocabula.*"

We have had before us a sufficient number of examples of scientific terms thus framed ; for they formed the first of three classes which we described in the First Aphorism. And we may again remark, that science, when she thus adopts terms which are in common use, always limits and fixes their meaning in a technical manner. We may also repeat here the warning

* *De Augm.*, Lib. iii. c. 4.

already given respecting terms of this kind, that they are peculiarly liable to mislead readers who do not take care to understand them in their technical instead of their common signification. *Force, momentum, inertia, impetus, vis viva,* are terms which are very useful, if we rigorously bear in mind the import which belongs to each of them in the best treatises on Mechanics; but if the reader content himself with conjecturing their meaning from the context, his knowledge will be confused and worthless.

In the application of this Third Aphorism, other rules are to be attended to, which 1 add.

Aphorism IV.

When common words are appropriated as technical terms, their meaning and relations in common use should be retained as far as can conveniently be done.

I WILL state an example in which this rule seems to be applicable. Mr. Davies Gilbert* has recently proposed the term *efficiency* to designate the work which a machine, according to the force exerted upon it, is capable of doing; the work being measured by the weight raised, and the space through which it is raised, jointly. The usual term employed among engineers for the work which a machine actually does, measured in the way just stated, is *duty.* But as there appears to be a little incongruity in calling that work *efficiency* which the machine *ought* to do, when we call that work *duty* which it really does, I have proposed to term these two quantities *theoretical efficiency* and *practical efficiency,* or *theoretical duty* and *practical duty.*

Since common words are often vague in their meaning, I add as a necessary accompaniment to the Third Aphorism the following :—

* *Phil. Trans.* 1827, p. 25.

Aphorism V.

When common words are appropriated as technical terms, their meaning may be modified, and must be rigorously fixed.

This is stated by Bacon in the above extract: " to retain the ancient terms, though I sometimes *alter the uses and definitions.*" The scientific use of the term is in all cases much more precise than the common use. The loose notions of *velocity* and *force* for instance, which are sufficient for the usual purposes of language, require to be fixed by exact measures when these are made terms in the science of Mechanics.

This scientific fixation of the meaning of words is to be looked upon as a matter of convention, although it is in reality often an inevitable result of the progress of science. *Momentum* is conventionally defined to be the product of the numbers expressing the weight and the velocity ; but then, it could be of no use in expressing the laws of motion if it were defined otherwise.

Hence it is no valid objection to a scientific term that the word in common language does not mean exactly the same as in its common use. It is no sufficient reason against the use of the term *acid* for a class of bodies, that all the substances belonging to this class are not sour. We have seen that a *trapezium* is used in geometry for any four-sided figure, though originally it meant a figure with two opposite sides parallel and the two others equal. A certain stratum which lies below the chalk is termed by English geologists *the green sand.* It has sometimes been objected to this denomination, that the stratum has very frequently no tinge of green, and that it is often composed of lime with little or no sand. Yet the term is a good technical term in spite of these apparent improprieties ; so long as it is carefully applied to that stratum which is geologically equivalent to the greenish sandy bed to which the appellation was originally applied.

When it appeared that *geometry* would have to be employed as much at least about the heavens as the earth, Plato exclaimed against the folly of calling the science by such a name ; since the word signifies " earth-measuring ;" yet the word *geometry* has retained its place and answered its purpose perfectly well up to the present day.

But though the meaning of the term may be modified or extended, it must be rigorously fixed when it is appropriated to science. This process is most abundantly exemplified by the terminology of Natural History, and especially of Botany, in which each term has a most precise meaning assigned to it. Thus Linnæus established exact distinctions between *fasciculus, capitulum, racemus, thyrsus, paniculus, spica, amentum, corymbus, umbella, cyma, verticillus;* or, in the language of English Botanists, *a tuft, a head, a cluster, a bunch, a panicle, a spike, a catkin, a corymb, an umbel, a cyme, a whorl.* And it has since been laid down as a rule*, that each organ ought to have a separate and appropriate name; so that the term *leaf,* for instance, shall never be applied to *a leaflet, a bractea,* or *a sepal* of the calyx.

Botanists have not been content with fixing the meaning of their terms by verbal definition, but have also illustrated them by figures, which address the eye. Of these, as excellent modern examples, may be mentioned those which occur in the works of Mirbel†, and Lindley‡.

Aphorism VI.

When common words are appropriated as technical terms, this must be done so that they are not ambiguous in their application.

An example will explain this maxim. The conditions of a body, as a solid, a liquid, and an air, have been distinguished as different *forms* of the body. But the word *form,* as applied to bodies, has other meanings; so that if we were to inquire in *what form* water exists in a snow-cloud, it might be doubted whether the forms of crystallization were meant, or the different forms of ice, water, and vapour. Hence I have proposed§ to reject the term *form* in such cases, and to speak of the different *consistence* of a body in these conditions. The term *consistence* is usually applied to conditions between solid and fluid; and may without

---

* Decandolle, *Theor. El.,* 328. † *Elemens de Botanique.*
‡ *Elements of Botany.* § *Hist. Ind. Sci.,* iii.

effort be extended to those limiting conditions. And though it may appear more harsh to extend the term *consistence* to the state of air, it may be justified by what has been said in speaking of Aphorism V.

I may notice another example of the necessity of avoiding ambiguous words. A philosopher who makes method his study, would naturally be termed a *methodist*; but unluckily this word is already appropriated to a religious sect: and hence we could hardly venture to speak of Cæsalpinus, Ray, Morison, Rivinus, Tournefort, Linnæus, and their successors, as *botanical methodists*. Again, by this maxim, we are almost debarred from using the term *physician* for a cultivator of the science of physics, because it already signifies a practiser of physic. We might, perhaps, still use *physician* as the equivalent of the French *physicien*, in virtue of Aphorism V.; but probably it would be better to form a new word. Thus we may say, that while the Naturalist employs principally the ideas of resemblance and life, the *Physicist* proceeds upon the ideas of force, matter, and the properties of matter.

Whatever may be thought of this proposal, the maxim which it implies is frequently useful. It is this.

Aphorism VII.

It is better to form new words as technical terms, than to employ old ones in which the last three Aphorisms cannot be complied with.

The principal inconvenience attending the employment of new words constructed expressly for the use of science, is the difficulty of effectually introducing them. Readers will not readily take the trouble to learn the meaning of a word, in which the memory is not assisted by some obvious suggestion connected with the common use of language. When this difficulty is overcome, the new word is better than one merely appropriated; since it is more secure from vagueness and confusion. And in cases where the inconveniences belonging to a scientific use of common words become great and inevitable, a new word must be framed and introduced.

The Maxims which belong to the construction of such words will be stated hereafter; but I may notice an instance or two tending to show the necessity of the Maxim now before us.

The word *Force* has been appropriated in the science of Mechanics in two senses : as indicating the cause of motion ; and again, as expressing certain measures of the effects of this cause, in the phrases *accelerating force* and *moving force*. Hence we might have occasion to speak of the accelerating or moving force *of* a certain *force ;* for instance, if we were to say that the centre of force which governs the motions of the planets resides in the sun ; and that the accelerating force *of* this *force* varies only with the distance, but its moving force varies as the product of the mass of the sun and the planet. This is a harsh and incongruous mode of expression ; and might have been avoided, if, instead of *accelerating force* and *moving force*, single abstract terms had been introduced by Newton : if, for instance, he had said that the velocity generated in a second measures the *accelerativity* of the force which produces it, and the momentum produced in a second measures the *motivity* of the force.

The science which treats of heat has hitherto had no special designation : treatises upon it have generally been termed treatises *On Heat*. But this practice of employing the same term to denote the property and the science which treats of it, is awkward and often ambiguous. And it is further attended with this inconvenience, that we have no adjective derived from the name of the science, as we have in other cases, when we speak of acoustical experiments and optical theories. This inconvenience has led various persons to suggest names for the Science of Heat. M. Le Comte terms it *Thermology*. In the History of the Sciences, I have named it *Thermotics*, which appears to me to agree better with the analogy of the names of other corresponding sciences, *Acoustics* and *Optics*.

Electricity is in the same condition as heat ; having only one word to express the property and the science. M. Le Comte proposes *Electrology :* for the same reason as before, I should conceive *Electrics* more agreeable to analogy. The coincidence of the word with the plural of Electric would not give rise to ambiguity; for *Electrics*, taken as the name of a science, would be singular, like *Optics* and *Mechanics*. But a term offers itself to express *common* or *machine Electrics*, which appears worthy of admission, though involving a theoretical view. The received doctrine of the difference between voltaic and common electricity is, that in the former case the fluid must be considered as in motion, in the latter as at rest. The science which treats of the former class of subjects is commonly termed *Electrodynamics*, which obviously suggests the name *Electrostatics* for the latter.

The subject of the Tides is, in like manner, destitute of any name which designates the science concerned about it. I have ventured to employ the term *Tidology*, having been much engaged in tidological researches.

Many persons possess a peculiarity of vision, which disables them from distinguishing certain colours. On examining many such cases, we find that in all such persons the peculiarities are the same; all of them confounding scarlet with green, and pink with blue. Hence they form a class, which, for the convenience of physiologists and others, ought to have a fixed designation. Instead of calling them, as has usually been done, " persons having a peculiarity of vision," we might take a Greek term implying this meaning, and term them *Idiopts*.

But my business at present is not to speak of the selection of new terms when they are introduced, but to illustrate the maxim that the necessity for their introduction often arises. The construction of new terms will be treated of subsequently.

APHORISM VIII.

Terms must be constructed and appropriated so as to be fitted to enunciate simply and clearly true general propositions.

THIS Aphorism may be considered as the fundamental principle and supreme rule of all scientific terminology. It is asserted by Cuvier, speaking of a particular case. Thus he says* of

---

* *Regne Animal,* Introd. viii.

Gmelin, that by placing the lamantin in the genus of morses, and the siren in the genus of eels, he had rendered every general proposition respecting the organization of those genera impossible.

The maxim is true of words appropriated as well as invented, and applies equally to the mathematical, chemical, and classificatory sciences. With regard to most of these, and especially the two former classes, it has been abundantly exemplified already, in what has previously been said, and in the History of the Sciences. For we have there had to notice many technical terms, with the occasions of their introduction; and all these occasions have involved the intention of expressing in a convenient manner some truth or supposed truth. The terms of Astronomy were adopted for the purpose of stating and reasoning upon the relations of the celestial motions, according to the doctrine of the sphere, and the other laws which were discovered by astronomers. The few technical terms which belong to Mechanics, *force, velocity, momentum, inertia,* &c., were employed from the first with a view to the expression of the laws of motion and of rest; and were, in the end, limited so as truly and simply to express those laws when they were fully ascertained. In Chemistry, the term *phlogiston* was useful, as has been shown in the History, in classing together processes which really are of the same nature; and the nomenclature of the *oxygen* theory was still preferable, because it enabled the chemist to express a still greater number of general truths.

To the connexion here asserted, of theory and nomenclature, we have the testimony of the author of the oxygen theory. In the Preface to his *Chemistry*, Lavoisier says:—" Thus while I thought myself employed only in forming a Nomenclature, and while I proposed to myself nothing more than to improve the chemical language, my work transformed itself by degrees, without my being able to prevent it, into a Treatise on the Elements of Chemistry." And he then proceeds to show how this happened.

It is, however, mainly through the progress of Natural History in modern times, that philosophers have been led to see the importance and necessity of new terms in expressing new truths. Thus

Harvey, in the Preface to his work on Generation, says:—" Be not offended if in setting out the History of the Egg I make use of a new method, and sometimes of unusual terms. For as they which find out a new plantation and new shores call them by names of their own coining, which posterity afterwards accepts and receives, so those that find out new secrets have good title to their compellation. And here, methinks, I hear Galen advising: If we consent in the things, contend not about the words."

The Nomenclature which answers the purposes of Natural History is a systematic nomenclature, and will be further considered under the next Aphorism. But we may remark, that the Aphorism now before us governs the use of words, not in science only, but in common language also. Are we to apply the name *fish* to animals of the whale kind? The answer is determined by our present rule: we are to do so, or not, accordingly as we can best express true propositions. If we are speaking of the internal structure and physiology of the animal, we must not call them *fish*; for in these respects they deviate widely from fishes: they have warm blood, and produce and suckle their young as land quadrupeds do. But this would not prevent our speaking of the *whale-fishery*, and calling such animals *fish* on all occasion connected with this employment; for the relations thus arising depend upon the animal's living in the water, and being caught in a manner similar to other fishes. A plea that human laws which mention fish do not apply to whales, would be rejected at once by an intelligent judge.

Aphorism IX.

In the Classificatory Sciences, a systematic Nomenclature is necessary; and the System and the Nomenclature are each essential to the utility of the other.

The inconveniences arising from the want of a good Nomenclature were long felt in Botany, and are still felt in Mineralogy. The attempts to remedy them by *Synonymies* are very ineffective,

f 2

for such comparisons of synonymes do not supply a systematic nomenclature ; and such a one alone can enable us to state general
truths respecting the objects of which the classificatory sciences
treat.   The *system* and the *names* ought to be introduced together;
for the former is a collection of asserted analogies and resemblances, for which the latter provide simple and permanent expressions.   Hence it has repeatedly occurred in the progress of
Natural History, that good systems did not take root, or produce
any lasting effect among naturalists, because they were not accompanied by a corresponding nomenclature.   In this way, as we have
already noticed, the excellent botanical system of Cæsalpinus was
without immediate effect upon the science.   The work of Willoughby, as Cuvier says*, forms an epoch, and a happy epoch in
Ichthyology ; yet because Willoughby had no nomenclature of his
own, and no fixed names for his genera, his immediate influence
was not great.   Again, in speaking of Schlotheim's work containing representations of fossil vegetables, M. Adolphe Brongniart observes† that the figures and descriptions are so good, that
if the author had established a nomenclature for the objects he
describes, his work would have become the basis of all succeeding
labours on the same subject.

As additional examples of cases in which the improvement of
classification, in recent times, has led philosophers to propose new
names, I may mention the term *Pœcilite*, proposed by Mr. Conybeare to designate the group of strata which lies below the oolites
and lias, including the new red or variegated sandstone, with the
keuper above, and the magnesian limestone below it.   Again, the
transition districts of our island have recently been reduced to
system by Professor Sedgwick and Mr. Murchison ; and this step
has been marked by the terms *Cambrian* system, and *Silurian*
system, applied to the two great groups of formations which they
have respectively examined, and by several other names of the
subordinate members of these formations.

Thus system and nomenclature are each essential to the other.
Without nomenclature, the system is not permanently incor-

---

* *Hist. des Poissons*, Pref.   † *Prodrom. Veg. Foss.*, p. 3.

porated into the general body of knowledge, and made an instrument of future progress. Without system, the names cannot express general truths, and contain no reason why they should be employed in preference to any other names.

This has been generally acknowledged by the most philosophical naturalists of modern times. Thus Linnæus begins that part of his Botanical Philosophy in which Names are treated of, by stating that the foundation of botany is twofold, *Disposition* and *Denomination;* and he adds this Latin line,

Nomina si nescis perit et cognitio rerum.

And Cuvier, in the Preface to his *Animal Kingdom*, explains, in a very striking manner, how the attempt to connect zoology with anatomy led him, at the same time, to reform the classifications, and to correct the nomenclature of preceding zoologists.

I have stated that in mineralogy we are still destitute of a good nomenclature generally current. From what has now been said, it will be seen that it may be very far from easy to supply this defect, since we have, as yet, no generally received system of mineralogical classification. Till we know what are really different species of minerals, and in what larger groups these species can be arranged, so as to have common properties, we shall never obtain a permanent mineralogical nomenclature. Thus *Leucocyclite* and *Tesselite* are minerals previously confounded with apophyllite, which Sir John Herschel and Sir David Brewster distinguished by those names, in consequence of certain optical properties which they exhibit. But are these properties definite distinctions? and are there any external differences corresponding to them? If not, can we consider them as separate species? and if not separate species, ought they to have separate names? In like manner, we might ask if *Augite* and *Hornblende* are really the same species, as Gustavus Rose has maintained? if *Diallage* and *Hypersthene* are not definitely distinguished, which has been asserted by Kobell? Till such questions are settled, we cannot have a fixed nomenclature in mineralogy. What appears the best course to follow in the present state of the science, I shall consider when we come to speak of the form of technical terms.

I may, however, notice here that the main forms of systematic nomenclature are two:—terms which are produced by combining words of higher and lower generality, as the binary names, consisting of the name of the genus and the species, generally employed by natural historians since the time of Linnæus;—and terms in which some relation of things is indicated by a change in the form of the word, for example, an alteration of its termination, of which kind of nomenclature we have a conspicuous example in the modern chemistry.

Aphorism X.

New terms and changes of terms, which are not needed in order to express truth, are to be avoided.

As the Seventh Aphorism asserted that novelties in language may be and ought to be introduced, when they aid the enunciation of truths, we now declare that they are not admissible in any other case. New terms and new systems of terms are not to be introduced, for example, in virtue of their own neatness or symmetry, or other merits, if there is no occasion for their use.

I may mention, as an old example of a superfluous attempt of this kind, an occurrence in the history of astronomy. In 1628 John Bayer and Julius Schiller devised a *Cœlum Christianum*, in which the common names of the planets, &c., were replaced by those of Adam, Moses, and the Patriarchs. The twelve Signs became the twelve Apostles, and the constellations became sacred places and things. Peireskius, who had to pronounce upon the value of this proposal, praised the piety of the inventors, but did not approve, he said*, the design of perverting and confounding whatever of celestial information from the period of the earliest memory is found in books.

Nor are slight anomalies in the existing language of science sufficient ground for a change, if they do not seriously interfere with the expression of our knowledge. Thus Linnæus says† that a fair generic name is not to be exchanged for another though apter one: and‡ if we separate an old genus into several,

---

* Gassendi, *Vita Peireskii*, 300.   † *Phil. Bot.*, 246.   § *Ib.*, 247.

we must try to find names for them among the synonyms which describe the old genus. This maxim excludes the restoration of ancient names long disused, no less than the needless in vention of new ones. Linnæus lays down this rule*; and adds, that the botanists of the sixteenth century well nigh ruined botany by their anxiety to recover the ancient names of plants. In like manner Cuvier† laments it as a misfortune, that he has had to introduce many new names; and declares earnestly that he has taken great pains to preserve those of his predecessors.

The great bulk which the synonymy of botany and of mineralogy have attained, shows us that this maxim has not been universally attended to. In these cases, however, the multiplication of different names for the same kind of object has arisen in general from ignorance of the identity of it under different circumstances, or from the want of a system which might assign to it its proper place. But there are other instances, in which the multiplication of names has arisen not from defect, but from excess, of the spirit of system. The love which speculative men bear towards symmetry and completeness is constantly at work, to make them create systems of classification more regular and more perfect than can be verified by the facts: and as good systems are closely connected with a good nomenclature, systems thus erroneous and superfluous lead to a nomenclature which is prejudicial to science. For although such a nomenclature is finally expelled, when it is found not to aid us in expressing the true laws of nature, it may obtain some temporary sway, during which, and even afterwards, it may be a source of much confusion.

We have a conspicuous example of such a result in the geological nomenclature of Werner and his school. Thus it was assumed, in Werner's system, that his *First, Second,* and *Third Flötz Limestone,* his *Old* and *New Red Sandstone,* were universal formations; and geologists looked upon it as their business to detect these strata in other countries. Names were thus assigned to the rocks of various parts of Europe, which created immense perplexity before they were again ejected. The geological terms which now prevail, for instance, those of Smith, are for the most

* *Phil. Bot.,* 248.   † *Regne Anim.,* Pref. p. **xvi.**

part not systematic, but are borrowed from accidents, as localities, or popular names ; as *Oxford Clay* and *Cornbrash* ; and hence they are not liable to be thrust out on a change of system.   On the other hand we do not find sufficient reason to accept the system of names of strata proposed by Mr. Conybeare in the *Introduction to the Geology of England and Wales*, according to which the *Carboniferous Rocks* are the *Medial Order*,—having above them the *Supermedial Order* (*New Red Sand, Oolites and Chalk*), and above these the *Superior Order* (*Tertiary Rocks*); and again,—having below, the *Submedial Order* (the *Transition Rocks*), and the *Inferior Order* (*MicaSlate, Gneiss, Granite*).   For though these names have long been proposed, it does not appear that they are useful in enunciating geological truths.   We may, it would seem, pronounce the same judgment respecting the system of geological names proposed by M. Alexander Brongniart, in his *Tableau des Terrains qui composent l'écorce du Globe*.   He divides these strata into nine classes, which he terms *Terrains Alluviens, Lysiens, Pyrogenes, Clysmiens, Yzemiens, Hemilysiens, Agalysiens, Plutoniques, Vulcaniques*.   These classes are again variously subdivided : thus the Terrains Yzemiens are *Thalassiques, Pelagiques*, and *Abyssiques ;* and the Abyssiques are subdivided into *Lias, Keuper, Conchiliens, Pœciliens, Peneens, Rudimentaires, Entritiques, Houillers, Carbonifers* and *Gres Rouge Ancien*. Scarcely any amount of new truths would induce geologists to burthen themselves at once with this enormous system of new names : but in fact, it is evident that any portion of truth, which any author can have brought to light, may be conveyed by means of a much simpler apparatus.   Such a nomenclature carries its condemnation on its own face.

Nearly the same may be said of the systematic nomenclature proposed for mineralogy by Professor Mohs.   Even if all his Genera be really natural groups, (a doctrine which we can have no confidence in till they are confirmed by the evidence of chemistry,) there is no necessity to make so great a change in the received names of minerals.   His proceeding in this respect, so different from the temperance of Linnæus and Cuvier, has probably ensured a speedy oblivion to this part of his system.

In crystallography, on the other hand, in which Mohs's improvements have been very valuable, there are several terms introduced by him, as *rhombohedron, scalenohedron, hemihedral, systems* of crystallization, which will probably be a permanent portion of the language of science.

I may remark, in general, that the only persons who succeed in making great alterations in the language of science, are not those who make names arbitrarily and as an exercise of ingenuity, but those who have much new knowledge to communicate; so that the vehicle is commended to general reception by the value of what it contains. It is only eminent discoverers to whom the authority is conceded of introducing a new system of names; just as it is only the highest authority in the state which has the power of putting a new coinage in circulation.

I will here quote some judicious remarks of Mr. Howard, which fall partly under this Aphorism, and partly under some which follow. He had proposed, as names for the kinds of clouds, the following : *Cirrus, Cirrocumulus, Cirrostratus, Cumulostratus, Cumulus, Nimbus, Stratus.* In an abridgment of his views, given in the Supplement to the *Encyclopædia Britannica*, English names were proposed as the equivalents of these; *Curlcloud, Sondercloud, Wanecloud, Twaincloud, Stackencloud, Raincloud, Fallcloud.* Upon these Mr. Howard observes : " I mention these, in order to have the opportunity of saying that I do not adopt them. The names for the clouds which I deduced from the Latin, are but seven in number, and very easy to remember. They were intended as *arbitrary terms* for the *structure* of clouds, and the meaning of them was carefully fixed by a definition. The observer having once made himself master of this, was able to apply the term with correctness, after a little experience, to the subject under all its varieties of form, colour, or position. The new names, if meant to be another set of arbitrary terms, are superfluous ; if intended to convey in themselves an explanation in English, they fail in this, by applying to some part or circumstance only of the definition ; the *whole* of which must be kept in view to study the subject with success. To take for an example the first of the modifications. The term *cirrus* very readily takes an abstract meaning, equally applicable to the rectilinear as to the flexuous forms of the subject. But the name of *curl-cloud* will not, without some violence to its *obvious sense*, acquire this more extensive one: and will therefore be apt to mislead the reader rather than further his progress. Others of these names are as devoid of a meaning obvious to the English reader, as the Latin terms themselves. But the principal objection to English or any other local terms, remains to be stated. They take away from the nomenclature its general advantage of constituting, as far as it goes, an universal language, by means of which the intelligent of every country may convey to each other their ideas without the necessity of translation."

I here adduce these as examples of the arguments against changing an established nomenclature. As grounds of selecting a new one, they may be taken into account hereafter.

Aphorism XI.

Terms which imply theoretical views are admissible, as far as the theory is proved.

It is not unfrequently stated that the circumstances from which the names employed in science borrow their meaning, ought to be facts and not theories. But such a recommendation implies a belief that facts are rigorously distinguished from theories and directly opposed to them; which belief, we have repeatedly seen, is unfounded. When theories are firmly established, they become facts; and names founded on such theoretical views are unexceptionable. If we speak of the *minor axis* of Jupiter's *orbit*, or of his *density*, or of *the angle of refraction*, or *the length of an undulation* of red light, we assume certain theories; but inasmuch as the theories are now the inevitable interpretation of ascertained facts, we can have no better terms to designate the conceptions thus referred to. And hence the rule which we must follow is, not that our terms must involve no theory, but that they imply the theory only in that sense in which it is the interpretation of the facts.

For example, the term *polarization* of light was objected to, as involving a theory. Perhaps the term was at first suggested by conceiving light to consist of particles having poles turned in a particular manner. But among intelligent speculators, the notion of polarization soon reduced itself to the simple conception of opposite properties in opposite positions, which is a bare statement of the fact: and the term being understood to have this meaning, is a perfectly good term, and indeed the best which we can imagine for designating what is intended.

I need hardly add the caution, that names involving theoretical views not in accordance with facts are to be rejected. The following instances exemplify both the positive and the negative application of this maxim.

The distinction of *primary* and *secondary* rocks in geology was founded upon a theory; namely, that those which do not contain any organic remains were first deposited, and afterwards, those which contain plants and animals. But this theory was insecure from the first. The difficulty of making the separation which it implied, led to the introduction of a class of *transition* rocks. And the recent researches of geologists lead them to the conclusion, that those rocks which are termed *primary*, may be the newest, not the oldest, productions of nature.

In order to avoid this incongruity, other terms have been proposed as substitutes for these. Mr. Lyell remarks*, that granite, gneiss, and the like, form a class which should be designated by a common name; which name should not be of chronological import. He proposes *hypogene*, signifying "nether-formed;" and thus he adopts the theory that they have not assumed their present form and structure at the surface, but determines nothing of the period when they were produced.

These hypogene rocks, again, he divides into unstratified or *plutonic,* and altered, stratified, or *metamorphic;* the latter term implying the hypothesis that the stratified rocks to which it is applied have been altered, by the effect of fire or otherwise, since they were deposited. That fossiliferous strata, in some cases at least, have undergone such a change, is demonstrable from facts†.

The modern nomenclature of chemistry implies the oxygen

---

* *Princ. Geol.,* iv. 386.   † *Elem. Geol.,* p. 17.

theory of chemistry. Hence it has sometimes been objected to. Thus Davy, in speaking of the Lavoisierian nomenclature, makes the following remarks, which, however plausible they may sound, will be found to be utterly erroneous[*]. " Simplicity and precision ought to be the characteristics of a scientific nomenclature : words should signify *things*, or the *analogies* of things, and not *opinions*. . . . A substance in one age supposed to be simple, in another is proved to be compound, and *vice versâ*. A theoretical nomenclature is liable to continual alterations : *oxygenated muriatic acid* is as improper a term as *dephlogisticated marine acid*. Every school believes itself to be in the right : and if every school assumes to itself the liberty of altering the names of chemical substances in consequence of *new ideas* of their composition, there can be no permanency in the language of the science ; it must always be confused and uncertain. Bodies which are *similar* to each other should always be classed together ; and there is a presumption that their composition is *analogous*. *Metals, earths, alkalis*, are appropriate names for the bodies they represent, and independent of all speculation : whereas *oxides, sulphurets*, and *muriates* are terms founded upon opinions of the composition of bodies, some of which have been already found erroneous. The least dangerous mode of giving a systematic form to a language seems to be to signify the analogies of substances by some common sign affixed to the beginning or the termination of the word. Thus as the metals have been distinguished by a termination in *um*, as *aurum*, so their calciform or oxidated state might have been denoted by a termination in *a*, as *aura :* and no progress, however great, in the science could render it necessary that such a mode of appellation should be changed."

These remarks are founded upon distinctions which have no real existence. We cannot separate *things* from their *properties*, nor can we consider their properties and analogies in any other way than by having *opinions* about them. By contrasting *analogies* with *opinions*, it might appear as if the author maintained that there were certain analogies about which there was no room for erroneous opinions. Yet the analogies of chemical compounds,

* *Elements of Chem. Phil.*, p. 46.

are, in fact, those points which have been most the subject of difference of opinion, and on which the revolutions of theories have have most changed men's views. As an example of analogies which are still recognized under alterations of theory, the writer gives the relation of a metal to its oxide or calciform state. But this analogy of metallic oxides, as Red Copper or Iron Ore, to Calx, or burnt lime, is very far from being self-evident;—so far indeed, that the recognition of the analogy was a great step in chemical *theory*. The terms which he quotes, *oxygenated muriatic acid* (and the same may be said of *dephlogisticated marine acid*,) if improper, are so not because they involve theory, but because they involve false theory;—not because those who framed them did not endeavour to express analogies, but because they expressed analogies about which they were mistaken. Unconnected names, as *metals, earths, alkalis*, are good as the *basis* of a systematic nomenclature, but they are not substitutes for such a nomenclature. A systematic nomenclature is an instrument of great utility and power, as the modern history of chemistry has shown. It would be highly unphilosophical to reject the use of such an instrument, because, in the course of the revolutions of science, we may have to modify, or even to remodel it altogether. Its utility is not by that means destroyed. It has retained, transmitted, and enabled us to reason upon, the doctrines of the earlier theory, so far as they are true; and when this theory is absorbed into a more comprehensive one, (for this, and not its refutation, is the end of a theory so far as it is true,) the nomenclature is easily translated into that which the new theory introduces. We have seen, in the history of astronomy, how valuable the theory of *epicycles* was, in its time: the nomenclature of the relations of a planet's orbit, which that theory introduced, was one of Kepler's resources in discovering the *elliptical* theory; and, though now superseded, is still readily intelligible to astronomers.

This is not the place to discuss the reasons for the *form* of scientific terms; otherwise we might ask, in reference to the objections to the Lavoisierian nomenclature, if such forms as *aurum* and *aura* are good to represent the absence or presence of oxygen, why such forms as *sulphite* and *sulphate* are not equally good to represent the presence of what we may call a smaller or larger dose of oxygen, so long as the oxygen theory is admitted in its present form; and to indicate still the difference of the same substances, if under any change of theory it should come to be interpreted in a new manner.

But I do not now dwell upon such arguments, my object in this place being to show that terms involving theory are not only allowable, if understood so far as the theory is proved, but of great value, and indeed of indispensable use, in science. The objection to them is inconsistent with the objects of science. If, after all that has been done in chemistry or any other science, we have arrived at no solid knowledge, no permanent truth;—if all that we believe now may be proved to be false tomorrow;—then indeed our opinions and theories are corruptible elements, on which it would be unwise to rest any thing important, and which we might wish to exclude, even from our names. But if our knowledge has no more security than this, we can find no reason why we should wish to have names of things, since the names are needed mainly that we may reason upon and increase our knowledge such as it is. If we are condemned to endless alternations of varying opinions, then, no doubt, our theoretical terms may be a source of confusion; but then, where would be the advantage of their being otherwise? what would be the value of words which should express in a more precise manner opinions equally fleeting? It will perhaps be said, our terms must express facts, not theories: but of this distinction so applied we have repeatedly shown the futility. Theories firmly established are facts. Is it not a fact that the rusting of iron arises from the metal combining with the oxygen of the atmosphere? Is it not a fact that a combination of oxygen and hydrogen produces water? That our terms should express such facts, is precisely what we are here inculcating.

Our examination of the history of science has led us to a view very different from that which represents it as consisting in the succession of hostile opinions. It is, on the contrary, a progress, in which each step is recognized and employed in the succeeding one. Every theory, so far as it is true, (and all that have prevailed extensively and long, contain a large portion of truth,) is taken up into the theory which succeeds and seems to expel it. All the narrower inductions of the first are included in the more comprehensive generalizations of the second. And this is performed mainly by means of such terms as we are now considering;—terms involving the previous theory. It is by means of such terms, that the truths at first ascertained become so familiar and manageable, that they can be employed as elementary facts in the formation of higher inductions.

These principles must be applied also, though with great caution, and in a temperate manner, even to descriptive language. Thus the mode of describing the forms of crystals adopted by Werner and Romé de l'Isle was to consider an original form, from which other forms are derived by *truncations* of the edges and the angles. Haüy's method of describing the same forms, was to consider them as built up of rows of small solids, the angles being determined by the *decrements* of these rows. Both these methods of description involve hypothetical views ; and the last was intended to rest on a true physical theory of the constitution of crystals. Both hypotheses are doubtful or false : yet both these methods are good as modes of description: nor is Haüy's terminology vitiated, if we suppose (as in fact we must suppose in many instances,) that crystalline bodies are not really made up of such small solids. The mode of describing an octahedron of fluor spar, as derived from the cube, by decrements of one row on all the edges, would still be proper and useful as a description, whatever judgment we should form of the material structure of the body. But then, we must consider the solids which are thus introduced into the description as merely hypothetical geometrical forms, serving to determine the angles of the faces. It is in this way alone that Haüy's nomenclature can now be retained.

In like manner we may admit theoretical views into the descriptive phraseology of other parts of Natural History: and the theoretical terms will replace the obvious images, in proportion as the theory is generally accepted and familiarly applied. For example, in speaking of the Honeysuckle, we may say that the upper leaves are *perfoliate*, meaning that a single orbicular leaf is perforated by the stalk or threaded upon it. Here is an image which sufficiently conveys the notion of the form. But it is now generally recognized that this apparent single leaf is, in fact, two opposite leaves joined together at their bases. If this were doubted, it may be proved by comparing the upper leaves with the lower, which are really separate and opposite. Hence the term *connate* is applied to these conjoined opposite leaves, implying that they grow together; or they are called *connato-perfoliate*. Again; formerly the corolla was called *monopetalous* or *polypetalous*, as it consisted of one part or of several : but it is now agreed among botanists that those corollas which appear to consist of a single part, are, in fact, composed of several soldered together; hence the term *gamopetalous* is now employed (by Decandolle and his followers) instead of monopetalous *.

In this way the language of natural history not only expresses, but inevitably implies, general laws of nature; and words are thus fitted to aid the progress of knowledge in this, as in other provinces of science.

Aphorism XII.

If terms are systematically good, they are not to be rejected because they are etymologically inaccurate.

Terms belonging to a system are defined, not by the meaning of their radical words, but by their place in the system. That they should be appropriate in their signification, aids the processes of introducing and remembering them, and should therefore be carefully attended to by those who invent and establish them; but this once done, no objections founded upon their etymological import are of any material weight. We find no inconvenience in the circumstance that *geometry* means the measuring of the earth, that the name *porphyry* is applied to many rocks which have no fiery spots, as the word implies, and *oolite* to strata which have no roelike structure. In like manner, if the term *pœcilite*

* On this subject, see Illiger, *Versuch einer Systematischen Vollständigen Terminologie für das Thierreich und Pflanzenreich.* (1810.) Decandolle, *Theorie Elementaire de la Botanique.*

were already generally received, as the name of a certain group of strata, it would be no valid ground for quarreling with it, that this group was not always variegated in colour, or that other groups were equally variegated: although undoubtedly in *introducing* such a term, care should be taken to make it as distinctive as possible. It often happens, as we have seen, that by the natural progress of changes in language, a word is steadily confirmed in a sense quite different from its etymological import. But though we may accept such instances, we must not wantonly attempt to imitate them. I say, not wantonly: for if the progress of scientific identification compel us to follow any class of objects into circumstances where the derivation of the term is inapplicable, we may still consider the term as an unmeaning sound, or rather an historical symbol, expressing a certain member of our system. Thus if, in following the course of the *mountain* or *carboniferous* limestone, we find that in Ireland it does not form mountains nor contain coal, we should act unwisely in breaking down the nomenclature in which our systematic relations are already expressed, in order to gain, in a particular case, a propriety of language which has no scientific value.

All attempts to act upon the maxim opposite to this, and to make our scientific names properly descriptive of the objects, have failed and must fail. For the marks which really distinguish the natural classes of objects, are by no means obvious. The discovery of them is one of the most important steps in science; and when they are discovered, they are constantly liable to exceptions, because they do not contain the essential differences of the classes. The natural order *Umbellatæ*, in order to be a natural order, must contain some plants which have not umbels, as *Eryngium**. " In such cases," said Linnæus, " it is of small import what you *call* the order, if you take a proper series of plants, and give it some name which is clearly understood to apply to the plants you have associated." " I have," he adds, " followed the rule of borrowing the name *à fortiori*, from the principal feature."

The distinction of crystals into systems according to the degree of symmetry which obtains in them, has been explained elsewhere.

See *Hist. Ind. Sci.*, iii. 324.

Two of these systems, of which the relat on as to symmetry might be expressed by saying that one is *square pyramidal* and the other *oblong pyramidal*, or the first *square prismatic* and the second *oblong prismatic*, are termedby Mohs, the first, *Pyramidal*, and the second *Prismatic*. And it may be doubted whether it is worth while to invent other terms, though these are thus defective in characteristic significance. As an example of a needless rejection of old terms in virtue of a supposed impropriety in their meaning, I may mention the attempt made in the last edition of Haüy's Mineralogy, to substitute *autopside* and *heteropside* for *metallic* and *unmetallic*. It was supposed to be proved that all bodies have a metal for their basis ; and hence it was wished to avoid the term *unmetallic*. But the words *metallic* and *unmetallic* may mean that minerals *seem* metallic and unmetallic, just as well as if they contained the element *opside* to imply this seeming. The old names express all that the new express, and with more simplicity, and therefore should not be disturbed.

The maxim on which we are now insisting, that we are not to be too scrupulous about the etymology of scientific terms, may, at first sight, appear to be at variance with our Fourth Aphorism, that words used technically are to retain their common meaning as far as possible. But it must be recollected, that in the Fourth Aphorism we spoke of *common* words *appropriated* as technical terms ; we here speak of words *constructed* for scientific purposes. And although it is, perhaps, impossible to draw a broad line between these two classes of terms, still the rule of propriety may be stated thus : In technical terms, deviations from the usual meaning of words are bad in proportion as the words are more familiar in our own language. Thus we may apply the term *Cirrus* to a cloud composed of filaments, even if these filaments are straight ; but to call such a cloud a *Curl cloud* would be much more harsh.

Since the names of things, and of classes of things, when constructed so as to involve a description, are constantly liable to become bad, the natural classes shifting away from the descriptive marks thus prematurely and casually adopted, I venture to lay down the following maxim.

APHORISM XIII.

The fundamental terms of a system of Nomenclature may be conveniently borrowed from casual or arbitrary circumstances.

For instance, the names of plants, of minerals, and of geological strata, may be taken from the places where they occur conspicuously or in a distinct form; as *Parietaria, Parnassia, Chalcedony, Arragonite, Silurian* system, *Purbeck* limestone. These names may be considered as at first supplying standards of reference; for in order to ascertain whether any rock be *Purbeck* limestone, we might compare it with the rocks in the Isle of Purbeck. But this reference to a local standard is of authority only till the place of the object in the system, and its distinctive marks, are ascertained. It would not vitiate the above names, if it were found that the *Parnassia* does not grow on Parnassus; that *Chalcedony* is not found in Chalcedon; or even that *Arragonite* no longer occurs in Arragon; for it is now firmly established as a mineral species. Even in geology such a reference is arbitrary, and may be superseded, or at least modified, by a more systematic determination. *Alpine* limestone is no longer accepted as a satisfactory designation of a rock, now that we know the limestone of the Alps to be of various ages.

Again, names of persons, either casually connected with the object, or arbitrarily applied to it, may be employed as designations. This has been done most copiously in botany, as for example, *Nicotiana, Dahlia, Fuchsia, Jungermannia, Lonicera*. And Linnæus has laid down rules for restricting this mode of perpetuating the memory of men, in the names of plants. Those generic names, he says*, which have been constructed to preserve the memory of persons who have deserved well of botany, are to be religiously retained. This, he adds, is the sole and supreme reward of the botanist's labours, and must be carefully guarded and scrupulously bestowed, as an encouragement and an honour. Still more arbitrary are the terms borrowed from the names of the gods and goddesses, heroes and heroines of

* *Phil. Bot.*, 241.

g 2

ant quity, to designate new genera in those departments of natural history in which so many have been discovered in recent times as to weary out all attempts at descriptive nomenclature. Cuvier has countenanced this method. " I have had to frame many new names of genera and sub-genera," he says[*], " for the sub-genera which I have established wer⁰ so numerous and various, that the memory is not satisfied with numerical indications. These I have chosen either so as to indicate some character, or among the usual denominations, which I have latinized, or finally, after the example of Linnæus, among the names of mythology, which are in general agreeable to the ear, and which are far from being exhausted."

This mode of framing names from the names of persons to whom it was intended to do honour, has been employed also in the mathematical and chemical sciences; but such names have rarely obtained any permanence, except when they recorded an inventor or discoverer. Some of the constellations, indeed, have retained such appellations, as *Berenice's Hair*; and the new star which shone out in the time of Cæsar, would probably have retained the name given to it, of the *Julian Star*, if it had not disappeared again soon after. In the map of the Moon, almost all the parts have had such names imposed upon them by those who have constructed such maps, and these names have very properly been retained. But the names of new planets and satellites thus suggested have not been generally accepted; as the *Medicean* stars, the name employed by Galileo for the satellites of Jupiter, the *Georgium Sidus*, the appellation proposed by Herschel for Uranus when first discovered; Ceres *Ferdinandea*, the name which Piazzi wished to impose on the small planet Ceres. The names given to astronomical tables by the astronomers who constructed them have been most steadily adhered to, being indeed names of books, and not of natural objects. Thus there were the *Ilchanic*, the *Alphonsine*, the *Rudolphine*, the *Carolinian* Tables. Comets which have been ascertained to be periodical, have very properly had assigned to them the name of the person who established this point; and of these we have thus, *Halley's,*

* *Regne An.*, p. xvi.

Encke's, and *Gambart's Comets;* the latter is often unjustly called *Biela's* comet.

In the case of discoveries in science or inventions of apparatus, the name of the inventor is very properly employed as the designation. Thus we have the *Torricellian* Vacuum, the *Voltaic* Pile, *Fahrenheit's* Thermometer. And in the same manner with regard to laws of nature, we have *Kepler's* Laws, *Boyle* or *Mariotte's* law of the elasticity of air, *Huyghens's* law of double refraction, *Newton's* scale of colours. *Descartes'* law of refraction is an unjust appellation; for the discovery of the law of sines was made by Snell. In deductive mathematics, where the invention of a theorem is generally a more definite step than an induction, this mode of designation is more common, as *Demoivre's* Theorem, *Maclaurin's* Theorem, *Lagrange's* Theorem, *Eulerian* Integrals.

In the History of Science*, I have remarked that in the discovery of what is termed galvanism, Volta's office was of a higher and more philosophical kind than that of Galvani; and I have, on this account, urged the propriety of employing the term *voltaic,* rather than *galvanic* electricity. I may add that the elec tricity of the common machine is often placed in contrast with this, and appears to require an express name. Mr. Faraday calls it common, or machine electricity; but I think that *franklinic* electricity would form a more natural correspondence with *voltaic,* and would be well justified by Franklin's place in the history of that part of the subject.

Aphorism XIV.

In forming a Terminology, words may be invented when necessary, but they cannot be conveniently borrowed from casual or arbitrary circumstances.

It will be recollected that Terminology is a language employed for describing objects, Nomenclature, a body of names of the objects themselves. The *names,* as was stated in the last maxim, may be arbitrary; but the *descriptive terms* must be

* iii., 69,

borrowed from words of suitable meaning in the modern or the classical languages. Thus the whole terminology which Linnæus introduced into botany, is founded upon the received use of Latin words, although he defined their meaning so as to make it precise when it was not so, according to Aphorism V. But many of the terms were invented by him and other botanists, as *Perianth, Nectary, Pericarp;* so many, indeed, as to form, along with the others, a considerable language. Many of the terms which are now become familiar were originally invented by writers on botany. Thus the word *petal,* for one division of the corolla, was introduced by Fabius Columna. The term *sepal* was devised by Neckar to express each of the divisions of the calyx. And up to the most recent times, new denominations of parts and conditions of parts have been devised by botanists, when they found them necessary, in order to mark important differences or resemblances. Thus the general *receptacle* of the flower, as it is termed by Linnæus, or *torus,* by Salisbury, is continued into organs which carry the stamina and pistil, or the pistil alone, or the whole flower; this organ has hence been termed* *gonophore, carpophore,* and *anthophore,* in these cases.

In like manner when Cuvier had ascertained that the lower jaws of Saurians consisted always of six pieces having definite relations of form and position, he gave names to them, and termed them respectively the *dental,* the *angular,* the *coronoid,* the *articular,* the *complementary,* and the *opercular* bones.

In all these cases, the descriptive terms thus introduced have been significant in their derivation. An attempt to circulate a perfectly arbitrary word as a means of description would probably be unsuccessful. We have, indeed, some examples approaching to arbitrary designations, in the Wernerian names of colours, which are a part of the terminology of Natural History. Many of these names are borrowed from natural resemblances, as *Auricula purple, Apple green, Straw yellow;* but the names of others are taken from casual occurrences, mostly, however, such as were already recognized in common language, as *Prussian blue, Dutch orange, King's yellow.*

---

* Decandolle's *Th. El.,* 405.

The extension of arbitrary names in scientific terminology is by no means to be encouraged. I may mention a case in which it was very properly avoided. When Mr. Faraday's researches on Voltaic electricity had led him to perceive the great impropriety of the term *poles*, as applied to the apparatus, since the processes have not reference to any opposed points, but to two opposite directions of a path, he very suitably wished to substitute for the phrases *positive pole* and *negative pole* two words ending in *ode*, from ὅδος, a way. A person who did not see the value of our present maxim, that descriptive terms should be descriptive in their origin, might have proposed words perfectly arbitrary, as *Alphode* and *Betode*: or, if he wished to pay a tribute of respect to the discoverers in this department of science, *Galvanode* and *Voltaode*. But such words would very justly have been rejected by Mr. Faraday, and would hardly have obtained any general currency among men of science. *Zincode* and *Platinode*, terms derived from the metal which, in one modification of the apparatus, forms what was previously termed the pole, are to be avoided, because in their origin too much is casual; and they are not a good basis for derivative terms. The pole at which the zinc is, is the Anode or Cathode, according as it is associated with different metals. Either the *zincode* must sometimes mean the pole at which the Zinc is, and at other times that at which the Zinc is not, or else we must have as many names for poles as there are metals. *Anode* and *Cathode*, the terms which Mr. Faraday adopted, were free from these objections; for they refer to a natural standard of the direction of the voltaic current, in a manner which, though perhaps not obvious at first sight, is easily understood and retained. *A*node and *Cath*ode, the *rising* and the *setting* way, are the directions which correspond to east and west in that voltaic current to which we must ascribe terrestrial magnetism. And with these words it was easy to connect *anion* and *cathion*, to designate the opposite elements which are separated and liberated at the two *electrodes*.

The following Aphorisms respect the Form of Technical Terms:

By the *Form* of Terms, I mean their philological conditions;

as, for example, from what languages they may be borrowed, by what modes of inflexion they must be compounded, how their derivatives are to be formed, and the like. In this, as in other parts of the subject, I shall not lay down a system of rules, but shall propose a few maxims.

APHORISM XV.

The two main conditions of the Form of technical terms are, that they must be generally intelligible, and susceptible of such grammatical relations as their scientific use requires.

THESE conditions may at first appear somewhat vague, but it will be found that they are as definite as we could make them, without injuriously restricting ourselves. It will appear, moreover, that they have an important bearing upon most of the questions respecting the form of the words which come before us; and that if we can succeed in any case in reconciling the two conditions, we obtain terms which are practically good, whatever objections may be urged against them from other considerations.

1. The former condition, for instance, bears upon the question whether scientific terms are to be taken from the learned languages, Greek and Latin, or from our own. And the latter condition very materially affects the same question, since in English we have scarcely any power of inflecting our words; and therefore must have recourse to Greek or Latin in order to obtain terms which admit of grammatical modification. If we were content with the term *Heat* to express the science of heat, still it would be a bad technical term, for we cannot derive from it an adjective like *thermotical*. If *bed* or *layer* were an equally good term with *stratum*, we must still retain the latter, in order that we may use the derivative *stratification*, for which the English words cannot produce an equivalent substitute. We may retain the words *lime* and *flint*, but their adjectives for scientific purposes are not *limy* and *flinty*, but *calcareous* and *siliceous;* and hence we are able to form a compound, as *calcareo-siliceous*, which we could not do with indigenous

words. We might fix the phrases *bent back* and *broken* to mean (of optical rays) that they are *reflected* and *refracted*; but then we should have no means of speaking of the angles of *reflection* and *refraction*, of the *refractive* indices, and the like.

Thus one of the advantages of going to the Greek and Latin languages for the origin of our scientific terms is, that in this way we obtain words which admit of the formation of adjectives and abstract terms, of composition, and of other inflexions. Another advantage of such an origin is, that such terms, if well selected, are readily understood over the whole lettered world. For this reason, the descriptive language of science, of botany for instance, has been, for the most part, taken from the Latin; many of the terms of the mathematical and chemical sciences have been derived from the Greek; and when occasion occurs to construct a new term, it is generally to that language' that recourse is had. The advantage of such terms is, as has already been intimated, that they constitute an universal language, by means of which cultivated persons in every country may convey to each other their ideas without the need of translation.

On the other hand, the advantage of indigenous terms is, that so far as the language extends, they are intelligible much more clearly and vividly than those borrowed from any other source, as well as more easily manageable in the construction of sentences. In the descriptive language of botany, for example, in an English work, the terms *drooping*, *nodding*, *one-sided*, *twining*, *straggling*, appear better than *cernuous*, *nutant*, *secund*, *volubile*, *divaricate*. For though the latter terms may by habit become as intelligible as the former, they cannot become more so to any readers; and to most English readers they will give a far less distinct impression.

2. Since the advantage of indigenous over learned terms, or the contrary, depends upon the balance of the capacity of inflexion and composition on the one hand, against a ready and clear signi-ficance on the other, it is evident that the employment of scientific terms of the one class or of the other may very properly be ex-tremely different in different languages. The German possesses in a very eminent degree that power of composition and derivation,

which in English can hardly be exercised at all, in a formal manner. Hence German scientific writers use native terms to a far greater extent than do our own authors. The descriptive terminology of botany, and even the systematic nomenclature of chemistry, are represented by the Germans by means of German roots and inflexions. Thus the description of *Potentilla anserina*, in English botanists, is that it has *Leaves interruptedly pinnate, serrate, silky, stem creeping, stalks axillar, one-flowered.* Here we have words of Saxon and Latin origin mingled pretty equally. But the German description is entirely Teutonic. *Die Blume in Achsel; die Blätter unterbrochen gefiedert, die Blättchen scharf gesagt, die Stämme kriechend, die Bluthenstiele einblumig.* We could imitate this in our own language, by saying *brokenly-feathered, sharp-sawed;* by using *threed* for *ternate,* as the Germans employ *gedreit;* by saying *fingered-feathered* for *digitato-pinnate,* and the like. But the habit which we have, in common as well as in scientific language, of borrowing words from the Latin for new cases, would make such usages seem very harsh and pedantic.

We may add that, in consequence of these different practices in the two languages, it is a common habit of the German reader to impose a scientific definiteness upon a common word, such as our Fifth Aphorism requires; whereas the English reader expects rather that a word which is to have a technical sense shall be derived from the learned languages. *Die Kelch* and *die Blume* (the cup and the flower) easily assume the technical meaning of *calyx* and *corolla; die griffel* (the pencil) becomes *the pistil;* and a name is easily found for the *pollen,* the *anthers,* and the *stamens,* by calling them the dust, the dust-cases, and the dust-threads (*der staub, die staub-beutel* or *staub-fächer,* and *die staub-fäden*). This was formerly done in English to a greater extent than is now possible without confusion and pedantry. Thus, in Grew's book on the *Anatomy of Plants,* the calyx is called the *impalement,* and the sepals the *impalers;* the petals are called the *leaves of the flower;* the stamens with their anthers are the *seminiform attire.* But the English language, as to such matters, is now less flexible than it then was; partly in consequence of having adopted the Linnæan terminology almost entire, without any attempt to naturalise it. For any attempt at idiomatic description would interfere with the scientific language now generally received in this country. In Germany, on the other hand, those who wrote upon science in their own language imitated the Latin words which they found in foreign writers, instead of transferring new roots into their own language. Thus the *numerator* and *denominator* of a fraction they called the *namer* and the *counter* (*nenner* and *zähler*). This course they pursued even where the expression was erroneous. Thus that portion of the intestines which ancient anatomists called *duodenum*, because they falsely estimated its length at twelve inches, the Germans also term *zwölffingerdarm* (twelve-inch-gut), though this intestine in a whale is twenty feet long, and in a frog not above twenty lines. As another example of this process in German, we may take the word *muttersackbauchblatte*, the uterine peritonæum.

It is a remarkable evidence of this formative power of the German language, that it should have been able to produce an imitation of the systematic chemical nomenclature of the French school, so complete, that it is used in Germany as familiarly as the original system is in France and England. Thus Oxygen and Hydrogen are *Sauerstoff* and *Wafferstoff*; Azote is *Stickstoff* (suffocating matter) ; Sulphuric and Sulphurous Acid are *Schwefel-säure* and *Schwefelichte-säure*. The Sulphate and Sulphite of Baryta, and Sulphuret of Baryum, are *Schwefel-säure Baryterde*, *Schwefelichte-säure Baryterde*, and *Schwefel-baryum*. Carbonate of Iron is *Kohlen-säures Eisenoxydul* ; and we may observe that, in such cases, the German name is much more agreeable to anology than the English one; for the Protoxide of Iron, and not the Iron itself, is the base of the salt. And the German language has not only thus imitated the established nomenclature of chemistry, but has shown itself capable of supplying new forms to meet the demands which the progress of theory occasions. Thus the Hydracids are *Wasserstoff-säuren* ; and of these, the Hydriodic Acid is *Iodwasserstoff-säure*, and so of the rest. In like manner, the translator of Berzelius has found

German names for the sulpho-salts of that chemist; thus he has *Wasserstoffschwefliges Schwefel-lithium*, which would be (if we were to adopt his theoretical view,) hydro-sulphuret of sulphuret of lithium : and a like nomenclature for all other similar cases.

3. In English we have no power of imitating this process, and must take our technical phrases from some more flexible language, and generally from the Latin or Greek. We are indeed so much accustomed to do this, that except a word has its origin in one of these languages, it hardly seems to us a technical term; and thus by employing indigenous terms, even descriptive ones, we may, perhaps, lose in precision more than we gain in the vividness of the impression. Perhaps it may be better to say *cuneate, lunate, hastate, sagittate, reniform,* than *wedge-shaped, crescent-shaped, halbert-headed, arrow-headed, kidney-shaped. Ringent* and *personate* are better than any English words which we could substitute for them ; *labiate* is more precise than *lipped* would readily become. *Urceolate, trochlear,* are more compact than *pitcher-shaped, pulley-shaped;* and *infundibuliform, hypocrateriform,* though long words, are not more inconvenient than *funnel-shaped* and *salver-shaped.* In the same way it is better to speak (with Dr. Prichard*,) of *repent* and *progressive* animals, than of *creeping* and progressive : the two Latin terms make a better pair of correlatives.

4. But wherever we may draw the line between the proper use of English and Latin terms in descriptive phraseology, we shall find it advisable to borrow almost all other technical terms from the learned languages. We have seen this in considering the new terms introduced into various sciences in virtue of our Ninth Maxim. We may add as further examples the names of the various animals of which a knowledge has been acquired from the remains of them which exist in, various strata, and which have been reconstructed by Cuvier and his successors. Such are the *Palæotherium,* the *Anoplotherium,* the *Megatherium,* the *Dinotherium,* the *Chirotherium,* the *Megalichthys,* the *Mastodon,* the *Ichthyosaurus,* the *Plesiosaurus,* the *Pterodactylus.* To these others are every year added; as, for instance, very recently, the

---

* *Researches,* p. 69.

Toxodon, Zeuglodon, and *Phascolotherium* of Mr. Owen, and the *Thylacotherium* of M. Valenciennes. The names of species, as well as of genera, are thus formed from the Greek: as the Plesiosaurus *dolichodeirus,* (long-necked), Ichthyosaurus *platyodon* (broadtoothed), the Irish elk, termed Cervus *megaceros* (large-horned). But the descriptive specific names are also taken from the Latin, as Plesiosaurus *brevirostris, longirostris, crassirostris;* besides which there are arbitrary specific names, which we do not here consider. These names being all constructed at a period when naturalists were familiar with an artificial system, the standard language of which is Latin, have not been taken from modern language. But the names of living animals, and even of their classes, long ago formed in the common language of men, have been in part adopted in the systems of naturalists, agreeably to Aphorism Third. Hence the language of systems in natural history is mixed of ancient and modern languages. Thus Cuvier's divisions of the vertebrated animals are *Mammifères* (Latin), *Oiseaux, Reptiles, Poissons; Bimanes, Quadrumanes, Carnassieres, Rongeurs, Pachydermes* (Greek), *Ruminans* (Latin), *Cetacés* (Latin). In the subordinate divisions the distribution being more novel, the names are less idiomatic: thus the kinds of Reptiles are *Cheloniens, Sauriens, Ophidiens, Batriciens,* all which are of Greek origin. In like manner, Fish are divided into *Chondropterygiens, Malacopterygiens, Acanthopterygiens.* The unvertebrated animals are *Mollusques* or *Animaux articules,* and *Animaux rayonnées;* and the former are divided into six classes, according to the position of their foot; namely, *Cephalopodes, Pteropodes, Gasteropodes, Acephales, Brachiopodes, Cirrhopodes.*

In transferring these terms into English, when the term is new in French as well as English, we have little difficulty; for we may take nearly the same liberties in English which are taken in French; and hence we may say *mammifers* (rather *mammals*), *cetaceans* or *cetaces, batracians* (rather *batrachians*), using the words as substantives. But in other cases we must go back to the Latin: thus we say *radiate* animals, or *radiata* (rather *radials*), for *rayonnées.* These changes, however, rather refer to another Aphorism.

5. When new mineral species have been established in recent times, they have generally had arbitrary names assigned to them, derived from some person or places. In some instances, however, descriptive names have been selected; and then these have been generally taken from the Greek, as *Augite, Stilbite, Diaspore, Dichroite, Dioptase.* Several of these Greek names imposed by Haüy, refer to some circumstances, often fancifully selected, in his view of the crystallization of the substance, as *Epidote, Peridote, Pleonast.* Similar terms of Greek origin have been introduced by others, as *Orthite, Anorthite, Periklin.* Greek names founded on casual circumstances are less to be commended. Berzelius has termed a mineral *Eschynite*, from αἰσχύνη, *shame*, because it is, he conceives, a shame for chemists not to have separated its elements more distinctly than they did at first.

6. In Botany, the old names of genera of Greek origin are very numerous, and many of them are descriptive, as *Glycyrhiza* (γλυκὺς and ῥίζα, sweet root) liquorice, *Rhododendron* (rose tree), *Hæmatoxylon* (bloody wood), *Chrysocoma* (golden hair), *Alopecurus* (fox tail), and many more. In like manner there are names which derive a descriptive significance from the Latin, either adjectives, as *Impatiens, Gloriosa, Sagittaria*, or substantives irregularly formed, as *Tussilago* (à tussis domatione), *Urtica* (ab urendo tactu), *Salsola* (à salsedine). But these, though good names when they are established by tradition, are hardly to be imitated in naming new plants. In most instances, when this is to be done, arbitrary or local names have been selected, as *Strelitzia.*

7. In Chemistry, new substances have of late had names assigned them from Greek roots, as *Iodine*, from its violet colour, *Chlorine* from its green colour. In like manner fluorine has by the French chemists been called *Phthor*, from its destructive properties. So the new metals, *Chrome, Rhodium, Iridium, Osmium*, had names of Greek derivation descriptive of their properties. Some such terms, however, were borrowed from localities, as *Strontia, Yttria*, the names of new earths. Others have a mixed origin, as *Pyrogallic, Pyroacetic*, and *Pyroligneous* Spirit. In some cases the deviation has been extravagantly capricious.

Thus in the process for making Pyrogallic Acid, a certain sub-stance is left behind, from which M. Braconnot extracted an acid which he called *Ellagic* Acid, framing the root of the name by reading the word *Galle* backwards.

The new laws which the study of electro-chemistry brought into view, required a new terminology to express their conditions: and in this case, as we have observed in speaking of the Twelfth Maxim, arbitrary words are less suitable. Mr. Faraday very properly borrowed from the Greek his terms *Electrolyte, Electrode, Anode, Cathode, Anïon, Cathïon, Dilectric*. In the mechanico-chemical and mechanical sciences, however, new terms are less copiously required than in the sciences of classification, and when they are needed, they are generally determined by analogy from existing terms. *Thermo-electricity* and *Electro-dynamics* were terms which very naturally offered themselves; Nobili's *thermo-mul-tiplier*, Snow Harris's *unit-jar*, were almost equally obvious names. In such cases, it is generally possible to construct terms both compendious and descriptive, without introducing any new radical words.

8. The subject of crystallography has inevitably given rise to many new terms, since it brings under our notice a great number of new relations of a very definite but very complex form. Haüy attempted to find names for all the leading varieties of crystals, and for this purpose introduced a great number of new terms, founded on various analogies and allusions. Thus the forms of calc-spar are termed by him *primitive, equiaxe, inverse, metastatique, contrastante, imitable, birhomboidale, prismatique, apophane, uniternaire, bisunitaire, dodécaèdre, contractée, dilatée, sexduodecimale, bisalterne, binoternaire,* and many others. The want of uniformity in the origin and scheme of these denomina-tions would be no valid objection to them, if any general truth could be expressed by means of them : but the fact is, that there is no definite distinction of these forms. They pass into each other by insensible gradations, and the optical and physical pro-perties which they possess are common to all of them. And as a mere enunciation of laws of form, this terminology is insuffi-cient. Thus it does not at all convey the relation between the

bisalterne and the *binoternaire*, the former being a combination of the *metastatique* with the *prismatique*, the latter of the *metastique* with the *contrastante*: again, the *contrastante*, the *mixte*, the *cuboide*, the *contractée*, the *dilatée*, all contain faces generated by a common law, the index being respectively altered so as to be in these cases, 3, $\frac{3}{2}$, $\frac{4}{4}$, $\frac{2}{4}$, $\frac{6}{6}$; and this, which is the most important geometrical relation of these forms, is not at all recorded or indicated by the nomenclature. The fact is, that it is probably impossible, the subject of crystallography having become so complex as it now is, to devise a system of names which shall express the relations of form. Numerical symbols, such as those of Weiss or Naumann, or Professor Miller, are the proper ways of expressing these relations, and are the only good crystallographic terminology for cases in detail.

The terms used in expressing crystallographic laws have been for the most part taken from the Greek by all writers except some of the Germans. These, we have already stated, have constructed terms in their own language, as *zwei-und-ein gliedrig*, and the like.

In Optics we have some new terms connected with crystalline laws, as *uniaxal* and *biaxal* crystals, *optical axes*, which offered themselves without any effort on the part of the discoverers. In the whole history of the undulatory theory, very few innovations in language were found necessary, except to fix the sense of a few phrases, as *plane-polarized* light in opposition to *circularly-polarized*, and the like.

This is still more the case in Mechanics, Astronomy, and pure mathematics. In these sciences, several of the primary stages of generalization being already passed over, when any new steps are made, we have before us some analogy by which we may frame our new terms. Thus when the *plane of maximum areas* was discovered, it had not some new arbitrary denomination assigned it, but the name which obviously described it was fixed as a technical name.

The result of this survey of the scientific terms of recent formation seems to be this;—that indigenous terms may be employed in the descriptions of facts and phenomena as they at first present themselves; and in the first induction from these; but that when we come to generalize and theorize, terms borrowed from the learned languages are more readily fixed and made definite, and are also more easily connected with derivatives. Our native terms are more impressive, and at first more intelligible; but they may wander from their scientific meaning, and are capable of little inflexion. Words of classical origin are precise to the careful student, and capable of expressing, by their inflexions, the relations of general ideas; but they are unintelligible, even to the learned man, without express definition, and convey instruction only through an artificial and rare habit of thought.

Since in the balance between words of domestic and of foreign origin so much depends upon the possibility of inflexion and derivation, I shall consider a little more closely what are the limits and considerations which we have to take into account in reference to that subject.

Aphorism XVI.

In the composition and inflexion of technical terms, philological analogies are to be preserved if possible, but modified according to scientific convenience.

In the language employed or proposed by writers upon subjects of science, many combinations and forms of derivation occur, which would be rejected and condemned by those who are careful of the purity and correctness of language. Such anomalies are to be avoided as much as possible; but it is impossible to escape them altogether, if we are to have a scientific language which has any chance of being received into general use. It is better to admit compounds which are not philologically correct, than to invent many new words, all strange to the readers for whom they are intended: and in writing on science in our own language, it is not possible to avoid making additions to the vocabulary of common life; since science requires exact names for many things which common language has not named. And

although these new names should, as much as possible, be
constructed in conformity with the analogies of the language,
such extensions of analogy can hardly sound, to the gram-
marian's ear, otherwise than as solecisms.   But, as our maxim
indicates, the analogy of science is of more weight with us
than the analogy of language: and although anomalies in our
phraseology should be avoided as much as possible, innovations
must be permitted wherever a scientific language, easy to acquire,
and convenient to use, is unattainable without them.

I shall proceed to mention some of the transgressions of strict
philological rules, and some of the extensions of grammatical forms,
which the above conditions appear to render necessary.

1. The combination of different languages in the derivation
of words, though to be avoided in general, is in some cases admissible.

Such words are condemned by Quintilian and other gramma-
rians, under the name of *hybrids*, or things of a mixed race; as
biclinium, from *bis* and κλίνη; *epitogium*, from ἐπὶ and *toga*.
Nor are such terms to be unnecessarily introduced in science.
Whenever a homogeneous word can be formed and adopted with
the same ease and convenience as a hybrid, it is to be preferred.
Hence we must have *ichthyology*, not *piscology*, *entomology*, not
insectology, *insectivorous* not *insectophagous*.   In like manner, it
would be better to say *unoculus* than *monoculus*, though the
latter has the sanction of Linnæus, who was a purist in such
matters.   Dr. Turner, in his *Chemistry*, speaks of *protoxides* and
binoxides, which combination violates the rule for making the
materials of our terms as homogeneous as possible; *protoxide*
and *deutoxide* would be preferable, both on this and on other
accounts.

Yet this rule admits of exceptions. *Mineralogy*, with its
Greek termination, has for its root *minera*, a medieval Latin word
of Teutonic origin, and is preferable to *oryctology*.   *Terminology*
appears to be better than *glossology*: which according to its derivation would be rather the science of language in general than of
technical terms; and *horology*, from ὅρος, a term, would not be
immediately intelligible, even to Greek scholars; and is already

employed to indicate the science which treats of horologes, or time-pieces.

Indeed, the English reader is become quite familiar with the termination *ology*, the names of a large number of branches of science and learning having that form. This termination is at present rather apprehended as a formative affix in our own language, indicating a science, than as an element borrowed from a foreign language. Hence, when it is difficult or impossible to find a Greek term which clearly designates the subject of a science, it is allowable to employ some other, as in *Tidology*, the doctrine of the tides.

The same remark applies to some other Greek elements of scientific words: they are so familiar to us that in composition they are almost used as part of our own language. This naturalization has taken place very decidedly in the element *arch*, (ἀρχὸς, a leader,) as we see in *archbishop*, *archduke*. It is effected in a great degree for the preposition *anti*: thus we speak of *anti-slavery* societies, *anti-reformers*, *anti-bilious*, or *anti-acid*, medicines, without being conscious of any anomaly. The same is the case with the Latin preposition *præ* or *pre*, as appears from such words as *pre-engage*, *pre-arrange*, *pre-judge*, *pre-paid*; and in some measure with *pro*, for in colloquial language we speak of *pro-catholics* and *anti-catholics*. Also the preposition *ante* is similarly used, as *ante-nicene* fathers. The preposition *co*, abbreviated from *con*, and implying things to be simultaneous or connected, is firmly established as part of the language, as we see in *coexist*, *coheir*, *coordinate*; hence I have called those lines *cotidal* lines which pass through places where the high water of the tide occurs simultaneously.

2. As in the course of the mixture by which our language has been formed, we have thus lost all habitual consciousness of the difference of its ingredients (Greek, Latin, Norman, French, and Anglo-Saxon): we have also ceased to confine to each ingredient the mode of grammatical inflexion which originally belonged to it. Thus the termination *ive* belongs peculiarly to Latin adjectives, yet we say *sportive*, *talkative*. In like manner, *able* is added to words which are not Latin, as *eatable*, *drinkable*, *piti-*

able, enviable. Also the termination *al* and *ical* are used with various roots, as *loyal, royal, farcical, whimsical;* hence we may make the adjective *tidal* from tide. This ending, *al,* is also added to abstract terms in *ion,* as *occasional, provisional, intentional, national;* hence we may, if necessary, use such words as *educational, terminational.* The ending *ic* appears to be suited to proper names, as *Pindaric, Socratic, Platonic;* hence it may be used when scientific words are derived from proper names, as *Voltaic* or *Galvanic* electricity: to which I have proposed to add *Franklinic.*

In adopting scientific adjectives from the Latin, we have not much room for hesitation; for, in such cases, the habits of derivation from that language into our own are very constant; *ivus* becomes *ive,* as *decursive; inus* becomes *ine,* as in *ferine; atus* becomes *ate,* as *hastate;* and *us* often becomes *ous,* as *rufous; aris* becomes *ary,* as *axillary; ens* becomes *ent,* as *ringent.* And in adopting into our language, as scientific terms, words which in another language, the French for instance, have a Latin origin familiar to us, we cannot do better than form them as if they were derived directly from the Latin. Hence the French adjectives *cetacé, crustacé, testacé,* may become either *cetaceous, crustaceous, testaceous,* according to the analogy of *farinaceous, predaceous,* or else *cetacean, crustacean, testacean,* imitating the form of *patrician.* Since, as I shall soon have to notice, we require substantives as well as adjectives from these words, we must, at least for that use, take the forms last suggested.

In pursuance of the same remark, *rongeur* becomes *rodent,* and *edenté* would become *edentate;* but that this word is rejected on another account: the adjectives *bimane* and *quadrumane* are *bimanous* and *quadrumanous.*

3. There is not much difficulty in thus forming adjectives: but the purposes of Natural History require that we should have substantive words corresponding to these adjectives; and these cannot be obtained without some extension of the analogies of our language. We cannot in general use adjectives or participles as singular substantives. *The happy* or *the doomed* would, according to good English usage, signify those who are happy and those

who are doomed. Hence we could not speak of a particular scaled animal as *the squamate*, and still less could we call any such animal *a squamate*, or speak of *squamates* in the plural. Some of the forms of our adjectives, however, do admit of this substantive use. Thus we talk of *Europeans, plebeians, republicans;* of *divines* and *masculines;* of the *ultramontanes;* of *mordants* and *brilliants;* of *abstergents* and *emollients;* of *mercenaries* and *tributaries;* of *animals, manuals,* and *officials;* of *dissuasives* and *motives.* We cannot generally use in this way adjectives in *ous*, nor in *ate* (though *reprobates* is an exception), nor English participles, nor adjectives in which there is no termination imitating the Latin, as *happy, good.* Hence, if we have, for purposes of science, to convert adjectives into substantives, we ought to follow the form of examples like these, in which it has already appeared in fact, that such usage, though an innovation at first, may ultimately become a received part of the language.

By attention to this rule we may judge what expressions to select in cases where substantives are needed. I will take as an example the division of the mammalian animals into orders. These orders, according to Cuvier, are *Bimanes, Quadrumanes, Carnassiers, Rongeurs, Edentés, Ruminans, Pachydermes, Cetacés. Bimanes, Quadrumanes, Rodents, Ruminants,* are admissible as English substantives on the grounds just stated. *Cetaceous* could not be used substantively; but *Cetacean* in such a usage is sufficiently countenanced by such cases as we have mentioned, *patrician,* &c.; hence we adopt this form. We have no English word equivalent to the French *Carnassiers:* the English translator of Cuvier has not provided English words for his technical terms; but has formed a Latin word, *Carnaria,* to represent the French terms. From this we might readily form *Carnaries;* but it appears much better to take the Linnæan name *Feræ* as our root, from which we may take *Ferine,* substantive as well as adjective; and hence we call this order *Ferines.* The word for which it is most difficult to provide a proper representation, is *Edenté, Edentata:* for, as we have said, it would be very harsh to speak of the order as the *Edentates;* and if we were to abbreviate the word into *edent,* we should suggest a false analogy with *rodent*, for as *rodent* is *quod rodit*, that which gnaws, *edent* would be *quod edit*, that which eats. And even if we were to take *edent* as a substantive, we could hardly use it as an adjective: we should still have to say, for example, the *edentate* form of head. For these reasons it appears best to alter the form of the word, and to call the order the *Edentals*, which is quite allowable, both as adjective and substantive.

There are several other words in *ate* about which there is the same difficulty in providing substantive forms. Are we to speak of *Vertebrates?* or would it not be better, in agreement with what has been said above, to call these *Vertebrals*, and the opposite class *Invertebrals?*

There are similar difficulties with regard to the names of subordinate portions of zoological classification; thus the Ferines are divided by Cuvier into *Cheiroptéres, Insectivores, Carnivores ;* and these latter into *Plantigrades, Digitigrades, Amphibies, Marsupiaux.* There is not any great harshness in naturalizing these substantives as *Chiropters, Insectivores, Carnivores, Plantigrades, Digitigrades, Amphibians,* and *Marsupials.* The words *Carnivores* and *Insectivores* are better, because of more familiar origin, than Greek terms; otherwise we might, if necessary, speak of *Zoophagans* and *Entomophagans.*

It is only with certain familiar adjectival terminations, as *ous* and *ate,* that there is a difficulty in using the word as substantive. When this can be avoided, we readily accept the new word, as *Pachyderms,* and in like manner *Mollusks.*

If we examine the names of the Orders of Birds, we find that they are in Latin, *Predatores* or *Accipitres, Passeres, Scansores, Rasores* or *Gallinæ, Grallatores, Palmipedes* and *Anseres :* Cuvier's Orders are, *Oiseaux de Proie, Passereaux, Grimpeurs, Gallinacés, Echassiérs, Palmipedes.* These may be englished conveniently as *Predators, Passerines, Scansors, Gallinaceans,* (rather than *Rasors,) Grallators, Palmipedans. Scansors, Grallators,* and *Rasors* are better, as technical terms, than *Climbers, Waders,* and *Scratchers.* We might venture to anglicize the terminations of the names which Cuvier gives to the divisions of these Orders : thus the Predators are the *Diurnals* and the *Nocturnals ;* the Passerines are the *Dentirostres*, the *Fissirostres*, the *Conirostres*, the *Tenuirostres*, and the *Syndactyls*: the word *lustre* showing that the former termination is allowable. The Scansors are not subdivided, nor are the Gallinaceans. The Grallators are *Pressirostres*, *Cultrirostres*, and *Macrodactyls*. The Palmipedans are the *Plungers*, the *Longipens*, the *Totipalmes* and the *Lamellirostres*.

The next class of Vertebrals is the *Reptiles*, and these are either *Chelonians*, *Saurians*, *Ophidians*, or *Batrachians*. Cuvier writes *Batraciens*, but we prefer the spelling to which the Greek word directs us.

The next class is the *Fishes*, in which province Cuvier has himself been the great systematist, and has therefore had to devise many new terms. Many of these are of Greek or Latin origin, and can be anglicized by the analogies already pointed out, as *Chondropterygians*, *Malacopterygians*, *Lophobranchs*, *Plectognaths*, *Gymnodonts*, *Scleroderms*. *Discoboles* and *Apodes* may be English as well as French. There are other cases in which the author has formed the names of families, either by forming a word in *ides* from the name of a genus, as *Gadoides*, *Gobiöides*, or by gallicizing the Latin name of the genus, as *Salmones* from *Salmo*, *Clupes* from *Clupea*, *Esoces* from *Esox*, *Cyprins* from *Cyprinus*. In both these cases the best procedure seems to be to form the English substantive in *idan*, as *Gadoidans*, *Gobiöidans*, *Salmonidans*, *Clupeidans*, *Esocidans*, *Cyprinidans*. One of the orders of fishes, co-ordinate with the Chondropterygians and the Lophobranchs, is termed *Osseux* by Cuvier. It appears hardly worth while to invent a substantive word for this, when *Bony Fishes* is so simple a phrase, and may readily be understood as a technical name of a systematic order.

The Mollusks are the next class ; and these are divided into *Cephalopods*, *Gasteropods*, and the like. The Gasteropods are *Nudibranchs*, *Inferobranchs*, *Tectibranchs*, *Pectinibranchs*, *Scutibranchs*, and *Cyclobranchs*. In framing most of these terms Cuvier has made hybrids by a combination of a Latin word with *branchiæ*, which is the Greek name for the gills of a fish ; and has thus avoided loading the memory with words of an origin not obvious to most naturalists, as terms derived from the Greek would have been.   Another division of the Gasteropods is *Pulmonés*, which we must make *Pulmonians*.   In like manner the subdivisions of the Pectinibranchs are the *Trochoidans* and *Buccinoidans* (*Trochoïdes, Buccinoïdes*).   The *Acéphales*, another order of Mollusks, may be *Acephals* in English.

After these comes the third grand division of *Articulated Animals*, and these are *Annelidans, Crustaceans, Arachnidans,* and *Insects*.   I shall not dwell upon the names of these, as the form of English words which is to be selected must be sufficiently obvious from the preceding examples.

Finally, we have the fourth grand division of animals, the *Rayonnés*, or *Radiata;* which, for reasons already given, we may call *Radials*.   These are *Echinoderms, Intestinals, Acalephes* and *Polyps*.   The Polyps, which are composite animals in which many gelatinous individuals are connected so as to have a common life, have, in many cases, a more solid framework belonging to the common part of the animal.   This framework, of which coral is a special example, is termed in French *Polypier;* the word has been anglicized by the word *polypary*, after the analogy of *aviary* and *apiary*.   Thus Polyps are either *Polyps with Polyparies* or *Naked Polyps*.

Any common kind of Polyps has usually in the English language been called *Polypus*, the Greek termination being retained. This termination in *us*, however, whether Latin or Greek, is to be excluded from the English as much as possible, on account of the embarassment which it occasions in the formation of the plural.   For if we say *Polypi* the word ceases to be English, while *Polypuses* is harsh: and there is the additional inconvenience, that both these forms would indicate the plural of individuals rather than of classes.   If we were to say, " The Corallines are a Family of the *Polypuses with Polyparies*," it would not at once occur to the reader that the three last words formed a technical phrase.

This termination *us*, which must thus be excluded from the names of families, may be admitted in the designation of genera ; of animals, as *Nautilus, Echinus, Hippopotamus ;* and of plants, as *Crocus, Asparagus, Narcissus, Acanthus, Ranunculus, Fungus.*

The same form occurs in other technical words, as *Fucus, Mucus, Œsophagus, Hydrocephalus, Callus, Calculus, Uterus, Fœtus, Radius, Focus, Apparatus.* It is, however, advisable to retain this form only in cases where it is already firmly established in the language; for a more genuine English form is preferable. Hence we say, with Mr. Lyell, *Icthyosaur, Plesiosaur, Pterodactyl.* In like manner Mr. Owen anglicizes the termination *erium,* and speaks of the *Anoplothere* and *Paleothere.*

Since the wants of science thus demand adjectives which can be used also as substantive names of classes, this consideration may sometimes serve to determine our selection of new terms. Thus Mr. Lyell's names for the subdivisions of the tertiary strata, *Miocene, Pliocene,* can be used as substantives; but if such words as *Mioneous, Plioneous* had suggested themselves, they must have been rejected, though of equivalent signification, as not fulfilling this condition.

4. (1.) Abstract substantives can easily be formed from adjectives: from electric we have *electricity;* from galvanic, *galvanism;* from organic, *organization; velocity, levity, gravity,* are borrowed from Latin adjectives. *Caloric* is familiarly used for the matter of heat, though the form of the word is not supported by any obvious analogy.

(2.) It is quite intolerable to have words regularly formed in opposition to the analogy which their meaning offers; as when bodies are said to have conduct*ibility* or conduc*ibility* with regard to heat. The bodies are conduct*ive* and their property is conduct*ivity.*

(3.) The terminations *ize* (rather than *ise*), *ism,* and *ist* are applied to words of all origins: thus we have to *pulverize,* to *colonize, Witticism, Heathenism, Journalist, Tobacconist.* Hence we may make such words when they are wanted. As we cannot use physician for a cultivator of physics, I have called him a *physicist.* We need very much a name to describe a cultivator of science in general. I should incline to call him a *Scientist.* Thus we might say, that as an Artist is a Musician, Painter, or Poet, a Scientist is a Mathematician, Physicist, or Naturalist.

(4.) Connected with verbs in *ize,* we have abstract nouns in *ization*, as *polarization, crystallization,* These it appears proper to spell in English with *z* rather than *s;* governing our practice by the Greek verbal termination *ίζω* which we imitate. But we must observe that verbs and substantives in *yse,* (*analyse,*) belong to a different analogy, giving an abstract noun in *ysis* and an adjective *ytic* or *ytical;* (*analysis, analytic, analytical*). Hence *electrolyse* is more proper than *electrolyze.*

(5.) The names of many sciences end in *ics* after the analogy of *Mathematics, Metaphysics;* as *Optics, Mechanics.* But these in most other languages, as in our own formerly, have the singular form *Optice, l'Optique, Optik, Optick:* and though we now write *Optics,* we make such words of the singular number : " Newton's Opticks is an example." As, however, this connexion in new words is startling, as when we say, " Thermo-electrics is now much cultivated," it appears better to employ the singular form, after the analogy of *Logic* and *Rhetoric,* when we have words to construct. Hence we may call the science of languages *Linguistic,* as it is called by the best German writers, for instance, William von Humboldt.

5. In the derivation of English from Latin or Greek words, the changes of letters are to be governed by the rules which have generally prevailed in such cases. The Greek *οι* and *αι,* the Latin *oe* and *ae,* are all converted into a simple *e,* as in *E*conomy, Geod*e*sy, p*e*nal, C*e*sar. Hence, according to common usage, we should write ph*e*nomena, not ph*œ*nomena, paleontology, not pal*œ*ontology, mioc*e*ne not mioc*œ*ne, p*e*kilite not p*œ*kilite. But in order to keep more clearly in view the origin of our terms, it may be allowable to deviate from these rules of change, especially so long as the words are still new and unfamiliar. Dr. Buckland speaks of the *poikilitic,* not *pecilitic,* group of strata : *palæontology* is the spelling commonly adopted; and in imitation of this I have written *palætiology.* The diphthong *ει* was by the Latins changed into *i,* as in Arist*i*des; and hence this has been the usual form in English. Some recent authors indeed (Mr. Mitford for instance) write Arist*ei*des; but the former appears to be the more legitimate. Hence we write m*i*ocene, pl*i*ocene, not m*ei*ocene, pl*ei*ocene. The Greek *υ* becomes *y,* and *ου* becomes *u,* in

English as in Latin, as crystal, col*u*re. The consonants κ and χ become *c* and *ch* according to common usage. Hence we write *crystal*, not *chrystal*, batra*ch*ian not batracian, cryolite, not *ch*ryolite. As, however, the letter *c* before *e* and *i* differs from *k*, which is the sound we assign to the Greek κ, it may be allowable to use *k* in order to avoid this confusion. Thus, as we have seen, poi*k*ilite has been used, as well as pe*c*ilite. Even in common language some authors write *sk*eptic, which appears to be better than sceptic with our pronunciation, and is preferred by Dr. Johnson. For the same reason, namely to avoid confusion in the pronunciation, and also, in order to keep in view the connexion with *cathode*, the elements of an electrolyte which go to the anode and cathode respectively may be termed the anion and cat*h*ion; although the Greek would suggest cat*ï*on, (κατίον).

6. The example of chemistry has shown that we have in the terminations of words a resource of which great use may be made in indicating the relations of certain classes of objects: as sulphur*ous* and sulphur*ic* acids; sulph*ates*, sulph*ites*, and sulph*urets*. Since the introduction of the artifice by the Lavoisierian school, it has been extended to some new cases. Thus Chlor*ine*, Fluor*ine*, Brom*ine*, Iod*ine*, had their names put into that shape in consequence of their supposed analogy: and for the same reason have been termed Chlore, Phtore, Brome, Iode, by French chemists. In like manner, the names of metals in their Latin form have been made to end in *um*, as Osmium, Palladium; and hence it is better to say Platin*um*, Molybden*um*, than Platin*a*, Molybden*a*. It has been proposed to term the basis of Boracic acid Boron; and those who conceive that the basis of Silica has an analogy with Boron have proposed to term it Silic*on*, while those who look upon it as a metal would name it Silic*ium*. Selen*ium* was so named when it was supposed to be a metal: as its analogies are now acknowledged to be of another kind, it would be desirable, if the change were not too startling, to term it Sel*en*, as it is in German. Phosph*orus* in like manner might be Phosph*ur*, which would indicate its analogy with Sulph*ur*.

The resource which terminations offer has been applied in other cases. The names of many species of minerals end in *lite*,

or *ite*, as Stauro*lite*, Aug*ite*. Hence Adolphe Brongniart, in order
to form a name for a genus of fossil plants, has given this termination to the name of the recent genus which they nearly resemble, as Zam*ites* from Zamia, Lycopod*ites* from Lycopodium.

Names of different genera which differ in termination only
are properly condemned by Linnæus*; as Alsine, Alsinoides,
Alsinella, Alsinastrum; for there is no definite relation marked
by those terminations. Linnæus gives to such genera distinct
names, Alsine, Bufonia, Sagina, Elatine.

Terminations are well adapted to express definite systematic
relations, such as those of chemistry, but they must be employed
with a due regard to all the bearings of the system. Davy
proposed to denote the combinations of other substances with
chlorine by peculiar terminations; using *ane* for the smallest
proportion of Chlorine, and *anea* for the larger, as Cupr*ane*,
Cupr*anea*. In this nomenclature, common salt would be *Sodane*,
and Chloride of Nitrogen would be *Azotane*. This suggestion
never found favour. It was objected that it was contrary to the
Linnæan precept, that a specific name must not be united to a
generic as a termination. But this was not putting the matter
exactly on its right ground; for the rules of nomenclature of
natural history do not apply to chemistry; and the Linnæan rule
might with equal propriety have been adduced as a condemnation
of such terms as Sulphur*ous*, Sulphur*ic*. But Davy's terms were
bad; for it does not appear that Chlorine enters, as Oxygen does,
into so large a portion of chemical compounds, that its relations
afford a key to their nature, and may properly be made an
element in their names.

This resource, of terminations, has been abused, wherever it
has been used wantonly, or without a definite significance in the
variety. This is the case in M. Beudant's Mineralogy. Among
the names which he has given to new species, we find the following (besides many in *ite*), Scolexer*ose*, Opsim*ose*, Exanthel*ose*,
&c.; Diacr*ase*, Panab*ase*, Neopl*ase*; Neocl*ese*; Rhodo*ise*, Stibi-
con*ise*, &c.; Marcel*ine*, Wilhelm*ine*, &c.; Exit*ele*, and many
others. In addition to other objections which might be made

* *Phil. Bot.*, 231.

to these names, their variety is a material defect: for to make this variety depend on caprice alone, as in those cases it does, is to throw away a resource of which chemical nomenclature may teach us the value.

Aphorism XVII.

When alterations in technical terms become necessary, it is desirable that the new term should contain in its form some memorial of the old one.

We have excellent examples of the advantageous use of this maxim in Linnæus's reform of botanical nomenclature. His innovations were very extensive, but they were still moderated as much as possible, and connected in many ways with the names of plants then in use. He has himself given several rules of nomenclature, which tend to establish this connexion of the old and new in a reform. Thus he says, "Generic names which are current, and are not accompanied with harm to botany, should be tolerated*." "A passable generic name is not to be changed for another, though more apt†." New generic names are not to be framed so long as passable synonyms are at hand‡." "A generic name of one genus, except it be superfluous, is not to be transferred to another genus, though it suit the other better §." "If a received genus requires to be divided into several, the name which before included the whole, shall be applied to the most common and familiar kind ‖." And though he rejects all *generic* names which have not a Greek or Latin root¶, he is willing to make an exception in favour of those which from their form might be supposed to have such a root, though they are really borrowed from other languages, as *Thea*, which is the Greek for goddess ; *Coffea*, which might seem to come from a Greek word denoting silence (κωφὸς); *Cheiranthus*, which appears to mean hand-flower, but is really derived from the Arabic *Keiri :* and many others.

As we have already said, the attempt at a reformation of the

* *Philosophia Botanica*, Art. 242.   † P. 246.   ‡ P. 247.
  § P. 249.   ‖ P. 249.   ¶ P. 232.

nomenclature of Mineralogy made by Professor Mohs will probably not produce any permanent effect, on this account amongst
others, that it has not been conducted in this temperate mode ;
the innovations bear too large a proportion to the whole of the
names, and contain too little to remind us of the known appellations. Yet in some respects Professor Mohs has acted upon this
maxim. Thus he has called one of his classes *Spar*, because
Felspar belongs to it. I shall venture to offer a few suggestions
on this subject of mineralogical nomenclature.

It has already been remarked that the confusion and complexity
which prevail in this subject render a reform very desirable.
But it will be seen, from the reasons assigned under the Ninth
Aphorism, that no permanent system of names can be looked for,
till a sound system of classification be established. The best
mineralogical systems recently published, however, appear to converge to a common point ; and certain classes have been formed
which have both a natural-historical and a chemical significance.
These Classes, according to Naumann, whose arrangement appears
the best, are Hydrolytes, Haloids, Silicides, Oxides of Metals,
Metals, Sulphurides (Pyrites, Glances, and Blendes), and Anthracides. Now we find ;—that the Hydrolytes are all compounds,
such as are commonly termed *Salts ;*—that the Haloids are, many
of them, already called *Spars*, as *Calc Spar, Heavy Spar, Iron
Spar, Zinc Spar ;*—that the *Silicides*, the most numerous and
difficult class, are denoted for the most part, by single words,
many of which end in *ite ;*—that the other classes, or sub-classes,
Oxides, Pyrites, Glances, and *Blendes,* have commonly been so
termed ; as *Red Iron Oxide, Iron Pyrites, Zinc Blende ;*—while
pure metals have usually had the adjective Native prefixed,
as *Native Gold, Native Copper.* These obvious features of
the current names appear to afford us a basis for a systematic
nomenclature. The Salts and Spars might all have the word
salt or *spar* included in their name, as *Natron Salt, Glauber
Salt, Rock Salt ; Calc Spar, Bitter Spar* (Carbonate of Lime
and Magnesia), *Fluor Spar, Phosphor Spar* (Phosphate of
Lime), *Heavy Spar, Celestine Spar* (Sulphate of Strontian),
Chromic Lead Spar (Chromate of Lead) ; the *Silicides* might all

have the name constructed so as to be a single word ending in *ite*, as *Chabasite* (Chabasie), *Natrolite* (Mesotype), *Sommite* (Nepheline), *Pistacite* (Epidote) ; from this rule might be excepted the *Gems*, as *Topaz*, *Emerald*, *Corundum*, which might retain their old names. The Oxides, Pyrites, Glances, and Blendes, might be so termed ; thus we should have *Tungstic Iron Oxide* (usually called Tungstate of Iron), *Arsenical Iron Pyrites* (Mispickel), *Tetrahedral Copper Glance* (Fahlerz), *Quicksilver Blende* (Cinnabar), and the Metals might be termed *native*, as *Native Copper*, *Native Silver*.

Such a nomenclature would take in a very large proportion of commonly received appellations, especially if we were to select among the synonyms, as is proposed above in the case of *Glauber Salt*, *Bitter Spar*, *Sommite*, *Pistacite*, *Natrolite*. Hence it might be adopted without serious inconvenience. It would make the name convey information respecting the place of the mineral in the system ; and by imposing this condition, would limit the extreme caprice, both as to origin and form, which has hitherto been indulged in imposing mineralogical names.

The principle of a mineralogical nomenclature determined by the place of the species in the system, has been recognized by Mr. Beudant as well as Mr. Mohs. The former writer has proposed that we should say *Carbonate Calcaire*, *Carbonate Witherite*, *Sulphate Couperose*, *Silicate Stilbite*, *Silicate Chabasie*, and so on. But these are names in which the part added for the sake of the system is not incorporated with the common name, and would hardly make its way into common use.

We have already noticed Mr. Mohs's designations for two of the Systems of Crystallization, the *Pyramidal* and the *Prismatic*, as not characteristic. If it were thought advisable to re-form such a defect, this might be done by calling them the *Square Pyramidal* and the *Oblong Prismatic*, which terms, while they expressed the real distinction of the systems, would be intelligible at once to those acquainted with the Mohsian terminology.

I will mention another suggestion respecting the introduction of an improvement in scientific language. The term *Depolarization* was introduced, because it was believed that the effect of certain crystals, when polarized light was incident upon them in certain positions, was to destroy the peculiarity which polarization had produced.   But it is now well known that the effect of the second crystal in general is to divide the polarized ray of light into two rays, polarized in different planes.   Still this effect is often spoken of as *Depolarization,* no better term having been yet devised.   I have proposed and used the term *Dipolarization,* which well expresses what takes place, and so nearly resembles the older word, that it must sound familiar to those already acquainted with writings on this subject.

I may mention one term in another department of literature which it appears desirable to reform in the same manner.   The theory of the Fine Arts, or the philosophy which speculates concerning what is beautiful in painting, sculpture or architecture, and other arts, often requires to be spoken of in a single word. Baumgarten and other German writers have termed this province of speculation *Æsthetics;* ἀισθάνεσθαι, *to perceive,* being a word which appeared to them fit to designate the perception of beauty in particular.   Since, however, *æsthetics* would naturally denote the doctrine of perception ; since this doctrine requires a name ; since the term *æsthetics* has actually been applied to it by other German writers (as Kant) ; and since the essential point in the philosophy now spoken of is that it attends to beauty;—it appears desirable to change this name.   In pursuance of the maxim now before us, I should propose the term *Callæsthetics,* or rather (in agreement with what was said in page cxiv.) *Callæsthetic,* the science of the perception of beauty.

I may here notice a principle which may sometimes be allowed to influence us, in selecting one form rather than another for a technical term.   It is convenient to make correlative terms resemble each other in termination, even when the resemblance is only apparent ; thus we may speak of marine and *terrene* animals, rather than *terrestrial* or *tellurian.*   Dr. Prichard speaks of *carnivorous* and *phytiborous* insects ; preferring the latter term to *phytophagous,* on account of its sound, I suppose, as well as for other reasons.

THE

PHILOSOPHY

OF THE

INDUCTIVE SCIENCES.

PART I.

OF IDEAS.

QUÆ adhuc inventa sunt in Scientiis, ea hujusmodi sunt ut notionibus vulgaribus fere subjaceant : ut vero ad interiora et remotiora naturæ penetretur, necesse est ut tam NOTIONES quam AXIOMATA magis certâ et munitâ viâ à particularibus abstrahantur ; atque omnino melior et certior intellectûs adoperatio in usum veniat.

BACON, *Nov. Org.*, Lib. 1. Aphor. xviii.

BOOK I.

OF IDEAS IN GENERAL.

Chapter I.

INTRODUCTION.

The Philosophy of Science, if the phrase were to be understood in the comprehensive sense which most naturally offers itself to our thoughts, would imply nothing less than a complete insight into the essence and conditions of all real knowledge, and an exposition of the best methods for the discovery of new truths. We must narrow and lower this conception, in order to mould it into a form in which we may make it the immediate object of our labours with a good hope of success; yet still it may be a rational and useful undertaking, to endeavour to make some advance towards such a Philosophy, even according to the most ample conception of it which we can form. The present work has been written with a view of contributing, in some measure, however small it may be, towards such an undertaking.

But in this, as in every attempt to advance beyond the position which we at present occupy, our hope of success must depend mainly upon our being able to profit, to the fullest extent, by the progress already made. We may best hope to understand the nature and conditions of real knowledge, by studying the nature and conditions of the most certain and stable portions of knowledge which we already possess: and we are most likely to learn the best methods of discovering truth, by examining how truths, now universally recognised, have really been discovered. Now there do exist among us doctrines of solid and acknowledged certainty, and truths of which the discovery has been received with universal applause. These constitute what we commonly term *Sciences;* and of these bodies of exact and enduring knowledge, we have within our reach so large and varied a collection, that we may examine them, and the history of their formation, with a good prospect of deriving from the study such instruction as we seek. We may best hope to make some progress towards the Philosophy of Science, by employing ourselves upon THE PHILOSOPHY OF THE SCIENCES.

The sciences to which the name is most commonly and unhesitatingly given, are those which are concerned about the material world; whether they deal with the celestial bodies, as the sun and stars, or the earth and its products, or the elements; whether they consider the differences which prevail among such objects, or their origin, or their mutual operation. And in all these sciences it is familiarly understood and assumed, that their doctrines are obtained by a common process of collecting general truths from particular observed facts, which process is termed *Induction.* It is further assumed that both in these and in other provinces of knowledge, so long as this process is duly and legitimately performed, the results will be real substantial truth. And although this process, with the conditions under which it is legitimate, and the general laws of the formation of sciences, will hereafter be subjects of discussion in this work, I shall at present so far adopt the assumption of which I speak, as to give to the sciences from which our lessons are to be collected the name of *Inductive* sciences. And thus it is that I am led to designate my work as THE PHILOSOPHY OF THE INDUCTIVE SCIENCES.

The views respecting the nature and progress of knowledge, towards which we shall be directed by such a course of inquiry as I have pointed out, though derived from those portions of human knowledge which are more peculiarly and technically termed Sciences, will by no means be confined, in their bearing, to the domain of such sciences as deal with the material world, nor even to the whole range of sciences now existing. On the contrary, we shall be led to believe that the nature of truth is in all subjects the same, and that its discovery involves, in all cases, the like conditions. On one subject of human speculation after another, man's knowledge assumes that exact and substantial character which leads us to term it Science; and in all these cases, whether inert matter or living bodies, whether permanent relations or successive occurrences be the subject of our attention, we can point out certain universal characters which belong to truth, certain general laws which have regulated its progress among men. And we naturally expect that even when we extend our range of speculation wider still, when we contemplate the world within us as well as the world without us, when we consider the thoughts and actions of men as well as the motions and operations of unintelligent bodies, we shall still find some general analogies which belong to the essence of truth, and run through the whole intellectual universe. Hence we have reason to trust that a just philosophy of the sciences may throw light upon the nature and extent of our knowledge in every department of human speculation. By considering what is the real import of our acquisitions, where they are certain and definite, we may learn something respecting the difference between true knowledge and its precarious or illusory semblances; by examining the steps by which such acquisitions have been made, we may discover the conditions under which truth is to be obtained; by tracing the boundary-line between our knowledge and our ignorance, we may ascertain in some measure the extent of the powers of man's understanding.

But it may be said, in such a design there is nothing new; these are objects at which inquiring men have often before aimed. To determine the difference between real and imaginary knowledge, the conditions under which we arrive at truth, the range of the powers of the human mind, has been a favourite employment of speculative men from the earliest to the most recent times. To inquire into the original, certainty, and compass of man's knowledge, the limits of his capacity, the strength and weakness of his reason, has been the professed purpose of many of the most conspicuous and valued labours of the philosophers of all periods up to our own day. It may appear, therefore, that there is little necessity to add one more to these numerous essays; and little hope that any new attempt will make any very important addition to the stores of thought upon such questions, which have been accumulated by the profoundest and acutest thinkers of all ages.

To this I reply, that without at all disparaging the value or importance of the labours of those who have previously written respecting the foundations and conditions of human knowledge, it may still be possible to add something to what they have done. The writings of all great philosophers, up to our own time, form a series which is not yet terminated. The books and systems of philosophy which have, each in its own time, won the admiration of men, and exercised a powerful influence upon their thoughts, have had each its own part and functions in the intellectual history of the world; and other labours which shall succeed these may also have their proper office and useful effect. We may not be able to do much, and yet still it may be in our power to effect something. Perhaps the very advances made by former inquirers may have made it possible for us, at present, to advance still further. In the discovery of truth, in the developement of man's mental powers and privileges, each generation has its assigned part; and it is for us to endeavour to perform our portion of this perpetual task of our species. Although the terms which describe our undertaking may be the same which have often been employed by previous writers to express their purpose, yet our position is different from theirs, and thus the result may be different too. We have, as they had, to run our appropriate course of speculation with the exertion of our best powers; but our course lies in a more advanced part of the great line along which philosophy travels from age to age. However familiar and old, therefore, be the design of such a work as this, the execution may have, and if it be performed in a manner suitable to the time, will have, something that is new and not unimportant.

Indeed, it appears to be absolutely necessary, in order to check the prevalence of grave and pernicious error, that the doctrines which are taught concerning the foundations of human knowledge and the powers of the human mind, should be from time to time revised and corrected or extended. Erroneous and partial views are promulgated and accepted; one portion of the truth is insisted upon to the undue exclusion of another; or principles true in themselves are exaggerated till they produce on men's minds the effect of falsehood. When evils of this kind have grown to a serious height, a *reform* is requisite. The faults of the existing systems must be remedied by correcting what is wrong, and supplying what is wanting. In such cases, all the merits and excellencies of the labours of the preceding times do not supersede the necessity of putting forth new views suited to the emergency which has arrived. The new form which error has assumed makes it proper to endeavour to give a new and corresponding form to truth. Thus the mere progress of time, and the natural growth of opinion from one stage to another, leads to the production of new systems and forms of philosophy. It will be found, I think, that some of the doctrines now most widely prevalent respecting the foundations and nature of truth are of such a kind that a reform is needed. The present age seems, by many indications, to be called upon to seek a sounder philosophy of knowledge than is now current among us. To contribute towards such a philosophy is the object of the present work. The work is, therefore, like all works which take into account the most recent forms of speculative doctrine, invested with a certain degree of novelty in its aspect and import, by the mere time and circumstances of its appearance.

But, moreover, we can point out a very important peculiarity by which this work is, in its design, distinguished from preceding essays on like subjects; and this difference appears to be of such a kind as may well entitle us to expect some substantial addition to our knowledge as the result of our labours. The peculiarity of which I speak has already been announced;—it is this: that we purpose to collect our doctrines concerning the nature of knowledge, and the best mode of acquiring it, from a contemplation of the structure and history of those sciences (the material sciences), which are universally recognised as the clearest and surest examples of knowledge and of discovery. It is by surveying and studying the whole mass of such sciences, and the various steps of their progress, that we now hope to approach to the true Philosophy of Science.

Now this, I venture to say, is a new method of pursuing the philosophy of human knowledge. Those who

have hitherto endeavoured to explain the nature of knowledge, and the process of discovery, have, it is true, often
illustrated their views by adducing special examples of
truths which they conceived to be established, and by
referring to the mode of their establishment. But these
examples have, for the most part, been taken at random,
not selected according to any principle or system. Often
they have involved doctrines so precarious or so vague
that they confused rather than elucidated the subject;
and instead of a single difficulty,—What is the nature
of knowledge? these attempts at illustration introduced
two,—What was the true analysis of the doctrines thus
adduced? and,—Whether they might safely be taken as
types of real knowledge?

This has usually been the case when there have been
adduced, as standard examples of the formation of human
knowledge, doctrines belonging to supposed sciences other
than the material sciences;—doctrines, for example,
of political economy, or philology, or morals, or the philosophy of the fine arts. I am very far from thinking
that, in regard to such subjects, there are no important
truths hitherto established: but it would seem that those
truths which have been obtained in these provinces of
knowledge, have not yet been fixed by means of
distinct and permanent phraseology, and sanctioned
by universal reception, and formed into a connected
system, and traced through the steps of their gradual
discovery and establishment, so as to make them instructive examples of the nature and progress of truth
in general. Hereafter we trust to be able to show that
the progress of moral, and political, and philological, and
other knowledge, is governed by the same laws as that
of physical science. But since, at present, the former
class of subjects are full of controversy, doubt, and
obscurity, while the latter consist of undisputed truths

clearly understood and expressed, it may be considered
a wise procedure to make the latter class of doctrines
the basis of our speculations. And on the having taken
this course, is, in a great measure, my hope founded, of
obtaining valuable truths which have escaped preceding
inquirers.

But it may be said that many preceding writers on
the nature and progress of knowledge have taken their
examples abundantly from the physical sciences. It
would be easy to point out admirable works, which have
appeared during the present and former generations, in
which instances of discovery, borrowed from the physical
sciences, are introduced in a manner most happily
instructive. And to the works in which this has been
done, I gladly give my most cordial admiration. But at
the same time I may venture to remark that there still
remains a difference between my design and theirs: and
that I use the physical sciences as exemplifications of the
general progress of knowledge in a manner very materially different from the course which is followed in works
such as are now referred to. For the conclusions stated
in the present work, respecting knowledge and discovery,
are drawn from *a connected and systematic survey of the
whole range of physical science and its history;* whereas,
hitherto, philosophers have contented themselves with
adducing detached examples of scientific doctrines, drawn
from one or two departments of science. So long as we
select our examples in this arbitrary and limited manner,
we lose the best part of that philosophical instruction,
which the sciences are fitted to afford when we consider
them as all members of one series, and as governed by rules
which are the same for all. Mathematical and chemical
truths, physical and physiological doctrines, the sciences of
classification and of causation, must alike be taken into
our account, in order that we may learn what are the

general characters of real knowledge. When our conclusions assume so comprehensive a shape that they apply
to a range of subjects so vast and varied as these, we
may feel some confidence that they represent the genuine
form of universal and permanent truth. But if our
exemplification is of a narrower kind, it may easily
cramp and disturb our philosophy. We may, for instance,
render our views of truth and its evidence so rigid and
confined as to be quite worthless, by founding them too
much on the contemplation of mathematical truth. We
may overlook some of the most important steps in the
general course of discovery, by fixing our attention too
exclusively upon some one conspicuous group of discoveries, as, for instance, those of Newton. We may
misunderstand the nature of physiological discoveries, by
attempting to force an analogy between them and discoveries of mechanical laws, without attending to the
intermediate sciences which fill up the vast interval
between these extreme terms in the series of material
sciences. In these and in many other ways, a partial
and arbitrary reference to the material sciences in our
inquiry into human knowledge may mislead us; or at
least may fail to give us those wider views, and that
deeper insight, which should result from a systematic study
of the whole range of sciences with this particular object.

The design of the following work, then, is to form a
Philosophy of Science, by analysing the substance and
examining the progress of the existing body of the
sciences. As a preliminary to this undertaking, a survey
of the history of the sciences was necessary. This,
accordingly, I have already performed; and the result of
the labour thus undertaken has been laid before the
public as a *History of the Inductive Sciences*.

In that work I have endeavoured to trace the steps
by which men acquired each main portion of that know-

ledge on which they now look with so much confidence and satisfaction. The events which that history relates, the speculations and controversies which are there described, and discussions of the same kind, far more extensive, which are there omitted, must all be taken into our account at present, as the prominent and standard examples of the circumstances which attend the progress of knowledge. With so much of real historical fact before us, we may hope to avoid such views of the processes of the human mind as are too partial and limited, or too vague and loose, or too abstract and unsubstantial, to represent fitly the real forms of discovery and of truth.

Of former attempts, made with the same view of tracing the conditions of the progress of knowledge, that of Bacon is perhaps the most conspicuous: and his labours on this subject were opened by his book on the Advancement of Learning, which contains, among other matter, a survey of the then existing state of knowledge. But this review was undertaken rather with the object of ascertaining in what quarters future advances were to be hoped for, than of learning by what means they were to be made. His examination of the domain of human knowledge was conducted rather with the view of discovering what remained undone, than of finding out how so much had been done. Bacon's survey was made for the purpose of tracing the boundaries, rather than of detecting the principles of knowledge. " I will now attempt," he says*, "to make a general and faithful perambulation of learning, with an inquiry what parts thereof lie fresh and waste, and not improved and converted by the industry of man; to the end that such a plot made and recorded to memory, may both minister light to any public designation, and also serve to excite voluntary endeavours." Nor will it be foreign to our scheme also hereafter to

* *Advancement of Learning*, b. i. p. 74.

examine with a like purpose the frontier of man's intellectual estate. But the object of our perambulation in the first place, is not so much to determine the extent of the field, as the sources of its fertility. We would learn by what plan and rules of culture, conspiring with the native forces of the bounteous soil, those rich harvests have been produced which fill our garners. Bacon's maxims, on the other hand, respecting the mode in which he conceived that knowledge was thenceforth to be cultivated, have little reference to the failures, still less to the successes, which are recorded in his Review of the learning of his time. His precepts are connected with his historical views in a slight and unessential manner. His philosophy of the sciences is not collected from the sciences which are noticed in his survey. Nor, in truth, could this, at the time when he wrote, have easily been otherwise. At that period, scarce any branch of physics existed as a science, except astronomy. The rules which Bacon gives for the conduct of scientific researches are obtained, as it were, by divination, from the contemplation of subjects with regard to which no sciences as yet were. His instances of steps rightly or wrongly made in this path, are in a great measure cases of his own devising. He could not have exemplified his Aphorisms by references to treatises then extant, on the laws of nature; for the constant burden of his exhortation is, that men up to his time had almost universally followed an erroneous course. And however we may admire the sagacity with which he pointed the way along a better path, we have this great advantage over him;—that we can interrogate the many travellers who since his time have journeyed on this road. At the present day, when we have under our notice so many sciences, of such wide extent, so well established; a Philosophy of the Sciences ought, it must seem, to be founded, not upon conjecture, but upon an examination of many instances ;—should not consist of a few vague and unconnected maxims, difficult and doubtful in their application, but should form a system of which every part has been repeatedly confirmed and verified.

This accordingly it is the purpose of the present work to attempt. But I may further observe, that as my hope of making any progress in this undertaking is founded upon the design of keeping constantly in view the whole result of the past history and present condition of science, I have also been led to draw my lessons from my examples in a manner more systematic and regular, as appears to me, than has been done by preceding writers. Bacon, as I have just said, was led to his maxims for the promotion of knowledge by the sagacity of his own mind, with little or no aid from previous examples. Succeeding philosophers may often have gathered useful instruction from the instances of scientific truths and discoveries which they adduced, but their conclusions were drawn from their instances casually and arbitrarily. They took for their moral any which the story might suggest. But such a proceeding as this cannot suffice for us, whose aim is to obtain a consistent body of philosophy from a contemplation of the whole of Science and its History. For our purpose it is necessary to resolve scientific truths into their conditions and ingredients, in order that we may see in what manner each of these has been and is to be provided, in the cases which we may have to consider. This accordingly is necessarily the first part of our task : —*to analyse scientific truth into its elements.* This attempt will occupy the earlier portion of the present work ; and will necessarily be somewhat long, and perhaps, in many parts, abstruse and uninviting. The risk of such an inconvenience is inevitable ; for the inquiry brings before us many of the most dark and entangled questions in which men have at any time busied themselves. And even if these can now be made clearer and plainer than of yore, still they can be made so only by means of mental discipline and mental effort. Moreover this analysis of scientific truth into its elements contains much, both in its principles and in its results, different from the doctrines most generally prevalent among us in recent times : but on that very account this analysis is an essential part of the doctrines which I have now to lay before the reader : and I must therefore crave his indulgence towards any portion of it which may appear to him obscure or repulsive.

There is another circumstance which may tend to make the present work less pleasing than others on the same subject, in the nature of the examples of human knowledge to which I confine myself; all my instances being, as I have said, taken from the material sciences. For the truths belonging to these sciences are, for the most part, neither so familiar nor so interesting to the bulk of readers as those doctrines which belong to some other subjects. Every general proposition concerning politics or morals at once stirs up an interest in men's bosoms, which makes them listen with curiosity to the attempts to trace it to its origin and foundation. Every rule of art or language brings before the mind of culti- vated men subjects of familiar and agreeable thought, and is dwelt upon with pleasure for its own sake as well as on account of the philosophical lessons which it may convey. But the curiosity which regards the truths of physics or chemistry, or even of physiology and astro- nomy, is of a more limited and less animated kind. Hence, in the mode of inquiry which I have prescribed to myself, the examples which I have to adduce will not amuse and relieve the reader's mind as much as they might do, if I could allow myself to collect them from the whole field of human knowledge. They will have in them nothing to engage his fancy, or to warm his heart. I am compelled to detain the listener in the chilly air of the external world, in order that we may have the advantage of full daylight.

But although I cannot avoid this inconvenience, so far as it is one, I hope it will be recollected how great are the advantages which we obtain by this restriction. We-are thus enabled to draw all our conclusions from doctrines which are universally allowed to be eminently certain, clear, and definite. The portions of knowledge to which I refer are well known, and well established among men. Their names are familiar, their assertions uncontested. Astronomy and geology, mechanics and chemistry, optics and acoustics, botany and physiology, are each recognised as large and substantial collections of undoubted truths. Men are wont to dwell with pride and triumph on the acquisitions of knowledge which have been made in each of these provinces; and to speak with confidence of the certainty of their results. And all can easily learn in what repositories these treasures of human knowledge are to be found. When, therefore, we begin our inquiry from such examples, we proceed upon a solid foundation. With such a clear ground of confidence, we shall not be met with general assertions of the vagueness and un- certainty of human knowledge; with the question, what truth is and how we are to recognise it; with complaints concerning the hopelessness and unprofitableness of such researches. We have, at least, a definite problem before us. We have to examine the structure and scheme, not of a shapeless mass of incoherent materials, of which we doubt whether it be a ruin or a natural wilderness, but of a fair and lofty palace, still erect and tenanted, where hundreds of different apartments belong to a common plan, where every generation adds something to the extent and magnificence of the pile. The certainty and

the constant progress of science are things so unques-
tioned, that we are at least engaged in an intelligible
inquiry, when we are examining the grounds and nature
of that certainty, the causes and laws of that progress.

To this inquiry, then, we now proceed. And in
entering upon this task, however our plan or our principles may differ from those of the eminent philosophers
who have endeavoured, in our own or in former times, to
illustrate or enforce the philosophy of science, we most
willingly acknowledge them as in many things our
leaders and teachers. Each reform must involve its own
peculiar principles, and the result of our attempts, so far
as they lead to a result, must be, in some respects,
different from those of former works. But we may still
share with the great writers who have treated this
subject before us, their spirit of hope and trust, their
reverence for the dignity of the subject, their belief in
the vast powers and boundless destiny of man. And we
may once more venture to use the words of hopeful
exhortation, with which the greatest of those who have
trodden this path encouraged himself and his followers
when he set out upon his way.

"Concerning ourselves we speak not; but as touching
the matter which we have in hand, this we ask;—that
men deem it not to be the setting up an Opinion, but the
performing of a Work: and that they receive this as a
certainty; that we are not laying the foundations of any
sect or doctrine, but of the profit and dignity of mankind.
Furthermore, that being well disposed to what shall
advantage themselves, and putting off factions and prejudices, they take common counsel with us, to the end
that being by these our aids and appliances freed and
defended from wanderings and impediments, they may
lend their hands also to the labours which remain to be
performed: and yet further, that they be of good hope;

neither imagine to themselves this our Reform as something of infinite dimension, and beyond the grasp of mortal man, when in truth it is the end and true limit of infinite errour; and is by no means unmindful of the condition of mortality and humanity, not confiding that such a thing can be carried to its perfect close in the space of one single age, but assigning it as a task to a succession of generations."

<hr />

Chapter II.

OF FACTS AND THEORIES.

1. I REGRET very much that I must begin my discussion by questioning the validity of a distinction which is usually considered to be clear and plain. For my purpose is to establish distinctions, not to obliterate them; and with regard to such contrasts as are commonly recognised among men, it will generally be my business rather to point out their real import, and give them as much definiteness as possible, than to endeavour to involve them in doubt and confusion. And, indeed, though I am compelled at first to expose the obscurity of the supposed line which separates Fact and Theory, I shall afterwards have to show that the contrast which we mark by these terms does really involve an antithesis which is the foundation of the whole philosophy of knowledge.

Every one is familiar with the distinction of Fact and Theory as commonly understood. Facts offer themselves to our senses on every side: ingenious men have framed Theories, that is, modes of mental conception, by which the facts are interpreted, connected, and accounted for. Every moment offers us examples of the two. The day dawns; the sun's bright edge beams over the distant hills; that is the fact. The theory is that the earth's surface rolls round towards the sun, and thus brings him into view. The dew-drops hang on the blade and the leaf; their globular form is a fact. By our theory we see in this fact a mutual attraction of the minutest portion of the water which composes the drops. Each drop, as it hangs in the sunshine, has on its surface a bright spot which shifts as the beholder moves, and has behind it another bright speck which falls on some neighbouring object. These facts our theories make us contemplate as the reflected and refracted light of the sun. The plant thus hung with dew exhibits to us its leaves and flowers, but in our minds we compare it with other plants in which the leaves and flowers are more or less different; we consider these facts as indicating the relation of this particular plant to some wider family of the vegetable system, such as in our theory we have arranged it. Or if we are acquainted with the plants of other regions, we may see in the existence and features of such a plant the confirmation of a theory by which we look upon some portion of our vegetable population as strangers wandered hither from a distant land.

2. In all these cases, the distinction between the fact as it presents itself to our senses and the theoretical view, seems at first sight plain enough. Yet a little consideration may show us that this distinction is not in every case quite clear. Is it not a fact as well as a theory that we see the light reflected from the surface of a dew-drop, and do we not by our common language acknowledge it to be so? And is not the refraction of the light through the water as much a fact as its reflection from the surface? Does not the manner in which the drop hangs from the leaf show that it is a fact that the particles of water adhere to or attract each other? Is not

this as much a fact as the globular form of the drops, or
indeed more so, for the drops are not strictly globular?
That they are not so, we learn from theory, and thus
our theory corrects our facts. Is not the greater or less
resemblance of one plant to another a fact? and is not,
therefore, a classification, which is merely a collection of
such resemblances, a fact also? And if the doctrine of
the derivation of any particular plant from one region to
another be a true theory, is it not a fact on that very
account?

And with regard to the first mentioned of the above
cases, the theoretical motion of the earth, is not that also
a fact, if the theory be true? It may be said that the
theory contradicts the facts as noticed by our senses.
But that our senses may misinform us respecting facts,
we easily see. When we glide along smooth water in
a barge, our senses inform us that the shore moves away
from us; but we know the fact to be otherwise. And if
the motion of the barge be the fact in this case, is not
the motion of the earth, by which the sun's rising is produced, a fact no less?

Again, if it were said that *that* is a fact which our
senses perceive, the question must be asked, *whose* senses?
One man watches the stars all the night, and sees them
describe circles about the pole; another looks at them
carelessly and at intervals, and sees no circles. Is not
the diurnal circular motion of the stars a fact? Again,
a man may rightly apprehend the motion of the stars for
one night, but may not notice the motion of the moon
among the stars from night to night. Another man
notices this latter motion also: to him the moon's
monthly circuit through the heavens is a fact. And
again, to another observer, more vigilant, the annual
motion of the sun in the ecliptic is a fact just as much
as the monthly motion of the moon in her orbit. For

the only difference is, that the moon's light quenches only the smaller stars in her neighbourhood, while the sun obliterates all. And thus what is matter of theory to one observer is matter of fact to another.

Is it not, indeed, evident that a theory, if it be true, is on that very account a fact? All the great theories which have successively been established in the world, are now thought of as facts. Is not the motion of the earth round the sun a fact? Is not the elliptical form of the planets' orbits a fact? Is not the attraction of the sun upon the planets a fact? Is not the circulation of the blood a fact? The definite and multiple proportions of the elements of bodies, which make up what is commonly called the atomic theory, are not they facts?

Thus, the opposition of fact and theory—a contrast which at first appeared so broad and plain—as we examine it, becomes wavering, obscure, and doubtful. The line of demarcation is invisible; the application of the distinction full of difficulty. That which is a fact under one aspect is a theory under another. The most recondite theories, when firmly established, are accepted as facts; the simplest facts appear to involve something of the nature of theory.

3. But yet, in what has been said, something of a difference between fact and theory still remains apparent. It is only when theories are firmly established, and recognised as indisputably true, that they become facts, The view, originally theoretical, becomes at length so convincing, that it occurs to us as the most natural view, and then it is theoretical no longer. The interpretation of appearances, which was at first a novelty and an effort, becomes at last so familiar that we are not conscious of it; and then the distinction of theory and fact, in that instance, melts away. Theory is some interpretation of phenomena, or inference from them, which we make

by a conscious act of thought, adding some new form of conception to that which at first offers itself. And as the doubt, and the effort, and the consciousness of the mental act gradually depart, the theory is a theory no longer, but becomes a fact.

And thus, as we become more and more familiar with sound theoretical views, such views become to us as really facts as those which are most obvious to the senses. The astronomer, constantly observing the moon, and determining from her apparent her real motions, sees that she is drawn by the earth, as clearly as a common spectator sees the needle drawn by the magnet. That which is intellectual effort to others is unconscious habit in him. He sees the true motion in the apparent, and separates the compound course into the simple paths, with no more doubt than the voyager feels when, in judging of the course of a distant ship, he allows for the motion of his own vessel, and for his own movements as he walks the deck. And as this true motion of the paths of the earth and moon, which is to him an habitual and inevitable interpretation of their visible changes, is thus a fact, the mutual attraction of the two bodies, which is but a further interpretation, equally inevitable, of those motions, is also a fact to him : while to those less accustomed to such interpretations, and who, therefore, cannot apply them without a conscious act of thought, such a view of the case,· even when accepted as true, is more properly described as theory.

4. In this instance, the doctrine of which I have spoken,· the attraction which the earth exerts upon the moon, would be termed a theory by most persons ; because those to whom this is a familiar and simple inference from the phènomena are only a few accomplished astronomers. But, in other similar cases, many, or most persons, perform a similar act of interpretation, without being con-

scious of it. When we assert that the magnet draws the
needle, we see only the motion of the needle which occurs when the magnet is brought into its neighbourhood.
It is by an act of our own minds that we ascribe this
motion to a force. That in this case a force is exerted
upon the needle, such as we could by our volition exert,
is our unconscious interpretation of the phenomena, and
is hence received by us as a fact.

5. But it is not in such cases only that we interpret
phenomena in our own way, without being conscious of
what we do. We see a tree at a distance, and judge it
to be a chestnut or a lime ; yet this is only an inference
from the colour or form of the mass, according to preconceived classifications of our own. Our lives are full of
such unconscious interpretations. The farmer recognises
a good or bad soil; the artist a picture of a favourite
master ; the geologist a rock of a known locality, as we
recognise the faces and voices of our friends ; that is, by
judgments formed on what we see and hear ; but judgments in which we do not analyse the steps, or distinguish
the inference from the appearance. And in these mixtures of observation and inference, we speak of the
judgment thus formed, as a fact directly observed.

Even in the case in which our perceptions appear to
be most direct, and least to involve any interpretations of
our own,—in the simple process of seeing,—who does
not know how much we, by an act of the mind, add to
that which our senses receive ? Does any one fancy that
he sees a solid cube ? It is easy to show that the solidity
of the figure, the relative position of its faces and edges
to each other, are inferences of the spectator ; no more
conveyed to his conviction by the eye alone, than they
would be if he were looking at a painted representation of
a cube. The scene of nature is a picture without depth
of substance, no less than the scene of art ; and in the

one case as in the other, it is the mind which, by an act
of its own, discovers that colour and shape denote distance and solidity.   Most men are unconscious of this
perpetual habit of reading the language of the external
world, and translating as they read.   The draughtsman,
indeed, is compelled, for his purposes, to return back in
thought from the solid bodies which he has inferred, to
the shapes of surface which he really sees.   He knows
that there is a mask of theory over the whole face of
nature, if it be *theory* to infer more than we *see*.   But
other men, unaware of this masquerade, hold it to be a
fact that they see cubes and spheres, spacious apartments
and winding avenues.   And these things are facts to
them, because they are unconscious of the mental operation by which they have penetrated nature's disguise.

And thus we still have an intelligible distinction of
fact and theory, if we consider theory as a conscious, and
fact as an unconscious inference from the phenomena
which are presented to our senses.

6. Yet still the distinction thus stated is far from
being rigorous and permanent, as, in truth, we have already
seen that in practice it is very precarious and obscure.
The difference of conscious and unconscious acts is by no
means strongly marked.   Education, habit, the degree of
self-observation, the circumstances of the case, all serve
to make the person unconscious or conscious of mental
acts in innumerable degrees.   The draughtsman sees in
nature features and outlines which others do not see.
The practised astrologer sees the moon walk from house
to house in her path, as he sees his friend walk from house
to house in the street; the beginner in the study sees this
with conscious effort.   But one of these habits gradually
passes into the other.   The distinction of conscious and
unconscious acts of thought fades away as we examine it.
We may walk or talk, as well as see, without conscious

effort; yet walking and talking imply acts of thought, as we perceive when we walk on a rugged path, or talk in a foreign language. Hence if this greater or less consciousness of our own internal act be all that distinguishes fact from theory, we must allow that the distinction is still untenable. The boundary-line again melts away; the difference is unsubstantial; the opposition loses its significance as we examine it.

Still there appears to be something real in this antithesis, and we must return to the examination of it under another form.

CHAPTER III.

OF SENSATIONS AND IDEAS.

1. IT has appeared that facts as well as theories involve some act of the mind. But it is also clear that they must involve something else besides an act of the mind. If we must exercise an act of thought in order to see force exerted, or orbits described by bodies in motion, or even in order to see bodies in space, and to distinguish one kind of object from another, still the act of thought alone does not make these objects. There must be something besides, *on which* the thought is exerted. A colour, a form, a sound, are not produced by the mind, however they may be moulded, combined, and interpreted by our mental acts. A philosophical poet has spoken of

> .   .   .   . All the world
> Of eye and ear, both what they half create,
> And what perceive.

But it is clear that though they *half* create, they do not wholly create; there must be an external world of colour and sound to give impressions to the eye and ear, as well as internal powers by which we perceive what is offered to those organs. The mind is in some way passive as well as active: there are sensations as well as acts of thought; objects without, as well as faculties within.

2. Indeed this is so far generally acknowledged, that according to common apprehension, the mind is passive *rather* than active in acquiring the knowledge which it receives concerning the material world. Its sensations are generally considered as more evident than its operations. The world without is held to be more clearly real than the faculties within. That there is something different from ourselves, something external to us, something independent of us, something which no act of our minds can make or can destroy, is held by all men to be at least as evident as that our minds exert any effectual process in modifying and appropriating the impressions made upon them. Most persons are more likely to doubt whether the mind be always active in contemplating external objects, than whether it be always passive in perceiving them.

This question, however, we have already, in some measure, answered; for we have shown, that in many instances where we are at the time unconscious of what we do, we are combining, interpreting, reasoning from the appearances which we have before our eyes; and that without this operation we cannot know anything, nor even recognise any single body as existing in the space about us. This view of the process of perception will be further prosecuted hereafter; but, in the mean time, we have, it may be hoped, made it appear· that in his apprehension of the objects which nature presents to him, man is both active and passive: that he has both *Ideas* and *Sensations*.

3. I use the term *Idea* here to designate those inevitable general relations which are imposed upon our perceptions by acts of the mind, and which are different from anything which our senses directly offer to us. Thus

we see various shades, and colours, and shapes before us; but the *outlines* by which they are separated into distinct objects, the conception by which they are considered as *solid bodies*, at various distances from us; these elements are not ministered by the senses, but supplied by the mind itself. And in drawing the outlines of bodies, in placing them at different distances from us, the mind proceeds in accordance with certain necessary general relations which are involved in the Idea of *Space*. In like manner when, seeing the motions of a needle towards a magnet, we conceive an attractive force exerted and obeyed, we form this conception by referring these motions to the Idea of *Cause*.

Our sensations are constantly apprehended in subordination to such ideas as these. And ideas of this wide and comprehensive nature, such as space and time, number and figure, cause and resemblance, which are the source of an innumerable series of more limited conceptions, I term *Fundamental Ideas*; and I shall hereafter endeavour to enumerate and analyse some of the most important of them.

4. I am thus using the term *Idea* in a very wide sense. But yet this use of it is far more limited than that which occurs in common language. For I restrict its application to the relations and conditions which are imposed on our sensations through the activity of the mind; and thus I do not apply the term to any impressions made upon the mind in virtue of its passive nature merely. Whereas the term *idea* has often been used for almost all imaginable results of our passive and active powers combined. If we speak of an idea of any existing object, as for example, of St. Paul's cathedral, we denote by this use of the term, a combination of various recollected impressions of form and colour, as well as order and symmetry, and we thus include in the word a mixture of Sensations, as well as Ideas in the more exact sense which

I would assign to the term. But the word thus applied appears to answer no purpose of analysis; or at least not the purpose which we have here in view. The distinction of *Sensations* which the mind passively receives, and *Ideas* which it actively employs, is of the highest importance in order to the prosecution of our investigations. And in order that we may keep this difference steadily before us, I shall trust to be allowed the liberty of assigning to these terms, in these pages, this definite and constant sense. I must now further consider the distinction and independence of these two elements.

<hr />

CHAPTER IV.

OF THE DIFFERENCE AND OPPOSITION OF SENSATIONS AND IDEAS.

1. *Ideas and Sensations are distinct.*—Thus Ideas are the active, Sensations the passive element of our minds. But it may be urged, that it is impossible to make such a separation of our consciousness. There are, as we have already said, few cases, if any, in which the mind is entirely passive; some act of the mind accompanies the reception of our most tranquil perceptions. And on the other hand, it is clear that no act of the mind can be conceived without some impression previously made on the senses. Without the use of sight and touch, where would be our idea of space, or number, or resemblance, or cause? And thus, it may be said, ideas (in our sense of the term) without sensations, and sensations without ideas, are altogether idle and imaginary hypotheses. They can nowhere be found in reality. And a distinction where separation is impossible can be of no use or value.

To this we reply, that although it is impossible completely to separate, in any actual cases, sensations and

ideas, nevertheless the distinction is real, and the opposition of the two is a principle of the most essential importance in all philosophy. And this principle we must
endeavour to illustrate further.

The distinction has constantly exercised a very important influence on the speculations respecting the nature
of knowledge in which men have employed themselves in
all ages; and in the course of these speculations it has
been illustrated by means of various images. One of the
most ancient of these, and one still very instructive, is
that which speaks of the sensations as the *matter*, and
the ideas as the *form* of our knowledge : just as ivory
is the matter, and a cube the form, of a die. And this
comparison may at least show us how little force there is
in the objection just stated, that sensations and ideas are
not separable in fact, and therefore that their separation
in our reasonings can be of no service. For the same is
the case with respect to matter and form. These two
things cannot, by any means, be detached from each
other. The ivory must have some form; if not a cube,
a sphere, or some other. The cube, in order to be a cube,
must be of some material or other. A figure without
matter is merely a geometrical conception;—an idea.
Matter without figure is a mere abstract term;—a supposed union of sensible qualities which, so insulated from
others, cannot exist. Yet the distinction of matter and
form is real, for it is clear and plain as a subject of contemplation. And it is by no means useless. For the
speculations which treat of materials are very widely
separated from those which treat of figure. On each
subject there may be much to be said; and the two
subjects would be, through their whole extent, distinct.
The researches concerning the two may involve principles
as different, for example, as the principles of chemistry and
geometry. If, therefore, we were to refuse to consider the

matter and the form of bodies separately, because we
cannot exhibit matter and form separately, we should
shut the door to all philosophy on such subjects. And
the same is the case with the analogous instance of sensations and ideas.

2. *Ideas are not Transformed Sensations.*—In a certain
school of speculators there has existed a disposition to
derive all our ideas from our sensations, the term *idea*
having been used in its wider sense, so as to include all
modifications and limitations of our Fundamental Ideas.
The doctrines of this school have been summarily
expressed by saying that " every idea is a transformed
sensation." Now, even supposing this assertion to be
exactly true, we easily see, from what has been said, how
little we are likely to answer the ends of philosophy, by
putting forward such a maxim as one of primary importance. For we might say, in like manner, that every
statue is but a transformed block of marble, or every
edifice but a collection of transformed stones. But
what would these assertions avail us, if our object were to
trace the rules of art by which beautiful statues were
formed, or great works of architecture erected? The
question naturally occurs, What is the nature, the principle, the law of this transformation? In what faculty
resides the transforming power? What train of ideas
of beauty, and symmetry, and stability, in the mind of
the statuary or the architect, has produced those great
works which mankind look upon as among their most
valuable possessions;—the Apollo of the Belvedere, the
Parthenon, the Cathedral of Cologne? When this is what
we want to know, how are we helped by learning that the
Apollo is of Parian marble, or the Cathedral of basaltic
stone? We must know much more than this, in order
to acquire any insight into the principles of statuary or of
architecture. In like manner, in order that we may

make any progress in the philosophy of knowledge, which is our purpose, we must endeavour to learn something further respecting ideas than that they are transformed sensations, even if they were this.

But, in reality, the assertion that our ideas are transformed sensations, is erroneous as well as frivolous. For it conveys, and is intended to convey, the opinion that our sensations have one form which properly belongs to them; and that, in order to become ideas, they are converted into some other form. But the truth is, that our sensations, of themselves, without some act of the mind, such as involves what we have termed an idea, have no form. We cannot see one object without the idea of space; we cannot see two without the idea of resemblance or difference; and space and difference are not sensations. Thus, if we are to employ the metaphor of matter and form, which is implied in the expression to which I have referred, our sensations, from their first reception, have their form not *changed*, but *given* by our ideas. Without the relations of thought which we here term ideas, the sensations are matter without form. Matter without form cannot exist: and in like manner sensations cannot become perceptions of objects, without some formative power of the mind. By the very act of being received as perceptions, they have a formative power exercised upon them, the operation of which might be expressed by speaking of them, not as *transformed*, but simply as *formed;*—as invested with form, instead of being the mere formless material of perception. The word *inform*, according to its Latin etymology, at first implied this process by which matter is invested with form. Thus Virgil* speaks of the thunderbolt as *informed* by the

* Ferrum exercebant vasto Cyclopes in Antro
  Brontesque Steropesque et nudus membra Pyracmon;
  His informatum manibus, jam parte polita
  Fulmen erat.—*Æn.* viii. 424.

hands of Brontes, and Steropes, and Pyracmon. And
Dryden introduces the word in another place :—

> Let others better mould the running mass
> Of metals, or *inform* the breathing brass.

Even in this use of the word the form is something
superior to the brute matter, and gives it a new significance and purpose. And hence the term is again used
to denote the effect produced by an intelligent principle
of a still higher kind :—

> . . . . . . He *informed*
> This ill-shaped body with a daring soul.

And finally even the soul itself, in its original condition,
is looked upon as matter, when viewed with reference to
education and knowledge, by which it is afterwards
moulded; and hence these are, in our language, termed
information. If we confine ourselves to the first of these
three uses of the term, we may correct the erroneous
opinion of which we have just been speaking, and retain
the metaphor by which it is expressed, by saying, that
ideas are not *transformed*, but *informed* sensations.

3. *Subjective and Objective.*—There is another mode of
expressing the distinction of our sensations and our
ideas, which has been often used by writers on such
topics, although, in our own country, of late years, it has
not been familiar to general readers. According to the
technical language of ancient philosophy, any one's qualities and acts are attributes of which he is the *subject;*
and thus the mind is the subject to which its own ideas
and operations appertain. But these ideas are employed
upon external *objects*, and from external objects all his
sensations proceed. Hence that part of man's mental
occupation which springs from the faculties and operations
of his own mind is *subjective*, while that which flows in
upon him from the world external to him is *objective*.
And as in his contemplation of nature there is always

some act of thought which depends on himself, and some matter of thought which is independent of him, there is in every part of his knowledge a subjective and an objective element.

This phraseology is very familiar in the philosophical writers of Germany and France, and is not uncommon in every age of our own literature. But whether or no we think fit to adopt these terms, the opposition which they imply is one of essential and fundamental importance, in all our speculations concerning the nature of knowledge. We may express the opposition in what terms we please; we may speak, for instance, of internal and external sources of our knowledge; of the world within and the world without us; of man and nature; of ideas and experience; and of many other antitheses. But, in whatever way we denote the contrast of the subjective and objective part of our speculations, the distinction is real and solid, and we shall hereafter see how essential the principle of this contrast is, in order to express the laws of the successful prosecution of knowledge.

The combination of the two, of ideas and experience, is, as we shall see, necessary, in order to give us any knowledge of the external world, any insight into the laws of nature. Different persons, according to their mental habits and constitution, may be inclined to dwell by preference upon one or the other of these two elements. But no knowledge can exist without the practical union of the two, nor any philosophy without their speculative separation. It may, perhaps, interest the reader to see this combination and this opposition illustrated in the intercourse of two eminent men of genius of modern times, Göthe and Schiller.

Göthe himself gives us the account to which I refer, in his history of the progress of his speculations concerning the metamorphosis of plants; a mode of viewing their structure by which he explained, in a very striking and beautiful manner, the relations of the different parts of a plant to each other; as has been narrated in the *History of the Inductive Sciences.* Göthe felt a delight in the passive contemplation of nature, unmingled with the desire of reasoning and theorizing; a delight such as naturally belongs to those poets who merely embody the images which a fertile genius suggests, and do not mix with these pictures, judgments and reflections of their own. Schiller, on the other hand, both by his own strong feeling of the value of a moral purpose in poetry, and by his adoption of a system of metaphysics in which the subjective element was made very prominent, was well disposed to recognize fully the authority of ideas over external impressions.

Göthe for a time felt a degree of estrangement towards Schiller, arising from this contrariety in their views and characters. But on one occasion they fell into discussion on the study of natural history; and Göthe endeavoured to impress upon his companion his persuasion that nature was to be considered, not as composed of detached and incoherent parts, but as active and alive, and unfolding herself in each portion, in virtue of principles which pervade the whole. Schiller objected that no such view of the objects of natural history had been pointed out by observation, the only guide which the natural historians recommended; and was disposed on this account to think the whole of their study narrow and shallow. "Upon this," says Göthe, "I expounded to him, in as lively a way as I could, the metamorphosis of plants, drawing on paper for him, as I proceeded, a diagram to represent that general form of a plant which shows itself in so many and so various transformations. Schiller attended and understood; and, accepting the explanation, he said, 'This is not observation, but an idea.' I replied,"

adds Göthe, "with some degree of irritation; for the
point which separated us was most luminously marked by
this expression: but I smothered my vexation, and
merely said, 'I was happy to find that I had got ideas
without knowing it; nay, that I saw them before my
eyes.'" Göthe then goes on to say, that he had been
grieved to the very soul by maxims promulgated by
Schiller, that no observed fact ever could correspond with
an idea; since he himself loved best to wander in the
domain of external observation, he had been led to look
with repugnance and hostility upon anything which professed to depend upon ideas. "Yet," he observes, "it
occurred to me that if my observation was identical with
his idea, there must be some common ground on which
we might meet." They went on with their mutual explanations, and became intimate and lasting friends.
"And thus," adds the poet, "by means of that mighty
and interminable controversy between *object* and *subject*,
we two concluded an alliance which remained unbroken,
and produced much benefit to ourselves and others."

The general diagram of a plant, of which Göthe here
speaks, must have been a combination of lines and marks
expressing the relations of position and equivalence among
the elements of vegetable forms, by which so many of
their resemblances and differences may be explained.
Such a symbol is not an Idea in that general sense in
which we propose to use the term, but is a particular
modification of the general ideas of symmetry, developement, and the like; and we shall hereafter see, according
to the phraseology which we shall explain in the next
chapter, such a diagram might express the *ideal conception*
of a plant.

4. *Other modes of expressing this antithesis.*—Besides
this antithesis of subjective and objective, some of the
more recent schools of German metaphysics have ex-

pressed the same opposition in other ways. They have, for instance, divided the universe into the *Me* and the *Not-me* (*Ich* and *Nicht Ich*). Upon such attempts, we may observe, that the fundamental distinction between our own thoughts and the objects of our thoughts is of the highest consequence; but that, if this distinction be clearly understood and recognised, little appears to be gained by expressing it in any novel manner. The most weighty part of the philosopher's task is to analyse the operations of the mind, and for this purpose, it can aid us but little to call it, instead of the mind, the *subject*, or the *me*. Whenever it appears that our views can be enunciated more clearly by the use of such phraseology, we shall not scruple to avail ourselves of it; but we shall not think it necessary to dilate upon these different modes of expressing the same truth.

CHAPTER V.

OF IDEAL CONCEPTIONS.

1. By what·has been said, we are directed towards an analysis of our thoughts and our knowledge into two opposite elements—Sensations, and Ideas. The latter element will require further examination; and this must be the more carefully conducted, in consequence of the great vagueness and vacillation with which the term has commonly been used. The word *idea* is not unfrequently employed to designate those conceptions which the mind forms, and which it expresses by means of general terms; for example (taking, as our plan requires, instances from the sciences), an angle, a circle, a central force, a reflected or refracted ray, a neutral salt, a rose, a reptile. Or, again, we may employ this term *idea* to express cer-

tain wider fields of mental apprehension, each of which includes many such conceptions as the above; as when we speak of the idea of space, of time, of number, of cause, of composition, of resemblance, of symmetry, of organization. It will be necessary for our purpose to distinguish these two modes of thought. The latter I shall term Fundamental Ideas; and I shall, in the succeeding Books, enumerate and scrutinize such ideas in succession. The other class of notions I shall term Ideal Conceptions, for reasons which I shall soon state.

Each of the Fundamental Ideas supplies us with many Ideal Conceptions. Thus straight lines, angles, polygons, cubes, tangents, curvatures, and the like, are all modifications of the fundamental idea of *space*. In like manner, the fundamental idea of *cause* furnishes us with such conceptions as accelerating and moving force, pressure and inertia, attraction and repulsion. The fundamental idea of *resemblance* gives rise to the conceptions of class, genus, species; and when followed into futher detail, and developed by the suggestions of observation, this, along with other ideas, produces the conception of a particular genus or species, as a rose; and so on, in other cases.

2. Much perplexity and difference of opinion have prevailed among metaphysicians respecting these ideal conceptions. It has been a matter of long and intricate discussion, what is the object, or act, of thought, which is denoted by general terms. Some have held that we have in our minds a *real* idea, something of the nature of an image, which we signify by such terms;—that we have, in this sense, a general idea of an angle, a polygon, a central force, a crystal, a rose. Others have held that in using such terms there is merely an act of the mind marked by a *name;*—an act by which the mind collects and connects many impressions. These two views (that of the *Realists* and that of the *Nominalists*) have prevailed, with various

fluctuations and modifications, through all ages of philosophy. But that either opinion, in its extreme form, involves us in insuperable difficulties, is easily seen: and of late, both parties appear to be willing to adopt the word *conception* as expressing that which by such terms we intend. This word, indicating both an act of the mind by which unity is given to that which was previously scattered, and the result of the act abiding with us when the act is performed, partakes of both views, so far as each is true, and will most conveniently aid us in proceeding with our analysis.

3. But to the word *Conception* I join the adjective *Ideal*. For we have to use the term, not to describe the mental images of individual objects casually taken, but to denote those definite abstract conceptions which are the subjects of our general knowledge. These we can reason upon securely, precisely because they are modifications of our Fundamental Ideas; for these Ideas, as we shall hereafter show, contain the grounds of demonstrative truth. The Conception of a Circle is determined by relations involved in the Idea of space, and hence its properties can be certainly known. The Conception of mutual attraction involves necessary principles derived from the Idea of cause. The Conception of a crystalline arrangement of particles involves the Idea of symmetry; and the case is similar in other examples. Hence I term these *Ideal* Conceptions; intending by this designation to remind the reader that the unity which these conceptions give to the circumstances included in them, is not a casual or arbitrary unity, but is derived from the necessity of the case. There are ideal relations which necessarily form the foundation of our knowledge in each province of human thought; and these relations govern our conceptions at first, as well as determine the scientific truths which, by means of our conceptions once formed, we are able to enunciate.

4. Since the Ideal Conceptions, of which we here speak, are only modifications and limitations of the Fundamental Ideas themselves, the reader will not think it strange that sometimes it may not be easy to draw a line of distinction between Ideas and Conceptions, in the senses in which we have used the terms. The modification may be of so comprehensive a character that it may appear almost as extensive as the idea itself, and as well fitted to supply a foundation for general truths. Thus, we may doubt whether Number be a modification of the Idea of Time, or an independent Idea; and some persons may decide that the Idea of Number supplies us with principles which are the proper foundation of arithmetic, without any reference to the Idea of Time. In like manner, some may be of opinion that mechanical Force is a distinct Idea, distinguishable from the Idea of Cause, and capable of affording us those axioms on which the reasonings of the science of mechanics must rest. Now, with respect to doubts and ambiguities of this kind, we may observe, that it is of small moment to our view of the Philosophy of Science how they are decided. Whether Number and Force be called Ideas or Conceptions, they are fundamental so far as the sciences founded upon them are concerned; and they partake of the nature of ideas at least so far as this, that they are the sources of necessary truth, as we shall hereafter show. We shall analyse the truths of arithmetic and of mechanics, so as to see that they depend upon our necessary mode of apprehending number and force. Whether we can analyse these modes of apprehension still further, is another question;—not without interest in itself, but not affecting our previous analysis. Hence it will not be inconsistent with the general course of our speculations, if number, force, and the like very general modifications of our ideas, should occasionally, in these pages, be

themselves termed Ideas. To reduce our Fundamental
Ideas to the smallest possible number, rigorously independent of each other, is a problem which, perhaps, we
have not completely solved in the present work; but any
defect in the solution of this problem will by no means
affect our general reasonings.

5. It has been said by some writers*, that all conceptions, whether ideal, (that is, of such a general kind as
those just adduced,) or conceptions of particular objects,
are merely *states* or *feelings* of the mind. That these
conceptions all belong to the mind in some way, being
its creations or acts, or, if any one prefers the expression, its states or feelings, (although the latter terms appear
far less appropriate,) it is superfluous to assert or to deny.
But if it be meant, by saying that all conceptions are
merely feelings of the mind, to imply that this general
description of them supersedes or diminishes the necessity of examining minutely their differences, their properties, and the very curious and complex principles
which they involve, the opinion appears to be very
unphilosophical; and the phrase which suggests it is
likely only to mislead us. We shall, we trust, show
hereafter, that these acts or states of mind, by whatever
name they be called, contain in them very fertile and
varied elements of truth: and we are in no way forwarded in our pursuit of such elements, by being told
that all conceptions about which we can reason are
merely so many states of the mind. The question still
remains, what are the peculiarities of each of those
states? and to what conclusions do they entitle us to
proceed? When we say that the conceptions of straight
lines and circles are *merely* states of the mind, we rather
increase, than diminish, the difficulty of understanding

* Brown's *Lectures*, vol. ii.

how these states of mind, and no other, make the whole body of geometrical knowledge possible.

We must now endeavour to explain in what manner such Ideal Conceptions as those which we have pointed out, enter into the formation of our Knowledge.

CHAPTER VI.

OF INDUCTION.

1. WHEN we have become possessed of such ideal conceptions as those just described, cases frequently occur in which we can, by means of such conceptions, connect the facts which we learn from our senses, and thus obtain truths from materials supplied by experience. In such cases, the truth to which we are thus led is said to be collected from the observed facts, *by Induction*.

Thus Hipparchus, tracing the unequal motion of the sun among the stars, in different parts of the year, as learnt from observation, found that this inequality might be fitly represented by the conception of an *eccentric ;*—a circle in which the sun had an equable annual motion, the spectator not being situated in the centre of the circle. And thus he established, by Induction, the truth that the sun appears to move in such an eccentric. At a later period, Kepler, proceeding upon more exact observations, was able to show that, not a circle about an eccentric point, but an ellipse about the focus, was the conception which truly agreed with the motion of the earth, and of the other planets, about the sun. And thus the elliptical form of these orbits was established by Induction from many observed facts. Again, to take an example of another kind, the forms of flowers may have applied to them conceptions borrowed from the idea of symmetry of parts; and this symmetry may contain *three* similar portions, as in the lily and its tribe; or *five*, as in the wild rose, and many others. Now, it appears by observation of many particular cases, that these differences in the kind of symmetry of the flower are conjoined with differences in the seed: the tripartite symmetry prevailing in those seeds which have only one cotyledon, or lobe enveloping the embryo; and the quinquepartite symmetry in those seeds which have two cotyledons. Here, then, we have a truth concerning the laws of vegetable form, established by Induction*.

In these, and in all cases of induction, the ideal conception which the mind itself supplies is *superinduced* upon the facts as they are originally presented to observation. Before the inductive truth is detected, the facts are there, but they are many and unconnected. The conception which the discoverer applies to them gives them connexion and unity. Before Hipparchus, it was known that the motion of the sun was not equable; but it appeared to be irregular and lawless: all parts of the motion became regular and orderly, by the introduction of the conception of the eccentric. In the case of Hipparchus, we can only conjecture the nature of the efforts by which the conception was discovered and applied to the facts. But in Kepler's case we know from his own narrative how hard he struggled and laboured to find the right conception; how many conceptions he tried and rejected; what corrections and adjustments of his first guesses he afterwards introduced. In his case we see in the most conspicuous manner the philosopher impressing his own ideal conception upon the facts; the facts being exactly fitted to this conception, although no one before had detected such a fitness. And in like manner, in all other cases, the discovery of a truth by

* *Hist. Inductive Sciences*, iii. 338.

induction consists in finding a conception or combination of conceptions which agrees with, connects, and arranges the facts.

2. Such ideal conceptions or combinations of conceptions, superinduced upon the facts, and reducing them to rule and order, are *theories*. And thus we seem to have again brought before us, as a real and positive distinction, that separation of fact and theory, which, in the outset of our inquiry (Chap. II.), we found ourselves compelled to reject. For we are at present led to this result:—that a theory is a truth collected from facts by induction; that is, by superinducing upon the facts ideal conceptions such as they truly agree with.

Of the apparent contradiction thus brought before us, the explanation is this:—that what we commonly term *facts* involve an act of the mind of the same kind as that which we have described as *induction*, and thus do not in that respect differ essentially from *theory*. Thus we speak of the eccentric *theory* of the sun's motions, as collected by induction from the *facts* of his unequal motion at different times of the year. But these facts are themselves theories collected by induction. For they depend upon the conception of an ecliptic, or circle passing round the heavens, in which ecliptic the sun's motion at each time is to be inferred by referring his places to the stars. But this ecliptic and these modes of reference are manifestly creations of the mind. And notwithstanding this artificial mode of measuring the sun's motion, the motion itself is as much a fact as the moon's motion among the stars, which is visible to the eye. Nor is there essentially any difference even in the mode of perceiving these motions. For in our apprehension of the moon's motion among the stars, we assign to her a path and a velocity which are conceptions of our own minds, and no mere impressions upon the senses. And thus as theories are collected by induction from facts,

facts are collected by an induction of the same kind from other facts, and so on, till we approach to bare impressions upon the senses, which yet we can never quite divest of some conception or other. The act of the mind, by which it converts facts into theories, is of the same kind as that by which it converts impressions into facts. In both cases there is a new principle of unity introduced by the mind, an ideal connexion established: that which was many becomes one; that which was loose and law-less becomes connected and fixed by rule. And this is done by induction; or, as we have described this process, by superinducing upon the facts, as given by observation, the conception of our own minds.

3. It has already been noticed that there is in different cases a wide difference as to the degree in which we are conscious of this operation. In some cases we see the facts distinct and separate, before they are brought to-gether by the conception of our own minds. In other cases we never contemplate them thus detached, and can hardly conceive them under any other form than that which our conceptions give them. Yet it is easy to see that these two classes of cases pass into each other by in-sensible gradations. To take an example of this: if we had to decipher an ancient inscription, of which a few broken letters and imperfect marks only remained, we might possibly, by an intimate acquaintance with the lan-guage in which it was written, and with the usual forms of such inscriptions, and by the aid of great sagacity and perseverance, discover the meaning, so that no doubt should remain of the justness of our conjecture. In this case, we might with propriety assert the import of the legend to be obtained by induction from the few facts which were placed before us. If the inscription were entire and plainly legible, we should, without hesitation, assert it to be a fact that we had before our eyes the declaration, whatever it was, which the legend might con-

tain. Yet in the latter case, as well as in the former, it is plain that there is much which the mind itself supplies, in addition to the impressions which it receives; much which it brings, as well as that which it finds. In the one case, as in the other, the reader must be provided with knowledge of the letters and of the language; and, if not in the same degree as in the other case, yet no less necessarily, with attention and coherence of thought. If there be induction in the one case, it must exist, more obscurely, perhaps, but no less certainly, in the other also.

And thus it appears that, understanding the term *induction* in that comprehensive sense in which alone it is consistent with itself, it is requisite to give unity to a fact, no less than to give connexion to a theory; and the conclusion at which we formerly arrived, that fact and theory pass into each other by insensible degrees, is not disturbed, but confirmed and illustrated by our view of induction, as the act of superinducing upon the impressions of observation an ideal conception, by which they receive connexion and unity.

Chapter VII.

OF SUCCESSIVE GENERALIZATIONS.

1. Thus we are again led to the doctrine, that Fact and Theory have no essential difference, except in the degree of their certainty and familiarity. Theory, when it becomes firmly established and steadily lodged in the mind, becomes Fact; and thus, as our knowledge becomes more sure and more extensive, we are constantly transferring to the class of facts, opinions which were at first regarded as theories.

Now we have further to remark, that in the progress

of human knowledge respecting any branch of speculation, there may be *several* such steps in succession, each depending upon and including the preceding. The theoretical views which one generation of discoverers establishes, become the facts from which the next generation advances to new theories. As they rise from the particular to the general, they rise from what is general to what is more general. Each induction supplies the materials of fresh inductions; each generalization, with all that it embraces in its circle, may be found to be but one of many circles, comprehended within the circuit of some wider generalization.

This remark has already been made, and illustrated, in the *History of the Inductive Sciences**; and, in truth, the whole of the history of science is full of suggestions and exemplifications of this course of things. It may be convenient, however, to select a few instances which may further explain and confirm this view of the progress of scientific knowledge.

2. The most conspicuous instance of this succession is to be found in that science which has been progressive from the beginning of the world to our own times, and which exhibits by far the richest collection of successive discoveries: I mean astronomy. It is easy to see that each of these successive discoveries depended on those antecedently made, and that in each, the truths which were the highest point of the knowledge of one age were the fundamental basis of the efforts of the age which came next. Thus we find, in the days of Greek discovery, Hipparchus and Ptolemy combining and explaining the particular *facts* of the motion of the sun, moon, and planets, by means of the *theory* of epicycles and eccentrics;—a highly important step, which gave an intelligible connexion and rule to the motions of each of these luminaries. When

---

* *Hist. Inductive Sciences*, ii. 182.

these cycles and epicycles, thus truly representing the apparent motions of the heavenly bodies, had accumulated to an inconvenient amount, by the discovery of many inequalities in the observed motions, Copernicus showed that their effects might all be more simply included, by making the sun the centre of motion of the planets, instead of the earth. But in this new view he still retained the epicycles and eccentrics which governed the motion of each body. Tycho Brahe's observations, and Kepler's calculations, showed that, besides the vast number of facts which the epicyclical theory could account for, there were some which it would not exactly include, and Kepler was led to the persuasion that the planets move in ellipses. But this view of motion was at first conceived by Kepler as a modification of the conception of epicycles. On one occasion he blames himself for not sooner seeing that such a modification was possible. " What an absurdity on my part!" he cries*; " as if libration in the diameter of the epicycle might not come to the same thing as motion in the ellipse." But again; Kepler's *laws* of the elliptical motion of the planets were established; and these laws immediately became the *facts* on which the mathematicians had to found their mechanical theories. From these facts Newton, as we have related, proved that the central force of the sun retains the planets in their orbits, according to the law of the inverse square of the distance. The same *law* was shown to prevail in the gravitation of the earth. It was shown, too, by induction from the motions of Jupiter and Saturn, that the planets attract each other; by calculations from the figure of the earth, that the parts of the earth attract each other; and, by considering the course of the tides, that the sun and moon attract the waters of the ocean. And all these curious discoveries being established as *facts*, the subject

* *Hist. Inductive Sciences,* i. 428.

was ready for another step of generalization. By an unparalleled rapidity in the progress of discovery in this case, not only were all the inductions which we have first mentioned made by one individual, but the new advance, the higher flight, the closing victory, fell to the lot of the same extraordinary person.

The attraction of the sun upon the planets, of the moon upon the earth, of the planets on each other, of the parts of the earth on themselves, of the sun and moon upon the ocean;—all these truths, each of itself a great discovery, were included by Newton in the higher *generalization*, of the universal gravitation of matter, by which each particle is drawn to each other according to the law of the inverse square: and thus this long advance from discovery to discovery, from truths to truths, each justly admired when new, and then rightly used as old, was closed in a worthy and consistent manner, by a truth which is the most worthy admiration, because it includes all the researches of preceding ages of astronomy.

3. We may take another example of a succession of this kind from the history of a science, which, though it has made wonderful advances, has not yet reached its goal, as physical astronomy appears to have done, but seems to have before it a long prospect of future progress. I now refer to chemistry, in which I shall try to point out how the preceding discoveries afforded the materials of the succeeding; although this subordination and connexion is, in this case, less familiar to men's minds than in astronomy, and is, perhaps, more difficult to present in a clear and definite shape. Sylvius saw, in the facts which occur, when an acid and an alkali are brought together, the evidence that they neutralize each other. But cases of neutralization, and acidification, and many other effects of mixture of the ingredients of bodies, being thus viewed as *facts*, had an aspect of unity and law given them by

Geoffroy and Bergman*, who introduced the *conception* of the chemical affinity or elective attraction, by which certain elements select other elements, as if by preference. That combustion, whether a chemical union or a chemical separation of ingredients, is of the same nature with acidification, was the doctrine of Beccher and Stahl, and was soon established as a truth which must form a part of every succeeding physical theory. That the rules of affinity and chemical composition may include gaseous elements, was established by Black and Cavendish. And all these truths, thus brought to light by chemical discoverers,—affinity, the identity of acidification and combustion, the importance of gaseous elements,—along with all the facts respecting the weight of ingredients and compounds which the balance disclosed,—were taken up, connected, and included as *particulars* in the oxygen *theory* of Lavoisier. Again, the results of this theory, and the quantity of the several ingredients which entered into each compound—(such results, for the most part, being now no longer mere theoretical speculations, but recognised facts)—were the *particulars* from which Dalton derived that wide law of chemical combination which we term the atomic *theory*. And this law, soon generally accepted among chemists, is already in its turn become one of the *facts* included in Faraday's *theory* of the identity of chemical affinity and electric attraction.

It is unnecessary to give further exemplifications of this constant ascent from one step to a higher;—this perpetual conversion of true theories into the materials of other and wider theories. It will hereafter be our business to exhibit, in a more full and formal manner, the mode in which this principle determines the whole scheme and structure of all the most exact sciences. And thus, beginning with the facts of sense, we gradually climb to

* *Hist. Inductive Sciences*, iii. 112.

the highest forms of human knowledge, and obtain from experience and observation a vast collection of the most wide and elevated truths.

There are, however, truths of a very different kind, to which we must turn our attention, in order to pursue our researches respecting the nature and grounds of our knowledge. But before we do this, we must notice one more feature in that progress of science which we have already in part described.

CHAPTER VIII.

OF TECHNICAL TERMS.

1. It has already been stated that we gather know-ledge from the external world, when we are able to apply, to the facts which we observe, some ideal conception, which gives unity and connexion to multiplied and separate perceptions. We have also shown that our conceptions, thus verified by facts, may themselves be united and connected by a new bond of the same nature; and that man may thus have to pursue his way from truth to truth through a long progression of discoveries, each resting on the preceding, and rising above it.

It is now further to be noticed that each of these steps, in succession, is recorded, fixed, and made available, by some peculiar form of words; and such words, thus rendered precise in their meaning, and appropriated to the service of science, we may call Technical Terms. It is in a great measure by inventing such Terms that men not only best express the discoveries they have made, but also enable their followers to become so familiar with these discoveries, and to possess them so thoroughly, that they can readily use them in advancing to ulterior generalizations.

Most of our ideal conceptions are described by exact and constant words or phrases, such as those of which we here speak. We have already had occasion to employ many of these. Thus we have had instances of technical terms expressing geometrical conceptions, as *ellipsis*, *radius vector*, *axis*, *plane*, the proportion of the *inverse square*, and the like. Other terms have described mechanical conceptions, as *accelerating force* and *attraction*. Again, chemistry exhibits (as do all sciences) a series of terms which mark the steps of her progress. The views of the first real founders of the science are recorded by the terms which are still in use, *neutral salts*, *affinity*, and the like. The establishment of Dalton's theory has produced the use of the word *atom* in a peculiar sense, or of some other word, as *proportion*, in a sense equally technical. And Mr. Faraday has found it necessary, in order to expound his electro-chemical theory, to introduce such terms as *anode* and *cathode*, *anïon* and *cathïon*.

2. I need not adduce any further examples, for my object at present is only to point out the use and influence of such language: its rules and principles I shall hereafter try, in some measure, to fix. But what we have here to remark is, the extraordinary degree in which the progress of science is facilitated, by thus investing each new discovery with a compendious and steady form of expression. These terms soon become part of the current language of all who take an interest in speculation. However strange they may sound at first, they soon grow familiar in our ears, and are used without any effort or recollection of the difficulty they once involved. They become as common as the phrases which express our most frequent feelings and interests, while yet they have incomparably more precision than belongs to any terms which express feelings; and they carry with them, in their import, the results of deep and laborious trains of research. They convey the

mental treasures of one period to the generations that follow ; and laden with this, their precious freight, they sail safely across gulfs of time in which empires have suffered shipwreck, and the languages of common life have sunk into oblivion. We have still in constant circulation among us the terms which belong to the geometry, the astronomy, the zoology, the medicine of the Greeks, and the algebra and chemistry of the Arabians. And we can in an instant, by means of a few words, call to our own recollection, or convey to the apprehension of another person, phenomena and relations of phenomena in optics, mineralogy, chemistry, which are so complex and abstruse, that it might seem to require the utmost subtlety of the human mind to grasp them, even if that were made the sole object of its efforts. By this remarkable effect of technical language, we have the results of all the labours of past times not only always accessible, but so prepared that we may (provided we are careful in the use of our instrument) employ what is really useful and efficacious for the purpose of further success, without being in any way impeded or perplexed by the length and weight of the chain of past connexions which we drag along with us.

By such means,—by the use of the inductive process, and by the aid of technical terms,—man has been constantly advancing in the path of scientific truth. In a succeeding part of this work we shall endeavour to trace the general rules of this advance, and to lay down the maxims by which it may be most successfully guided and forwarded. But in order that we may do this to the best advantage, we must pursue still further the analysis of knowledge into its elements; and this will be our employment in the first part of the work.

Chapter IX.

OF NECESSARY AND CONTINGENT TRUTHS.

1. *Course of the Argument.*—Every advance in human knowledge consists, as we have seen, in adapting new ideal conceptions to ascertained facts, and thus in superinducing the form upon the matter, the active upon the passive processes of our minds. Every such step introduces into our knowledge an additional portion of the ideal element, and of those relations which flow from the nature of ideas. It is, therefore, important for our purpose to examine more closely this element, and to learn what the relations are which may thus come to form part of our knowledge. An inquiry into those ideas which form the foundations of our sciences;—into the reality, independence, extent, and principal heads of the knowledge which we thus acquire;—is a task on which we must now enter, and which will employ us for several of the succeeding Books.

In this inquiry our object will be to pass in review all the most important fundamental ideas which our sciences involve; and to prove more distinctly in reference to each, what we have already asserted with regard to all, that there are everywhere involved in our knowledge acts of the mind as well as impressions of sense; and that our knowledge derives, from these acts, a generality, certainty, and evidence which the senses could in no degree have supplied. But before I proceed to do this in particular cases, I will give some account of the argument in its general form.

We have already considered the separation of our knowledge into its two elements,—Impressions of Sense and Ideas,—as evidently indicated by this; that all knowledge possesses characters which neither of these elements alone could bestow.   Without our ideas, our sensations could have no connexion; without external impressions, our ideas would have no reality; and thus both ingredients of our knowledge must exist.   But there is another mode in which we may prove the distinct and independent existence of these two elements, namely, by considering that there are two large classes of truths which differ entirely from each other, and of which the difference arises from this, that the one class derives its nature from the one, and the other from the other, of these two elements.   These are what are technically termed *necessary* and *contingent* truths; truths of demonstration and truths of experience.   I shall first point out the difference of these two kinds of truths, which difference is briefly this, that the former are true universally and necessarily, the latter, only learnt from experience, and limited by experience.   I shall show that upon various subjects we possess truths of the former kind; that the universality and necessity which distinguish them can by no means be derived from experience; that these characters do in reality flow from the ideas which these truths involve; and that when their necessity is exhibited in the way of logical demonstration, it is found to depend upon certain fundamental principles, (Definitions and Axioms,) which may thus be considered as expressing, in some measure, the essential characters of our ideas. These fundamental principles I shall afterwards proceed to discuss and to exhibit in each of the principal departments of science.

2. *Of Necessary Truths.*—Necessary truths are those in which we not only learn that the proposition *is* true, but see that it *must be* true; in which the negation of the truth is not only false, but impossible; in which we cannot, even by an effort of imagination, or in a supposition, conceive the reverse of that which is asserted.

That there are such truths cannot be doubted. We may take, for example, all relations of number. Three and Two added together make Five. We cannot conceive it to be otherwise. We cannot, by any freak of thought, imagine Three and Two to make Seven.

It may be said that this assertion merely expresses what we mean by our words; that it is a matter of definition; that the proposition is an identical one.

But this is by no means so. The definition of Five is not Three and Two, but Four and One. How does it appear that Three and Two is the same number as Four and One? It is evident that it is so; but *why* is it evident?—not because the proposition is identical; for if that were the reason, all numerical propositions must be evident for the same reason. If it be a matter of definition that 3 and 2 make 5, it must be a matter of definition that 39 and 27 make 66. But who will say that the definition of 66 is 39 and 27? Yet the magnitude of the numbers can make no difference in the ground of the truth. How do we know that the product of 13 and 17 is 4 less than the product of 15 and 15? We see that it is so, if we perform certain operations by the rules of arithmetic; but how do we know the truth of the rules of arithmetic? If we divide 123375 by 987 according to the process taught us at school, how are we assured that the result is correct, and that the number 125 thus obtained is really the number of times one number is contained in the other?

The correctness of the rule, it may be replied, can be rigorously demonstrated. It can be shown that the process must inevitably give the true quotient.

Certainly this can be shown to be the case. And precisely because it *can* be shown that the result must be true, we have here an example of a necessary truth; and this truth, it appears, is not *therefore* necessary because it is itself evidently identical, however it may be possible to prove it by reducing it to evidently identical propositions. And the same is the case with all other numerical propositions; for, as we have said, the nature of all of them is the same.

Here, then, we have instances of truths which are not only true, but demonstrably and necessarily true. Now such truths are, in this respect at least, altogether different from truths, which, however certain they may be, are learnt to be so only by the evidence of observation, interpreted, as observation must be interpreted, by our own mental faculties. There is no difficulty in finding examples of these merely observed truths. We find that sugar dissolves in water, and forms a transparent fluid, but no one will say that we can see any reason beforehand why the result *must* be so. We find that all animals which chew the cud also have the divided hoof; but could any one have predicted that this would be universally the case? or supposing the truth of the rule to be known, can any one say that he cannot conceive the facts as occurring otherwise? Water expands when it crystallizes, some other substances contract in the same circumstances; but can any one know that this will be so otherwise than by observation? We have here propositions *rigorously* true, (we will assume,) but can any one say they are *necessarily* true? These, and the great mass of the doctrines established by induction, are actual, but so far as we can see, accidental laws; results determined by some unknown selection, not demonstrable consequences of the essence of things, inevitable and perceived to be inevitable. According to the phraseology which has been frequently used by philosophical writers, they are *contingent*, not necessary truths.

It is requisite to insist upon this opposition, because no insight can be obtained into the true nature of knowledge, and the mode of arriving at it, by any one who does not clearly appreciate the distinction. The separation of truths which are learnt by observation, and truths which can be seen to be true by a pure act of thought, is one of the first and most essential steps in our examination of the nature of truth, and the mode of its discovery. If any one does not clearly comprehend this distinction of necessary and contingent truths, he will not be able to go along with us in our researches into the foundations of human knowledge; nor, indeed, to pursue with success any speculation on the subject. But, in fact, this distinction is one that can hardly fail to be at once understood. It is insisted upon by almost all the best modern, as well as ancient, metaphysicians*, as of primary importance. And if any person does not fully apprehend, at first, the different kinds of truth thus pointed out, let him study, to some extent, those sciences which have necessary truth for their subject, as geometry, or the properties of numbers, so as to obtain a familiar acquaintance with such truth; and he will then hardly fail to see how different the evidence of the propositions which occur in these sciences, is from the evidence of the facts which are merely learnt from experience. That the year goes through its course in 365 days, can only be known by observation of the sun or stars: that 365 days is 52 weeks and a day, it requires no experience, but only a little thought to perceive. That bees build their cells in the form of hexagons, we cannot know without looking at them; that regular hexagons may be arranged so as to fill space, may be proved with the utmost rigour, even if there were not in existence such a thing as a material hexagon.

I have taken examples of necessary truths from the properties of number and space; but such truths exist no less in other subjects, although the discipline of

* Aristotle, Dr. Whately, Dugald Stewart, &c.

thought which is requisite to perceive them distinctly, may not be so usual among men with regard to the sciences of mechanics and hydrostatics, as it is with regard to the sciences of geometry and arithmetic. Yet every one may perceive that there are such truths in mechanics. If I press the table with my hand, the table presses my hand with an equal force : here is a self-evident and necessary truth. In any machine, constructed in whatever manner to increase the force which I can exert, it is certain that what I gain in force I must lose in the velocity which I communicate. This is not a contingent truth, borrowed from and limited by observation; for a man of sound mechanical views applies it with like confidence, however novel be the construction of the machine. When I come to speak of the ideas which are involved in our mechanical knowledge, I may, perhaps, be able to bring more clearly into view the necessary truth of general propositions on such subjects. That reaction is equal and opposite to action is as necessarily true as that two straight lines cannot inclose a space; it is as impossible theoretically to make a perpetual motion by mere mechanism as to make the diagonal of a square commensurable with the side.

The existence of these two kinds of truth, necessary and contingent, and their separate nature, being established or allowed, we proceed onwards with the argument.

Necessary truths must be *universal* truths. If any property belong to a right-angled triangle *necessarily*, it must belong to *all* right-angled triangles. And it shall be proved in the following Chapter, that truths possessing these two characters, of Necessity and Universality, cannot possibly be the mere results of experience.

Chapter X.

OF EXPERIENCE.

1. I HERE employ the term Experience in a more definite and limited sense than it possesses in common usage; for I restrict it to matters belonging to the domain of science. In such cases, the knowledge which we acquire, by means of experience, is of a clear and precise nature; and the passions and feelings and interests, which make the lessons of experience in practical matters so difficult to read aright, no longer disturb and confuse us. We may, therefore, hope, by attending to such cases, to learn what efficacy experience really has, in the discovery of truth.

That from *experience* (including intentional experience, or *observation*,) we obtain much knowledge which is highly important, and which could not be procured from any other source, is abundantly clear. We have already taken several examples of such kuowledge. We know by experience that animals which ruminate are cloven-hoofed; and we know this in no other manner. We know, in like manner, that all the planets and their satellites revolve round the sun from west to east. It has been found by experience that all meteoric stones contain chrome. Many similar portions of our knowledge might be mentioned.

Now what we have here to remark is this;—that in no case can experience prove a proposition to be *necessarily* or *universally* true. However many instances we may have observed of the truth of a proposition, yet if it be merely observation, there is nothing to assure us that the next case shall not be an exception to the rule. If it be strictly true that every ruminant animal yet known has cloven hoofs, we still cannot be sure that some creature will not hereafter be discovered which has the first of these attributes without having the other. When the planets and their satellites, as far as Saturn, had been all found to move round the sun in one direction, it was still possible that there might be other·such bodies not obeying this rule; and, accordingly, when the satellites of Uranus were detected, they appeared to offer an exception of this kind. Even in the mathematical sciences, we have examples of such rules suggested by experience, and also of their precariousness. However far they may have been tested, we cannot depend upon their correctness, except we see some reason for the rule. For instance, various rules have been given, for the purpose of pointing out prime numbers; that is, those which cannot be divided by any other number. We may try, as an example of such a rule, this one—any odd power of the number two, diminished by one. Thus the third power of two, diminished by one, is seven; the fifth power, diminished by one, is thirty-one; the seventh power so diminished is one hundred and twenty-seven. All these are prime numbers: and we might be led to suppose that the rule is universal. But the next example shows us the fallaciousness of such a belief. The ninth power of two, diminished by one, is five hundred and eleven, which is not a prime, being divisible by seven.

Experience must always consist of a limited nnmber of observations. And, however numerous these may be, they can show nothing with regard to the infinite number of cases in which the experiment has not been made. Experience being thus unable to prove a fact to be universal, is, as will readily be seen, still more incapable of proving a truth to be necessary. Experience cannot, indeed, offer the smallest ground for the necessity of a proposition. She can observe and record what has happened ; but she cannot find, in any case, or in any accumulation of cases, any reason for what *must* happen. She may see objects side by side ; but she cannot see a reason why they must ever be side by side. She finds certain events to occur in succession ; but the succession supplies, in its occurrence, no reason for its recurrence. She contemplates external objects ; but she cannot detect any internal bond, which indissolubly connects the future with the past, the possible with the real. To learn a proposition by experience, and to see it to be necessarily true, are two altogether different processes of thought.

2. But it may be said, that we do learn by means of observation and experience many universal truths; indeed, all the general truths of which science consists. Is not the doctrine of universal gravitation learnt by experience? Are not the laws of motion, the properties of light, the general principles of chemistry so learnt? How, with these examples before us, can we say that experience teaches no universal truths?

To this we reply, that these truths can only be known to be general, not universal, if they depend upon experience alone. Experience cannot bestow that universality which she herself cannot have, and that necessity of which she has no comprehension. If these doctrines are universally true, this universality flows from the *ideas* which we apply to our experience, and which are, as we have seen, the real sources of necessary truth. How far these ideas can communicate their universality and necessity to the results of experience, it will hereafter be our business to consider. It will then appear, that when the mind collects from observation truths of a wide and comprehensive kind, which approach to the simplicity and universality of the truths of pure science ; she gives them this character by throwing upon them the light of her own Fundamental Ideas.

But the truths which we discover by observation of the external world, even when most strikingly simple and universal, are not necessary truths. Is the doctrine of universal gravitation necessarily true? It was doubted by Clairaut (so far as it refers to the moon), when the progression of the apogee in fact appeared to be twice as great as the theory admitted. It has been doubted, even more recently, with respect to the planets, their mutual perturbations appearing to indicate a deviation from the law. It is doubted still, by some persons, with respect to the double stars. But suppose all these doubts to be banished, and the law to be universal; is it then proved to be necessary? Manifestly not: the very existence of these doubts proves that it is not so. For the doubts were dissipated by reference to observation and calculation, not by reasoning on the nature of the law. Clairaut's difficulty was removed by a more exact calculation of the effect of the sun's force on the motion of the apogee. The suggestion of Bessel, that the intensity of gravitation might be different for different planets, was found to be unnecessary, when Professor Airy gave a more accurate determination of the mass of Jupiter. And the question whether the extension of the law of the inverse square to the double stars be true, (one of the most remarkable questions now before the scientific world,) must be answered, not by any speculations concerning what the laws of attraction must necessarily be, but by carefully determining the laws of the motion of these curious objects, by means of the observations such as those which Sir John Herschel has collected for that purpose, by his unexampled survey of both hemispheres of the sky. And since the extent of this truth is thus to be determined by reference to observed facts, it is clear that no mere accumulation of them can make its universality certain, or its necessity apparent.

Thus no knowledge of the necessity of any truths can result from the observation of what really happens. This being clearly understood, we are led to an important inquiry.

The characters of universality and necessity in the truths which form part of our knowledge, can never be derived from the experience by which so large a part of our knowledge is obtained. But since, as we have seen, we really do possess a large body of truths which are necessary, and because necessary, therefore universal, the question still recurs, from what source these characters of universality and necessity are derived.

The answer to this question we will attempt to give in the next chapter.

CHAPTER XI.

OF THE GROUNDS OF NECESSARY TRUTHS.

1. To the question just stated, I reply, that the necessity and universality of the truths which form a part of our knowledge, are derived from the *Fundamental Ideas* which those truths involve. These ideas entirely shape and circumscribe our knowledge; they regulate the active operations of our minds, without which our passive sensations do not become knowledge. They govern these operations, according to rules which are not only fixed and permanent, but which may be expressed in plain and definite terms; and these rules, when thus expressed, may be made the basis of demonstrations by which the necessary relations imparted to our knowledge by our ideas may be

traced to their consequences in the most remote ramifications of scientific truth.

These enunciations of the necessary and evident conditions imposed upon our knowledge by the fundamental ideas which it involves, are termed *Axioms*. Thus the Axioms of Geometry express the necessary conditions which result from the idea of space; the Axioms of Mechanics express the necessary conditions which flow from the ideas of force and motion; and so on.

2. It will be the office of several of the succeeding Books of this work to establish and illustrate in detail what I have thus stated in general terms: I shall there pass in review many of the most important fundamental ideas on which the existing body of our science depends; and I shall endeavour to show, for each such idea in succession, that knowledge involves an active as well as a passive element; that it is not possible without an act of the mind, regulated by certain laws. I shall further attempt to enumerate some of the principal fundamental relations which each idea thus introduces into our thoughts, and to express them by means of definitions and axioms, and other suitable forms.

I will only add a remark or two to illustrate further this view of the ideal grounds of our knowledge.

3. To persons familiar with any of the demonstrative sciences, it will be apparent that if we state all the Definitions and Axioms which are employed in the demonstrations, we state the whole basis on which those reasonings rest. For the whole process of demonstrative or deductive reasoning in any science, (as in geometry, for instance,) consists entirely in combining some of these first principles so as to obtain the simplest propositions of the science; then combining these so as to obtain other propositions of greater complexity; and so on, till we advance to the most recondite demonstrable truths; these

last, however intricate and unexpected, still involving no principles except the original definitions and axioms. Thus, by combining the definition of a triangle, and of equal lines and equal angles, namely, that they are such as when applied to each other, coincide, with the axiom respecting straight lines (that two such lines cannot inclose a space,) we demonstrate the equality of triangles, under certain assumed conditions. Again, by combining this result with the definition of parallelograms, and with the axiom that if equals be taken from equals the wholes are equal, we prove the equality of parallelograms between the same parallels and upon the same base. From this proposition, again, we prove the equality of the square on the hypotenuse of a triangle to the squares on the two sides containing the right angle. But in all this there is nothing contained which is not rigorously the result of our geometrical definitions and axioms. All the rest of our treatises of geometry consists only of terms and phrases of reasoning, the object of which is to connect those first principles, and to exhibit the effects of their combination in the shape of demonstration.

4. This combination of first principles takes place according to the forms and rules of *Logic*. All the steps of the demonstration may be stated in the shape in which logicians are accustomed to exhibit processes of reasoning in order to show their conclusiveness, that is, in *Syllogisms*. Thus our geometrical reasonings might be resolved into such steps as the following:—

All straight lines drawn from the centre of a circle to its circumference are equal:

But the straight lines A B, A C, are drawn from the centre of a circle to its circumference:

Therefore the straight lines A B, A C, are equal.

Each step of geometrical, and all other demonstrative reasoning, may be resolved into three such clauses as these; and these three clauses are termed respectively, the *major premiss*, the *minor premiss*, and the *conclusion;* or, more briefly, the *major*, the *minor*, and the *conclusion*.

The principle which justifies the reasoning when exhibited in this syllogistic form, is this:—that a truth which can be asserted as generally, or rather as universally true, can be asserted as true also in each particular case. The *minor* only asserts a certain particular case to be an example of such conditions as are spoken of in the *major;* and hence the conclusion, which is true of the major by supposition, is true of the minor by consequence; and thus we proceed from syllogism to syllogism, in each one employing some general truth in some particular instance. Any proof which occurs in geometry, or any other science of demonstration, may thus be reduced to a series of processes, in each of which we pass from some general proposition to the narrower and more special propositions which it includes. And this process of deriving truths by the mere combination of general principles, applied in particular hypothetical cases, is called *deduction;* being opposed to *induction*, in which, as we have seen, a new general principle is introduced at every step.

5. Now we have to remark that, this being so, however far we follow such deductive reasoning, we can never have in our conclusion any truth which is not virtually included in the original principles from which the reasoning started. For since at any step we merely take out of a general proposition something included in it, while at the preceding step we have taken this general proposition out of one more general, and so on perpetually, it is manifest that our last result was really included in the principle or principles with which we began. I say principles, because, although our logical conclusion can only exhibit the legitimate issue of our first principles, it may, never-

theless, contain the result of the combination of several such principles, and may thus assume a great degree of complexity, and may appear so far removed from the parent truths, as to betray at first sight hardly any relationship with them. Thus the proposition which has already been quoted respecting the squares on the sides of a right-angled triangle, contains the results of many elementary principles; as the definitions of parallels, triangle, and square; the axioms respecting straight lines, and respecting parallels; and, perhaps, others. The conclusion is complicated by containing the effects of the combination of all these elements; but it contains nothing, and can contain nothing, but such elements and their combinations.

This doctrine, that logical reasoning produces no new truths, but only unfolds and brings into view those truths which were, in effect, contained in the first principles of the reasoning, is assented to by almost all who, in modern times, have attended to the science of logic. Such a view is admitted both by those who defend, and by those who depreciate the value of logic. " Whatever is established by reasoning, must have been contained and virtually asserted in the premises* " " The only truth which such propositions can possess consists in conformity to the original principles."

In this manner the whole substance of our geometry is reduced to the definitions and axioms which we employ in our elementary reasonings; and in like manner we reduce the demonstrative truths of any other science to the definitions and axioms which we there employ.

6. But in reference to this subject, it has sometimes been said that demonstrative sciences do in reality depend upon Definitions only; and that no additional kind of

---

* WHATELEY's *Logic*, pp. 237, 238.

principle, such as we have supposed Axioms to be, is absolutely required. It has been asserted that in geometry, for example, the source of the necessary truth of our propositions is this, that they depend upon definitions alone, and consequently merely state the identity of the same thing under different aspects.

That in the sciences which admit of demonstration, as geometry, mechanics, and the like, axioms as well as definitions are needed, in order to express the grounds of our necessary convictions, must be shown hereafter by an examination of each of these sciences in particular. But that the propositions of these sciences, those of geometry for example, do not merely assert the identity of the same thing, will, I think, be generally allowed, if we consider the assertions which we are enabled to make. When we declare that " a straight line is the shortest distance between two points," is this merely an identical proposition? the definition of a straight line in another form? Not so: the definition of a straight line involves the notion of form only, and does not contain anything about magnitude; consequently, it cannot contain anything equivalent to " shortest." Thus the propositions of geometry are not merely identical propositions; nor have we in their general character anything to countenance the assertion, that they are the results of definitions alone. And when we come to examine this and other sciences more closely, we shall find that axioms, such as are usually in our treatises made the fundamental principles of our demonstrations, neither have ever been, nor can be, dispensed with. Axioms, as well as definitions, are in all cases requisite, in order properly to exhibit the grounds of necessary truth.

7. Thus the real logical basis of every body of demonstrated truths are the Definitions and Axioms which are the first principles of the reasonings. But when we are

arrived at this point, the question further occurs, what is
the ground of the truth of these Axioms? It is not the
logical but the philosophical, not the formal but the real
foundation of necessary truth, which we are seeking.
Hence this inquiry, What is the ground of the axioms of
geometry, of mechanics, and of any other demonstrable
science, necessarily comes before us.

The answer which we are led to give, by the view
which we have taken of the nature of knowledge, has
already been stated. The ground of the axioms belonging to each science is the *idea* which the axiom involves.
The ground of the axioms of geometry is the *idea of
space:* the ground of the axioms of mechanics is the
idea of force, of *action* and *reaction,* and the like. And
hence these ideas are Fundamental Ideas; and since
they are thus the foundations, not only of demonstration
but of truth, an examination into their real import and
nature is of the greatest consequence to our purpose.

8. Not only the Axioms, but the Definitions which
form the basis of our reasonings, depend upon our Fundamental Ideas. And the definitions are not arbitrary definitions, but are determined by a necessity no less rigorous
than the axioms themselves. We could not think of
geometrical truths without conceiving a circle; and we
could not reason concerning such truths without defining
a circle in some mode equivalent to that which is commonly adopted. The definitions of parallels, of right
angles, and the like, are quite as necessarily prescribed
by the nature of the case, as the axioms which these definitions bring with them. Indeed we may substitute one
of these kinds of principles for another. We cannot
always put a definition in the place of an axiom; but we
may always find an axiom which shall take the place of
a definition. If we assume a proper axiom respecting
straight lines, we need no definition of a straight line.

But in whatever shape the principle appear, as definition or as axiom, it has about it nothing casual or arbitrary, but is determined to be what it is, as to its import, by the most rigorous necessity, growing out of the Idea of Space.

7. These principles,—definitions, and axioms,—thus exhibiting the primary developements of a fundamental idea, do in fact express the idea, so far as its expression in words forms part of our science. They are different views of the same body of truth; and though each principle, by itself, exhibits only one aspect of this body, taken together they convey a sufficient conception of it for our purposes. The idea itself cannot be fixed in words; but these various lines of truth proceeding from it, suggest sufficiently to a fitly-prepared mind, the place where the idea resides, its nature, and its efficacy.

It is true that these principles,—our elementary definitions and axioms,—even taken altogether, express the idea incompletely. Thus the definitions and axioms of geometry, as they are stated in our elementary works, do not fully express the idea of space as it exists in our minds. For, in addition to these, other axioms, independent of these, and no less evident, can be stated; and are in fact stated when we come to the higher geometry. Such, for instance, is the axiom of Archimedes—that a curve line which joins two points is less than a broken line which joins the same points and includes the curve. And thus the idea is disclosed but not fully revealed, imparted but not transfused, by the use we make of it in science. When we have taken from the fountain so much as serves our purpose, there still remains behind a deep well of truth, which we have not exhausted, and which we may easily believe to be inexhaustible.

Chapter XII.

THE FUNDAMENTAL IDEAS ARE NOT DERIVED FROM EXPERIENCE.

1. By the course of speculation contained in the last three Chapters, we are again led to the conclusion which we have already stated, that our knowledge contains an ideal element, and that this element is not derived from experience. For we have seen that there are propositions which are known to be necessarily true; and that such knowledge is not, and cannot be, obtained by mere observation of actual facts. It has been shown, also, that these necessary truths are the results of certain fundamental ideas, such as those of space, number, and the like. Hence it follows inevitably that these ideas and others of the same kind are not derived from experience. For these ideas possess a power of infusing into their developements that very necessity which experience can in no way bestow. This power they do not borrow from the external world, but possess by their own nature. Thus we unfold out of the idea of space the propositions of geometry, which are plainly truths of the most rigorous necessity and universality. But if the idea of space were merely collected from observation of the external world, it could never enable or entitle us to assert such propositions: it could never authorize us to say that not merely some lines, but *all* lines, not only have, but *must* have, those properties which geometry teaches. Geometry in every proposition speaks a language which experience never dares to utter; and indeed of which she but half comprehends the meaning. Experience sees that the assertions are true, but she sees not how profound and absolute is their truth. She unhesitatingly assents to the laws which geometry delivers, but she does not pre-

tend to see the origin of their obligation. She is always ready to acknowledge the sway of pure scientific principles as a matter of fact, but she does not dream of offering her opinion on their authority as a matter of right; still less can she justly claim to be herself the source of that authority.

David Hume asserted *, that we are incapable of seeing in any of the appearances which the world presents anything of necessary connexion; and hence he inferred that our knowledge cannot extend to any such connexion. It will be seen from what we have said that we assent to his remark as to the fact, but we differ from him altogether in the consequence to be drawn from it. Our inference from Hume's observation is, not the truth of his conclusion, but the falsehood of his premises;—not that, therefore, we can know nothing of natural connexion, but that, therefore, we have some other source of knowledge than experience:—not that we can have no idea of connexion or causation, because, in his language, it cannot be the copy of an impression; bnt that since we have such an idea, our ideas are not the copies of our impressions.

Since it thus appears that our fundamental ideas are not acquired from the external world by our senses, but have some separate and independent origin, it is important for us to examine their nature and properties, as they exist in themselves, and this it will be our business to do through a portion of the following pages. But it may be proper first to notice one or two objections which may possibly occur.

3. It may be said that without the use of our senses, of sight and touch, for instance, we should never have any idea of space; that this idea, therefore, may properly be said to be derived from those senses. And to this I reply

* *Essays*, vol. ii. p. 70.

by referring to a parallel instance. Without light we should have no perception of visible figure; yet the power of perceiving visible figure cannot be said to be derived from the light, but resides in the structure of the eye. If we had never seen objects in the light, we should be quite unaware that we possessed a power of vision; yet we should not possess it the less on that account. If we had never exercised the senses of sight and touch (if we can conceive such a state of human existence) we know not that we should be conscious of an idea of space. But the light reveals to us at the same time the existence of external objects and our own power of seeing. And in a very similar manner, the exercise of our senses discloses to us, at the same time, the external world, and our own ideas of space, time, and other conditions, without which the external world can neither be observed nor conceived. That light is necessary to vision, does not, in any degree, supersede the importance of a separate examination of the laws of our visual powers, if we would understand the nature of our own bodily faculties and the extent of the information they can give us. In like manner, the fact that intercourse with the external world is necessary for the conscious employment of our ideas, does not make it the less essential for us to examine those ideas in their most intimate structure, in order that we may understand the grounds and limits of our knowledge. Even before we see a single object, we have a faculty of vision; and in like manner, if we can suppose a man who has never contemplated an object in space or time, we must still assume him to have the faculties of entertaining the ideas of space and time, which faculties are called into play on the very first occasion of the use of the senses.

4. In answer to such remarks as the above, it has sometimes been said that to assume separate faculties in

the mind for so many different processes of thought, is to give a mere verbal explanation, since we learn nothing concerning our idea of space by being told that we have a faculty of forming such an idea. It has been said that this course of explanation leads to an endless multiplication of elements in man's nature, without any advantage to our knowledge of his true constitution. We may, it is said, assert man to have a faculty of walking, of standing, of breathing, of speaking; but what, it is asked, is gained by such assertions? To this I reply, that we undoubtedly have such faculties as those just named; that it is by no means unimportant to consider them; and that the main question in such cases is, whether they are separate and independent faculties, or complex and derivative ones; and, if the latter be the case, what are the simple and original faculties by the combination of which the others are produced. In walking, standing, breathing, for instance, a great part of the operation can be reduced to one single faculty; the voluntary exercise of our muscles. But in breathing this does not appear to be the whole of the process. The operation is, in part at least, involuntary; and it has been held that there is a certain sympathetic action of the nerves, in addition to the voluntary agency which they transmit, which is essential to the function. To determine whether or no this sympathetic faculty is real and distinct, and if so, what are its laws and limits, is certainly a highly philosophical inquiry, and well deserving the attention which has been bestowed upon it by eminent physiologists. And just of the same nature are the inquiries with respect to man's intellectual constitution, on which we propose to enter. For instance, man has a faculty of apprehending time, and a faculty of reckoning numbers; are these distinct, or is one faculty derived from the other? To analyse the various combinations of our ideas and observations into the original faculties which they involve; to show that these faculties are original, and not capable of further analysis; to point out the characters which mark these faculties and lead to the most important features of our knowledge;—these are the kind of researches on which we have now to enter, and these, we trust, will be found to be far from idle or useless parts of our plan. If we succeed in such attempts, it will appear that it is by no means a frivolous or superfluous step to distinguish separate faculties in the mind. If we do not learn much by being told that we have a faculty of forming the idea of space, we at least, by such a commencement, circumscribe a certain portion of the field of our investigations, which, we shall afterwards endeavour to show, requires and rewards a special examination. And though we shall thus have to separate the domain of our philosophy into many provinces, these are, as we trust it will appear, neither arbitrarily assigned, nor vague in their limits, nor infinite in number.

CHAPTER XIII.

OF THE PHILOSOPHY OF THE SCIENCES.

WE proceed, in the ensuing Books, to the closer examination of a considerable number of those Fundamental Ideas on which the sciences, hitherto most successfully cultivated, are founded. In this task, our objects will be to explain and analyse such Ideas so as to bring into view the Definitions and Axioms or other forms, in which we may clothe the conditions to which our speculative knowledge is subjected. I shall also try to prove, for some of these Ideas in particular, what has been already urged respecting them in general, that they are not derived from observation, but necessarily impose their conditions upon that knowledge of which observation supplies the materials. I shall further, in some cases, endeavour to trace the history of these Ideas as they have successively come into notice in the progress of science; the gradual developement by which they have arrived at their due purity and clearness; and, as a necessary part of such a history, I shall give a view of some of the principal controversies which have taken place with regard to each portion of knowledge.

An exposition and discussion of the Fundamental Ideas of each Science may, with great propriety, be termed the PHILOSOPHY of such science. These ideas contain in themselves the elements of those truths which the science discovers and enunciates; and in the progress of the sciences, both in the world at large and in the mind of each individual student, the most important steps consist in apprehending these ideas clearly, and in bringing them into accordance with the observed facts. I shall, therefore, in a series of Books, treat of the Philosophy of the Pure Sciences, the Philosophy of the Mechanical Sciences, the Philosophy of Chemistry, and the like, and shall analyse and examine the ideas which these sciences respectively involve.

In this undertaking, inevitably somewhat long, and involving many deep and subtle discussions, I shall take, as a chart of the country before me, by which my course is to be guided, the scheme of the sciences which I was led to form by travelling over the history of each in order*. Each of the sciences of which I then narrated the progress, depends upon several of the Fundamental Ideas of which I have to speak: some of these Ideas are peculiar to one field of speculation, others are common to more. A previous enumeration of Ideas thus collected

* *History of the Inductive Sciences.*

may serve both to show the course and limits of this part of our plan, and the variety of interest which it offers.

I shall, then, successively, have to speak of the ideas which are the foundation of geometry and arithmetic, (and which also regulate all sciences depending upon these, as astronomy and mechanics;) namely, the ideas of *space, time,* and *number :*

Of the ideas on which the mechanical sciences (as mechanics, hydrostatics, physical astronomy) more peculiarly rest; the ideas of *force* and *matter,* or rather the idea of *cause,* which is the basis of these :

Of the ideas which the secondary mechanical sciences (acoustics, optics, and thermotics) involve; namely, the ideas of the *externality* of objects, and of the *media* by which we perceive their qualities :

Of the ideas which are the basis of mechanico-chemical and chemical science, *polarity, chemical affinity,* and *substance;* and the idea of *symmetry,* a necessary part of the philosophy of crystallography :

Of the ideas on which the classificatory sciences proceed (mineralogy, botany, and zoology); namely, the ideas of *resemblance,* and of its gradations, and of *natural affinity :*

Finally, of those ideas on which the physiological sciences are founded; the ideas of separate vital powers, such as *assimilation* and *irritability ;* and the idea of *final cause.*

We have, besides these, the Palætiological sciences, which proceed mainly on the conception of *historical causation.*

It is plain that when we have proceeded so far as this, we have advanced to the verge of those speculations which have to do with mind as well as body. The extension of our philosophy to such a field, if it can be justly so extended, will be one of the most important results of our researches; but on that very account we must fully study the lessons which we learn in those fields of speculation where our doctrines are most secure, before we venture into a region where our principles will appear to be more precarious, and where they are inevitably less precise.

We now proceed to the examination of the above ideas, and to such essays towards the philosophy of each science as this course of investigation may suggest.

BOOK II.

THE PHILOSOPHY OF THE PURE SCIENCES.

CHAPTER I.

OF THE PURE SCIENCES.

1. ALL external objects and events which we can contemplate are viewed as having relations of Space, Time, and Number; and are subject to the general conditions which these Ideas impose, as well as to the particular laws which belong to each class of objects and occurrences. The special laws of nature, considered under the various aspects which constitute the different sciences, are obtained by a mixed reference to experience and to the fundamental ideas of each science. But besides the sciences thus formed by the aid of special experience, the conditions which flow from those more comprehensive ideas first mentioned, space, time, and number, constitute a body of science, applicable to objects and changes of all kinds, and deduced without recurrence being had to any observation in particular. These sciences, thus unfolded out of ideas alone, unmixed with any reference to the phenomena of matter, are hence termed *pure* sciences. The principal sciences of this class are geometry, theoretical arithmetic, and algebra considered in its most general sense, as the investigation of the relations of space and number by means of general symbols.

2. These pure sciences were not included in our survey of the history of the sciences, because they are not *inductive* sciences. Their progress has not consisted in collecting laws from phenomena, true theories from observed facts, and more general from more limited laws; but in tracing the consequences of the ideas themselves, and in detecting the most general and intimate analogies and connexions which prevail among such conceptions as are derivable from the ideas. These sciences have no principles besides definitions and axioms, and no process of proof but *deduction;* this process, however, assuming here a most remarkable character; and exhibiting a combination of simplicity and complexity, of rigour and generality, quite unparalleled in other subjects.

3. The universality of the truths, and the rigour of the demonstrations of these pure sciences, attracted attention in the earliest times; and it was perceived that they offered an exercise and a discipline of the intellectual faculties, in a form peculiarly free from admixture of extraneous elements. They were strenuously cultivated by the Greeks, both with a view to such a discipline, and from the love of speculative truth which prevailed among that people: and the name *mathematics,* by which they are designated, indicates this their character of *disciplinal* studies.

4. As has already been said, the ideas which these sciences involve extend to all the objects and changes which we observe in the external world; and hence the consideration of mathematical relations forms a large portion of many of the sciences which treat of the phenomena and laws of external nature, as astronomy, optics, and mechanics. Such sciences are hence often termed *mixed mathematics,* the relations of space and number being, in these branches of knowledge, combined with principles collected from special observation;

while geometry, algebra, and the like subjects, which involve no result of experience, are called *pure mathematics*.

5. Space, time, and number, may be conceived as *forms* by which the knowledge derived from our sensations is moulded, and which are independent of the differences in the *matter* of our knowledge, arising from the sensations themselves. Hence the sciences which have these ideas for their subject may be termed *formal sciences*. In this point of view, they are distinguished from sciences in which, besides these mere formal laws by which appearances are corrected, we endeavour to apply to the phenomena the idea of cause, or some of the other ideas which penetrate further into the principles of nature. We have thus, in the History, distinguished Formal Astronomy and Formal Optics from Physical Astronomy and Physical Optics.

We now proceed to our examination of the ideas which constitute the foundation of these formal or pure mathematical sciences, beginning with the idea of space.

CHAPTER II.

OF THE IDEA OF SPACE.

1. BY speaking of space as an Idea, I intend to imply, as has already been stated, that the apprehension of objects as existing in space, and of the relations of position, &c., which thus prevail among them, is not a consequence of experience, but a result of a peculiar constitution and activity of the mind, which is independent of all experience in its origin, though constantly combined with experience in its exercise.

That the idea of space is thus independent of experience, has already been pointed out in speaking of ideas in general: but it may be useful to illustrate the doctrine further in this particular case.

I assert, then, that space is not a notion obtained by experience. Experience gives us information concerning things without us: but our apprehending them *as* without us, takes for granted their existence in space. Experience acquaints us what are the form, position, magnitude of particular objects: but that they *have* form, position, magnitude, presupposes that they are in space. We cannot derive from appearances, by the way of observation, the habit of representing things to ourselves as in space; for no single act of observation is possible any otherwise than by beginning with such a representation, and conceiving objects as already existing in space.

2. That our mode of representing space to ourselves is not derived from experience, is clear also from this:— that through this mode of representation we arrive at propositions which are rigorously universal and necessary. Propositions of such a kind could not possibly be obtained from experience; for experience can only teach us by a limited number of examples, and therefore can never securely establish a universal proposition: and again, experience can only inform us that anything is so, and can never prove that it must be so. That two sides of a triangle are greater than the third is a universal and necessary geometrical truth: it is true of all triangles; it is true in such a way that the contrary cannot be conceived. Experience could not prove such a proposition. And experience has not proved it; for perhaps no man ever made the trial as a means of removing doubts: and no trial could, in fact, add in the smallest degree to the certainty of this truth. To seek for proof of geometrical propositions by an appeal to observation proves nothing in reality, except that the person who has recourse to such grounds has no due apprehension of the nature of geo-

metrical demonstration. We have heard of persons who
convinced themselves by measurement that the geometrical rule respecting the squares on the sides of a right-
angled triangle was true: but these were persons whose
minds had been engrossed by practical habits, and in
whom the speculative developement of the idea of space
had been stifled by other employments. The practical
trial of the rule may illustrate, but cannot prove it.
The rule will of course be confirmed by such trial, because
what is true in general is true in particular: but it cannot
be proved from any number of trials, for no accumulation
of particular cases makes up a universal case. To all
persons who can see the force of any proof, the geometrical rule above referred to is as evident, and its evidence
as independent of experience, as the assertion that sixteen
and nine make twenty-five. At the same time the truth
of the geometrical rule is quite independent of numerical
truths, and results from the relations of space alone.
This could not be if our apprehension of the relations of
space were the fruit of experience: for experience has no
element from which such truth and such proof could
arise.

3. Thus the existence of necessary truths, such as
those of geometry, proves that the idea of space from
which they flow, is not derived from experience. Such
truths are inconceivable on the supposition of their being
collected from observation; for the impressions of sense
include no evidence of necessity. But we can readily
understand the necessary character of such truths, if we
conceive that there are certain necessary conditions under
which alone the mind receives the impressions of sense.
Since these conditions reside in the constitution of the
mind, and apply to every perception of an object to which
the mind can attain, we easily see that their rules must
include, not only all that has been, but all that can be,

matter of experience. Our sensations can each convey
no information except about itself; each can contain no
trace of another additional sensation; and thus no relation and connexion between two sensations can be given
by the sensations themselves. But the mode in which
the mind perceives these impressions as objects, may and
will introduce necessary relations among them: and thus
by conceiving the idea of space to be a condition of perception in the mind, we can conceive the existence of
necessary truths, which apply to all perceived objects.

4. If we consider the impressions of sense as the
mere materials of our experience, such materials may
be accumulated in any quantity and in any order. But
if we suppose that this matter has a certain form given
it, in the act of being accepted by the mind, we can
understand how it is that these materials are subject to
inevitable rules;—how nothing can be perceived exempt
from the relations which belong to such a form. And
since there are such truths applicable to our experience, and arising from the nature of space, we may
thus consider space as a *form* which the materials given
by experience necessarily assume in the mind; as an
arrangement derived from the perceiving mind, and not
from the sensations alone.

5. Thus this phrase,—that space is a *form* belonging
to our perceptive power,—may be employed to express
that we cannot perceive objects as in space, without an
operation of the mind as well as of the senses—without
active as well as passive faculties. This phrase, however, is not necessary to the exposition of our doctrines.
Whether we call the conception of space a condition of
perception, a form of perception, or an idea, or by any
other term, it is something originally inherent in the mind
perceiving, and not in the objects perceived. And it is
because the apprehension of all objects is thus subjected

to certain mental conditions, forms or ideas, that our knowledge involves certain inviolable relations and necessary truths. The principles of such truths, so far as they regard space, are derived from the idea of space, and we must endeavour to exhibit such principles in their general form. But before we do this, we may notice some of the conditions which belong not to our Ideas in general, but to this Idea of Space in particular.

CHAPTER III.

OF SOME PECULIARITIES OF THE IDEA OF SPACE.

1. SOME of the Ideas which we shall have to examine involve conceptions of certain relations of objects, as the idea of Cause and of Likeness; and may appear to be suggested by experience, enabling us to *abstract* this general relation from particular cases. But it will be seen that Space is not such a general conception of a relation. For we do not speak of *Spaces* as we speak of Causes and Likenesses, but of space. And when we speak of spaces, we understand by the expression, parts of one and the same identical everywhere extended Space. We conceive a universal space; which is not made up of these partial spaces as its component parts, for it would remain if these were taken away; and these cannot be conceived without presupposing absolute space. Absolute space is essentially one; and the complication which exists in it, and the conception of various spaces, depends merely upon boundaries. Space must, therefore, be, as we have said, not a general conception abstracted from particulars, but a universal mode of representation, altogether independent of experience.

2. Space is infinite. We represent it to ourselves as

an infinitely great magnitude. Such an idea as that of
Likeness or Cause, is, no doubt, found in an infinite
number of particular cases, and so far includes these
cases. But these ideas do not include an infinite number
of cases as parts of an infinite whole. When we say
that all bodies and partial spaces exist *in* infinite space,
we use an expression which is not applied in the same
sense to any cases except those of space and time.

3. What is here said may appear to be a denial of
the real existence of space. It must be observed, however, that we do not deny, but distinctly assert, the
existence of space as a real and necessary condition of all
objects perceived ; and that we not only allow that
objects are seen external to us, but we found upon the
fact of their being so seen, our view of the nature of
space. If, however, it be said that we deny the reality
of space as an object or thing, this is true. Nor does it
appear easy to maintain that space exists as a thing,
when it is considered that this thing is infinite in all its
dimensions ; and, moreover, that it is a thing, which,
being nothing in itself, exists only that other things may
exist in it. And those who maintain the real existence
of space, must also maintain the real existence of time in
the same sense. Now two infinite things, thus really
existing, and yet existing only as other things exist in
them, are notions so extravagant that we are driven to
some other mode of explaining the state of the matter.

4. Thus space is not an object of which we perceive
the properties, but a form of our perception ; not a thing
which affects our senses, but an idea to which we conform the impressions of sense. And its peculiarities
appear to depend upon this, that it is not only a form of
sensation, but of *intuition ;* that in reference to space,
we not only perceive but *contemplate* objects. We see
objects in space, side by side, exterior to each other ;

space, and objects in so far as they occupy space, have parts exterior to other parts; and have the whole thus made up by the juxtaposition of parts. This mode of apprehension belongs only to the ideas of space and time. Space and time are made up of parts, but cause and likeness are not apprehended as made up of parts. And the term intuition (in its rigorous sense) is applicable only to that mode of contemplation in which we thus look at objects as made up of parts, and apprehend the relations of those parts at the same time and by the same act by which we apprehend the objects themselves.

5. As we have said, space limited by *boundaries* gives rise to various conceptions which we have often to consider. Thus limited, space assumes *form* or *figure;* and the variety of conceptions thus brought under our notice is infinite. We have every possible form of line, straight line, and curve; and of curves an endless number;—circles, parabolas, hyperbolas, spirals, helices. We have plane surfaces of various shapes,—parallelograms, polygons, ellipses; and we have solid figures,—cubes, cones, cylinders, spheres, spheroids, and so on. All these have their various properties, depending on the relations of their boundaries; and the investigation of their properties forms the business of the science of geometry.

6. Space has three dimensions, or directions in which it may be measured; it cannot have more or fewer. The simplest measurement is that of a straight line, which has length alone. A surface has both length and breadth: and solid space has length, breadth, and thickness or depth. The origin of such a difference of dimensions will be seen if we reflect that each portion of space has a boundary, and is extended both *in* the direction in which its boundary extends, and also in a direction *from* its boundary; for otherwise it would not be a boundary,

A point has no dimensions. A line has but one dimension,—the distance from its boundary, or its *length*. A plane, bounded by a straight line, has the dimension which belongs to this line, and also has another dimension arising from the distance of its parts from this boundary line; and this may be called *breadth*. A solid, bounded by a plane, has the dimensions which this plane has; and has also a third dimension, which we may call *height* or *depth*, as we consider the solid extended above or below the plane; or *thickness*, if we omit all consideration of up and down. And no space can have any dimensions which are not resoluble into these three.

We may now proceed to consider the mode in which the idea of space is employed in the formation of geometry.

CHAPTER IV.

OF THE DEFINITIONS AND AXIOMS WHICH
RELATE TO SPACE.

1. THE relations of space have been apprehended with peculiar distinctness and clearness from the very first unfolding of man's speculative powers. This was a consequence of the circumstance which we have just noticed, that the simplest of these relations, and those on which the others depend, are seen by intuition. Hence, as soon as men were led to speculate concerning the relations of space, they assumed just principles, and obtained true results. It is said that the science of *geometry* had its origin in Egypt, before the dawn of the Greek philosophy: but the knowledge of the early Egyptians (exclusive of their mythology) appears to have been purely practical; and, probably, their geometry consisted only in some maxims of *land-measuring*, which

is what the term implies. The Greeks of the time of Plato, had, however, not only possessed themselves of many of the most remarkable elementary theorems of the science; but had, in several instances, reached the boundary of the science in its elementary form; as when they proposed to themselves the problems of doubling the cube and squaring the circle.

But the deduction of these theorems by a systematic process, and the primary exhibition of the simplest principles involved in the idea of space, which such a deduction requires, did not take place, so far as we are aware, till a period somewhat later. The *Elements of Geometry* of Euclid, in which this task was performed, are to this day the standard work on the subject: the author of this work taught mathematics with great applause at Alexandria, in the reign of Ptolemy Lagus, about 280 years before Christ. The principles which Euclid makes the basis of his system have been very little simplified since his time; and all the essays and controversies which bear upon these principles, have had a reference to the form in which they are stated by him.

2. *Definitions.*—The first principles of Euclid's geometry are, as the first principles of any system of geometry must be, definitions and axioms respecting the various ideal conceptions which he introduces; as straight lines, parallel lines, angles, circles, and the like. But it is to be observed that these definitions and axioms are very far from being arbitrary hypotheses and assumptions. They have their origin in the idea of space, and are merely modes of exhibiting that idea in such a manner as to make it afford grounds of deductive reasoning. The axioms are necessary consequences of the conceptions respecting which they are asserted; and the defitions are no less necessary limitations of conceptions; not requisite in order to arrive at this or that consequence;

but necessary in order that it may be possible to draw any consequences, and to establish any general truths.

For example, if we rest the end of one straight staff upon the middle of another straight staff, and move the first staff into various positions, we, by so doing, alter the angles which the first staff makes with the other to the right hand and to the left. But if we place the staff in that special position in which these two angles are equal, each of them is a right angle, according to Euclid; and this is the definition of a right angle, except that Euclid employs the abstract conception of straight lines, instead of speaking, as we have done, of staves. But this selection of the case in which the two angles are equal is not a mere act of caprice; as it might have been if he had selected a case in which these angles are unequal in any proportion. For the consequences which can be drawn concerning the cases of unequal angles, do not lead to general truths, without some reference to that peculiar case in which the angles are equal: and thus it becomes necessary to single out and define that special .case, marking it by a special phrase. And this definition not only gives complete and distinct knowledge what a right angle is, to any one who can form the conception of an angle in general; but also supplies a principle from which all the properties of right angles may be deduced.

3. *Axioms.*—With regard to other conceptions also, as circles, squares, and the like, it is possible to lay down definitions which are a sufficient basis for our reasoning, so far as such figures are concerned. But, besides these definitions, it has been found necessary to introduce certain axioms among the fundamental principles of geometry. These are of the simplest character; for instance, that two straight lines cannot cut each other in more than one point, and an axiom concerning parallel lines. Like

the definitions, these axioms flow from the Idea of Space, and present that idea under various aspects. They are different from the definitions; nor can the definitions be made to take the place of the axioms in the reasoning by which elementary geometrical properties are established. For example, the definition of parallel straight lines is, that they are such as, however far continued, can never meet: but, in order to reason concerning such lines, we must further adopt some axiom respecting them: for example, we may very conveniently take this axiom; that two straight lines which cut one another are not both of them parallel to a third straight line*. The definition and the axiom are seen to be inseparably connected by our intuition of the properties of space; but the axiom cannot be proved from the definition, by any rigorous deductive demonstration. And if we were to take any other definition of two parallel straight lines, (as that they are both perpendicular to a third straight line,) we should still, at some point or other of our progress, fall in with the same difficulty of demonstratively establishing their properties without some further assumption.

4. Thus the elementary properties of figures, which are the basis of our geometry, are necessary results of our Idea of Space; and are connected with each other by the nature of that idea, and not merely by our hypotheses and constructions. Definitions and axioms must be combined, in order to express this idea so far as the purposes of demonstrative reasoning require. These verbal enunciations of the results of the idea cannot be made to depend on each other by logical consequence; but have a mutual dependence of a more intimate kind, which words cannot fully convey. It is not possible to resolve these truths into certain hypotheses, of which all the rest shall be the necessary logical consequence. The necessity is

---

* This axiom is simpler and more convenient than that of Euclid. It is employed by the late Professor Playfair in his *Geometry*.

not hypothetical, but intuitive.   The axioms require not
to be granted, but to be seen.   If any one were to assent
to them without seeing them to be true, his assent would
be of no avail for purposes of reasoning : for he would be
also unable to see in what cases they might be applied.
The clear possession of the Idea of Space is the first requisite for all geometrical reasoning; and this clearness of
idea may be tested by examining whether the axioms
offer themselves to the mind as evident.

5. The necessity of ideas added to sensations, in order
to produce knowledge, has often been overlooked or
denied in modern times.   The ground of necessary truth
which ideas supply being thus lost, it was conceived that
there still remained a ground of necessity in definitions ;—
that we might have necessary truths, by asserting especially what the definition implicitly involved in general.   It
was held, also, that this was the case in geometry :—that
all the properties of a circle, for instance, were implicitly
contained in the definition of a circle.   That this alone is
not the ground of the necessity of the truths which regard
the circle,—that we could not in this way unfold a definition into proportions, without possessing an intuition of
the relations to which the definition led,—has already been
shown.   But the insufficiency of the above account of the
grounds of necessary geometrical truth appeared in another way also.   It was found impossible to lay down a
system of definitions out of which alone the whole of
geometrical truth could be evolved.   It was found that
axioms could not be superseded.   No definition of a
straight line could be given which rendered the axiom
concerning straight lines superfluous.   And thus it appeared that the source of geometrical truths was not
definition alone; and we find in this result a confirmation
of the doctrine which we are here urging, that this source
of truth is to be found in the form or conditions of our
perception;—in the idea which we unavoidably combine

with the impressions of sense;—in the activity, and not in the passivity of the mind*.

6. This will appear further when we come to consider the mode in which we exercise our observation upon the relations of space. But we may, in the first place, make a remark which tends to show the connexion between our conception of a straight line, and the axiom which is made the foundation of our reasonings concerning space. The axiom is this;—that two straight lines, which have both their ends joined, cannot have the intervening parts separated so as to inclose a space. The necessity of this axiom is of exactly the same kind as the necessity of the definition of a right angle, of which we have already spoken. For as the line standing on another makes *right angles* when it makes the angles on the two sides of it equal; so a line is a *straight line* when it makes the two portions of space, on the two sides of it, similar. And as there is only a single position of the line first mentioned, which can make the angles equal, so there is only a single form of a line which can make the spaces near the line similar on one side and on the other: and therefore there cannot be two straight lines, such as the axiom describes, which, between the same limits, give two different boundaries to space thus separated. And thus we see a reason for the axiom. Perhaps this view may be further elucidated if we take a leaf of paper, double it, and crease the folded edge. We shall thus obtain a straight line at the folded edge; and this line divides the surface of the paper, as it was originally spread out, into two similar spaces. And that these

---

* I formerly stated views similar to these in some "Remarks" appended to a work which I termed *The Mechanical Euclid*, published in 1837. These Remarks, so far as they bear upon the question here discussed, were noticed and controverted in No. 135 of the *Edinburgh Review*. As an examination of the reviewer's objections may serve further to illustrate the subject, I shall annex to this chapter an answer to the article to which I have referred.

spaces are similar so far as the fold which separates them is concerned, appears from this;—that these two parts coincide when the paper is doubled. And thus a fold in a sheet of paper at the same time illustrates the definition of a straight line according to the above view, and confirms the axiom that two such lines cannot enclose a space.

If the separation of the two parts of space were made by any other than a straight line; if, for instance, the paper were cut by a concave line; then on turning one of the parts over, it is easy to see that the edge of one part being concave one way, and the edge of the other part concave the other way, these two lines might enclose a space. And each of them would divide the whole space into two portions which were not similar; for one portion would have a concave edge, and the other a convex edge. Between any two points there might be innumerable lines drawn, some convex one way and some convex the other way; but the straight line is the line which is not convex either one way or the other; it is the single medium standard from which the others may deviate in opposite directions.

Such considerations as these show sufficiently that the singleness of the straight line which connects any two points is a result of our fundamental conceptions of space. But yet the above conceptions of the similar form of the two parts of space on the two sides of a line, and of the form of a line which is intermediate among all other forms, are of so vague a nature, that they cannot fitly be made the basis of our elementary geometry ; and they are far more conveniently replaced, as they have been in almost all treatises of geometry, by the axiom that two straight lines cannot inclose a space.

7. But we may remark that in what precedes we have considered space only under one of its aspects:—as a plane. The sheet of paper which we assumed in order

to illustrate the nature of a straight line, was supposed to be perfectly *plane* or *flat*: for otherwise, by folding it, we might obtain a line not straight. Now this assumption of a plane appears to take for granted that very conception of a straight line which the sheet was employed to illustrate; for the definition of a plane given in the Elements of Geometry is, that it is a surface on which lie all straight lines drawn from one point of the surface to another. And thus the explanation above given of the nature of a straight line,—that it divides a plane space into similar portions on each side,—appears to be imperfect or nugatory.

And to this we reply, that the explanation must be rendered complete and valid by deriving the conception of a plane from considerations of the same kind as those which we employed for a straight line. Any portion of solid space may be divided into two portions by surfaces passing through any given line or boundaries. And these surfaces may be convex either on one side or on the other, and they admit of innumerable changes from being convex on one side to being convex on the other in any degree. So long as the surface is convex either way, the two portions of space which it separates are not similar, one having a convex and the other a concave boundary. But there is a certain intermediate position of the surface in which the two portions of space which it divides have their boundaries exactly similar. In this position the surface is neither convex nor concave, but plane. And thus a plane surface is determined by this condition of its being that single surface which is the intermediate form among all convex and concave surfaces by which solid space can be divided, and of its separating such space into two portions, of which the boundaries, though they are the same surface in two opposite positions, are exactly similar.

Thus a plane is the simplest and most symmetrical boundary by which a solid can be divided; and a straight line is the simplest and most symmetrical boundary by which a plane can be separated. These conceptions are obtained by considering the boundaries of an interminable space capable of imaginary division in every direction. And as a limited space may be separated into two parts by a plane, and a plane again separated into two parts by a straight line, so a line is divided into two portions by a point, which is the common boundary of the two portions; the end of the one and the beginning of the other portion having itself no magnitude, form, or parts.

8. The geometrical properties of planes and solids are deducible from the first principles of the Elements, without any new axioms; the definition of a plane above quoted,—that all straight lines joining its points lie in the plane,—being a sufficient basis for all reasoning upon these subjects. And thus the views which we have presented of the nature of space being verbally expressed by means of certain definitions and axioms, become the groundwork of a long series of deductive reasoning, by which is established a very large and curious collection of truths, namely, the whole science of elementary plane and solid geometry.

This science is one of indispensable use and constant reference to every student of the laws of nature; for the relations of space and number are the alphabet in which those laws are written. But besides the interest and importance of this kind which geometry possesses, it has a great and peculiar value for all who wish to understand the foundations of human knowledge, and the methods by which it is acquired. For the student of geometry acquires, with a degree of insight and clearness which the unmathematical reader can but feebly imagine, a conviction that there are necessary truths, many of them of a very complex and striking character; and that a few of the most simple and self-evident truths which it is possible for the mind of man to apprehend, may, by systematic deduction, lead to the most remote and unexpected results.

In pursuing such philosophical researches as that in which we are now engaged, it is of great advantage to the speculator to have cultivated to some extent the study of geometry; since by this study he may become fully aware of such features in human knowledge as those which we have mentioned. By the aid of the lesson thus learned from the contemplation of geometrical truths, we have been endeavouring to establish those further doctrines;—that these truths are but different aspects of the same Fundamental Idea, and that the ground of the necessity which these truths possess reside in the Idea from which they flow, this Idea not being a derivative result of experience, but its primary rule. When the reader has obtained a clear and satisfactory view of these doctrines, so far as they are applicable to our knowledge concerning space, he has, we may trust, overcome the main difficulty which will occur in following the course of the speculations now presented to him. He is then prepared to go forwards with us; to see over how wide a field the same doctrines are applicable; and how rich and various a harvest of knowledge springs from these seemingly scanty principles.

But before we quit the subject now under our consideration, we shall endeavour to answer some objections which have been made to the views here presented; and shall attempt to illustrate further the active powers which we have ascribed to the mind.

CHAPTER V.

OF SOME OBJECTIONS WHICH HAVE BEEN MADE TO THE DOCTRINES STATED IN THE PREVIOUS CHAPTER*.

THE *Edinburgh Review*, No. CXXXV., contains a critique on a work termed *The Mechanical Euclid*, in which opinions were delivered to nearly the same effect as some of those stated in the last chapter, and in Chapter XI. of the First Book. Although I believe that there are no arguments used by the reviewer to which the answers will not suggest themselves in the mind of any one who has read with attention what has been said in the preceding chapters (except, perhaps, one or two remarks which have reference to mechanical ideas), it may serve to illustrate the subject if I reply to the objections directly, taking them as the reviewer has stated them.

1. I had dissented from Stewart's assertion that mathematical truth is hypothetical, or depends upon arbitrary definitions; since we understand by an hypothesis a supposition, not only which we may make, but may abstain from making, or may replace by a different supposition;

---

* In order to render the present chapter more intelligible, it may be proper to state briefly the arguments which gave occasion to the review. After noticing Stewart's assertions, that the certainty of mathematical reasoning arises from its depending upon definitions, and that mathematical truth is hypothetical; I urged,—that no one has yet been able to construct a system of mathematical truths by the aid of definition alone; that a definition would not be admissible or applicable except it agreed with a distinct conception in the mind; that the definitions which we employ in mathematics are not arbitrary or hypothetical, but necessary definitions; that if Stewart had taken as his examples of axioms the peculiar geometrical axioms, his assertions would have been obviously erroneous, and that the real foundation of the truths of mathematics is the Idea of Space, which may be expressed (for purposes of demonstration) partly by definitions and partly by axioms.

whereas the definitions and hypotheses of geometry are necessarily such as they are, and cannot be altered or excluded. The reviewer (p. 84), informs us that he understands Stewart, when he speaks of hypotheses and definitions being the foundation of geometry, to speak of the hypothesis that real objects correspond to our geometrical definitions. "*If* a crystal be an exact hexahedron, the geometrical properties of the hexahedron may be predicated of that crystal." To this I reply, that such hypotheses as this are the grounds of our applications of geometrical truths to real objects, but can in no way be said to be the foundation of the truths themselves; that I do not think that the sense which the reviewer gives was Stewart's meaning; but that if it was, this view of the use of mathematics does not at all affect the question which both he and I proposed to discuss, which was, the ground of mathematical certainty. I may add, that whethei a crystal be an exact hexahedron, is a matter of observation and measurement, not of definition. I think the reader can have no difficulty in seeing how little my doctrine is affected by the connexion on which the reviewer thus insists. I have asserted that the proposition which affirms the square on the diagonal of a rectangle to be equal to the squares on two sides does not rest upon arbitrary hypotheses; the objector answers, that the proposition that the square on the diagonal *of this page* is equal to the squares on the sides, depends upon the arbitrary hypothesis that the page is a rectangle. Even if this fact were a matter of arbitrary hypothesis, what could it have to do with the general geometrical proposition? How could a single fact, observed or hypothetical, affect a universal and necessary truth, which would be equally true if the fact were false? If there be nothing arbitrary or hypothetical in geometry till we come to such steps in its application, it is plain that the truths themselves are not hypothetical, which is the question for us to decide.

2. The reviewer then (p. 85,) considers the doctrine that axioms as well as definitions are the foundations of geometry; and here he strangely narrows and confuses the discussion by making himself the advocate of Stewart, instead of arguing the question itself. I had asserted that some axioms are necessary as the foundations of mathematical reasoning, in addition to the definitions. If Stewart did not intend to discuss this question, I had no concern with what he had said about axioms. But I had every reason to believe that this was the question which Stewart did intend to discuss. I conceive there is no doubt that he intended to give an opinion upon the grounds of mathematical reasoning in general. For he begins his discussion (*Elements*, vol. ii., p. 38,) by contesting Reid's opinion on this subject, which is stated generally; and he refers again to the same subject, asserting in general terms, that the first principles of mathematics are not axioms but definitions. If, then, afterwards, he made his proof narrower than his assertion;—if having declared that no axioms are necessary, he afterwards limited himself to showing that seven out of twelve of Euclid's axioms are barren truisms, it was no concern of mine to contest this assertion, which left my thesis untouched. I had asserted that the proper geometrical axioms (that two straight lines cannot inclose a space, and the axiom about parallel lines) are indispensable in geometry. What account the reviewer gives of these axioms we shall soon see; but if Stewart allowed them to be axioms necessary to geometrical reasoning, he overturned his own assertion as to the foundations of such reasoning; and if he said nothing decisive about these axioms, which are the points on which the battle must turn, he left his assertion altogether unproved; nor was it neces-

sary for me to pursue the war into a barren and unimportant corner, when the metropolis was surrendered. The reviewer's exultation that I have not contested the first seven axioms is an amusing example of the self-complacent zeal of advocacy.

3. But let us turn to the material point: the proper geometrical axioms. What is the reviewer's account of these? Which side of the alternative does he adopt? Do they depend upon the definitions, and is he prepared to show the dependence? Or are they superfluous, and can he erect the structure of geometry without their aid? One of these two courses, it would seem, he must take. For we both begin by asserting the excellence of geometry as an example of demonstrated truth. It is precisely this attribute which gives an interest to our present inquiry. How, then, does the reviewer explain this excellence on his views? How does he reckon the foundation courses of the edifice which we agree in considering as a perfect example of intellectual building?

I presume I may take, as his answer to this question, his hypothetical statement of what Stewart would have said, (p. 87,) on the supposition that there had been, among the foundations of geometry, self-evident indemonstrable truths: although it is certainly strange that the reviewer should not venture to make up his mind as to the truth or falsehood of this supposition. If there were such truths they would be, he says, "legitimate filiations" of the definitions. They would be involved in the definitions. And again he speaks of the foundation of the geometrical doctrine of parallels as a flaw, and as a truth which requires, but has not received demonstration. And yet again, he tells us that each of these supposed axioms (Euclid's twelfth, for instance), is "merely an indication of the point at which geometry fails to perform that which it undertakes to perform" (p. 91); and that in reality her truths are not yet demonstrated. The amount of this is, that the geometrical axioms are to be held to be *legitimate filiations* of the definitions, because though certainly true, they cannot be proved from the definitions; that they are involved in the definitions, although they cannot be evolved out of them; and that rather than admit that they have any other origin than the definitions, we are to proclaim that geometry has failed to perform what she undertakes to perform.

To this I reply that I cannot understand what is meant by "legitimate filiations" of principles, if the phrase do not mean consequences of such principles established by rigorous and formal demonstration; that the reviewer, if he claims any real signification for his phrase, must substantiate the meaning of it by such a demonstration; he must establish his "legitimate filiation" by a genealogical table in a satisfactory form. When this cannot be done, to assert, notwithstanding, that the propositions are involved in the definitions, is a mere begging the question; and to excuse this defect by saying that geometry fails to perform what she has promised, is to calumniate the character of that science which we profess to make our standard, rather than abandon an arbitrary and unproved assertion respecting the real grounds of her excellence. I add, further, that if the doctrine of parallel lines, or any other geometrical doctrine of which we see the truth, with the most perfect insight of its necessity, have not hitherto received demonstration to the satisfaction of any school of reasoners, the defect must arise from their erroneous views of the nature of demonstrations, and the grounds of mathematical certainty.

4. I conceive, then, that the reviewer has failed altogether to disprove the doctrine that the axioms of geometry are necessary as a part of the foundations of the science. I had asserted further that these axioms supply

what the definitions leave deficient; and that they, along with definitions, serve to present the idea of space under such aspects that we can reason logically concerning it. To this the reviewer opposes (p. 96) the common opinion that a perfect definition is a complete explanation of a name, and that the test of its perfection is, that we may substitute the definition for the name wherever it occurs. I reply, that my doctrine, that a definition expresses a part, but not the whole, of the essential characters of an idea, is certainly at variance with an opinion sometimes maintained, that a definition merely explains a word, and should explain it so fully that it may always replace it. The error of this common opinion may, I think, be shown from considerations such as these;—that if we undertake to explain one word by several, we may be called upon, on the same ground, to explain each of these several by others, and that in this way we can reach no limit nor resting-place: that in point of fact, it is not found to lead to clearness, but to obscurity, when in the discussion of general principles, we thus substitute definitions for single terms; that even if this be done, we cannot reason without conceiving what the terms mean; and that, in doing this, the relations of our conceptions, and not the arbitrary equivalence of two forms of expression, are the foundations of our reasoning.

5. The reviewer conceives that some of the so-called axioms are really definitions. The axiom, that " magnitudes which coincide with each other, that is, which fill the same space, are equal," is a definition of geometrical *equality*: the axiom, that " the whole is greater than its part," is a definition of *whole* and *part*. But surely there are very serious objections to this view. It would seem more natural to say, if the former axiom is a definition of the word *equal*, that the latter is a definition of the word *greater*. And how can one short phrase define two

terms? If I say, "the heat of summer is greater than the heat of winter," does this assertion define anything, though the proposition is perfectly intelligible and distinct? I think, then, that this attempt to reduce these axioms to definitions is quite untenable.

6. I have stated that a definition can be of no use, except we can conceive the possibility and truth of the property connected with it; and that if we do conceive this, we may rightly begin our reasonings by stating the property as an axiom; which Euclid does, in the case of straight lines and of parallels. The reviewer inquires, (p. 92,) whether I am prepared to extend this doctrine to the case of circles, for which the reasoning is usually rested upon the definition; whether I would replace this definition by an axiom, asserting the possibility of such a circle. To this I might reply, that it is not at all incumbent upon me to assent to such a change; for I have all along stated that it is indifferent whether the fundamental properties from which we reason be exhibited as definitions or as axioms, provided their necessity be clearly seen. But I am ready to declare that I think the form of our geometry would be not at all the worse, if, instead of the usual definition of a circle,—" that it is a figure contained by one line, which is called the circumference, and which is such, that all straight lines drawn from a certain point within the circumference are equal to one another,"— we were to substitute an axiom and a definition, as follows:—

Axiom. If a line be drawn so as to be at every point equally distant from a certain point, this line will return into itself, or will be *one* line including a space.

Definitions. The space is called a *circle*, the line the *circumference*, and the point the *centre*.

And this being done, it would be true, as the reviewer remarks, that geometry cannot stir *one* step without resting on an axiom. And I do not at all hesitate to say, that the above axiom, expressed or understood, is no less necessary than the definition, and is tacitly assumed in every proposition into which circles enter.

7. I have, I think, now disposed of the principal objections which bear upon the proper axioms of geometry. The principles which are stated as the first seven axioms of Euclid's *Elements*, need not, as I have said, be here discussed. They are principles which refer, not to Space in particular, but to Quantity in general: such, for instance, as these; "If equals be added to equals the wholes are equal;"—"If equals be taken from equals the remainders are equal." But I will make an observation or two upon them before I proceed.

Both Locke and Stewart have spoken of these axioms as barren truisms: as propositions from which it is not possible to deduce a single inference: and the reviewer asserts that they are not first principles, but laws of thought. (p. 88.) To this last expression I am willing to assent; but I would add, that not only these, but all the principles which express the fundamental conditions of our knowledge, may with equal propriety be termed laws of thought; for these principles depend upon our ideas, and regulate the active operations of the mind, by which coherence and connexion are given to its passive impressions. But the assertion that no conclusions can be drawn from simple axioms, or laws of human thought which regard quantity, is by no means true. The whole of arithmetic,—for instance, the rules for the multiplication and division of large numbers, for finding a common measure, and, in short, a vast body of theory respecting numbers,—rests upon no other foundation than such axioms as have been just noticed, that if equals be added to equals the wholes will be equal. And even when Locke's assertion, that from these axioms no truths can be deduced, is modified by Stewart and the reviewer, and limited to *geometrical* truths, it is hardly tenable (although, in fact, it matters little to our argument whether it is or no). For the greater part of the Seventh Book of Euclid's *Elements*, (on Commensurable and Incommensurable Quantities,) and the Fifth Book, (on Proportion,) depend upon these axioms, with the addition only of the definition or axiom (for it may be stated either way) which expresses the idea of proportionality in numbers. So that the attempt to disprove the necessity and use of axioms, as principles of reasoning, fails even when we take those instances which the opponents consider as the more manifestly favourable to their doctrine.

8. But perhaps the question may have already suggested itself to the reader's mind, of what use can it be formally to state such principles as these, (for example, that if equals be added to equals the wholes are equal,) since, whether stated or no, they will be assumed in our reasoning? And how can such principles be said to be necessary, when our proof proceeds equally well without any reference to them? And the answer is, that it is precisely because these are the common principles of reasoning, which we naturally employ without specially contemplating them, that they require to be separated from the other steps and formally stated, when we *analyse* the demonstrations which we have obtained. In every mental process many principles are combined and abbreviated, and thus in some measure concealed and obscured. In analysing these processes the combination must be resolved, and the abbreviation expanded, and thus the appearance is presented of a pedantic and superfluous formality. But that which is superfluous for proof, is necessary for the analysis of proof. In order to exhibit the conditions of demonstration distinctly, they must be exhibited formally. In the same manner, in

demonstration we do not usually express every step in the form of a syllogism, but we see the grounds of the conclusiveness of a demonstration, by resolving it into syllogisms. Neither axioms nor syllogisms are necessary for conviction; but they are necessary to display the conditions under which conviction becomes inevitable. The application of a single one of the axioms just spoken of is so minute a step in the proof, that it appears pedantic to give it a marked place; but the very essence of demonstration consists in this, that it is composed of an indissoluble succession of such minute steps. The admirable circumstance is, that by the accumulation of such apparently imperceptible advances, we can in the end make so vast and so sure a progress. The completeness of the analysis of our knowledge appears in the smallness of the elements into which it is thus resolved. The minuteness of any of these elements of truth, of axioms for instance, does not prevent their being as essential as others which are more obvious. And any attempt to assume one kind of element only when the course of our analysis brings before us two or more kinds, is altogether unphilosophical. Axioms and definitions are the proximate constituent principles of our demonstrations; and the intimate bond which connects together a definition and an axiom on the same subject is not truly expressed by asserting the latter to be derived from the former. This bond of connexion exists in the mind of the reasoner, in his conception of that to which both definition and axiom refer, and consequently in the general Fundamental Idea of which that conception is a modification.

CHAPTER VI.

OF THE PERCEPTION OF SPACE.

1. ACCORDING to the views above explained, certain of the impressions of our senses convey to us the perception of objects as existing in space; inasmuch as by the constitution of our minds we cannot receive those impressions otherwise than in a certain form, involving such a manner of existence. But the question deserves to be asked, *What* are the impressions of sense by which we thus become acquainted with space and its relations? And as we have seen that this idea of space implies an act of the mind as well as an impression on the sense, what manifestations do we find of this activity in our observation of the external world?

It is evident that sight and touch are the senses by which the relations of space are perceived, principally or entirely. It does not appear that an odour, or a feeling of warmth or cold, would, independently of experience, suggest to us the conception of a space surrounding us. But when we *see* objects, we see that they are extended and occupy space; when we *touch* them, we feel that they are in a space in which we also are. We have before our eyes any object, for instance, a board covered with geometrical diagrams; and we distinctly perceive, by vision, those lines of which the relations are the subjects of our mathematical reasoning. Again, we see before us a solid object, a cubical box for instance; we see that it is within reach; we stretch out the hand and perceive by the touch that it has sides, edges, corners, which we had already perceived by vision.

2. Probably most persons do not generally apprehend that there is any material difference in these two cases;

that there are any different acts of mind concerned in perceiving by sight a mathematical diagram upon paper, and a solid cube lying on a table. Yet it is not difficult to show that, in the latter case at least, the perception of the shape of the object is not immediate. A very little attention teaches us that there is an act of judgment as well as a mere impression of sense requisite, in order that we may see any solid object. For there is no visible appearance which is inseparably connected with solidity. If a picture of a cube be rightly drawn in perspective and skilfully shaded, the impression upon the sense is the same as if it were a real cube. The picture may be mistaken for a solid object. But it is clear that in this case, the solidity is given to the object by an act of mental judgment. All that is seen is outline and shade, figures and colours on a flat board. The solid angles and edges, the relation of the faces of the figure by which they form a cube, is a matter of inference. This, which is evident in the case of the pictured cube, is true in all vision whatever. We see a scene before us on which are various figures and colours, but the eye cannot see more. It sees length and breadth, but no third dimension. In order to know that there are solids, we must infer as well as see. And this we do readily and constantly; so familiarly, indeed, that we do not perceive the operation. Yet we may detect this latent process in many ways; for instance, by attending to cases in which the habit of drawing such inferences misleads us. Most persons have experienced this delusion in looking at a scene in a theatre, and especially that kind of scene which is called a diorama, when the interior of a building is represented. In these cases, the perspective representations of the various members of the architecture and decoration impress us almost irresistibly with the conviction that we have before us a space of great extent and complex form, instead of a flat painted canvass. Here, at least, the

space is our own creation; but it is manifestly created
by the same act of thought as if we were really in the
palace or the cathedral of which the halls and aisles thus
seem to inclose us.   And the act by which we thus
create space of three dimensions out of visible extent
of length and breadth, is constantly and imperceptibly
going on.   We are perpetually interpreting in this
manner the language of the visible world.   From the
appearances of things which we directly see, we are constantly inferring that which we cannot directly see, their
distance from us, and the position of their parts.

3. The characters which we thus interpret are various.
They are, for instance, the visible forms, colours, and
shades of their parts, understood according to the maxims
of perspective; (for of perspective every one has a practical knowledge, as every one has of grammar;) the
effort by which we fix both our eyes on the same object,
and adjust each eye to distinct vision; and the like.
The right interpretation of the information which such
circumstances give us respecting the true forms and
distances of things, is gradually learned; the lesson being
begun in our earliest infancy, and inculcated upon us
every hour during which we use our eyes.   The completeness with which the lesson is mastered is truly
admirable; for we forget that our conclusion is obtained
indirectly, and mistake a judgment on evidence for an
intuitive perception.   We see the breadth of the street,
as clearly and readily as we see the house on the other
side of it; and we see the house to be square, however
obliquely it be presented to us.   This, however, by no
means throws any doubt or difficulty on the doctrine
that in all these cases we do interpret and infer.   The
rapidity of the process, and the unconsciousness of the
effort, are not more remarkable in this case than they are
when we understand the meaning of the speech which
we hear, or of the book which we read.   In these latter

cases we merely hear noises or see black marks; but we make, out of these elements, thought and feeling, without being aware of the act by which we do so. And by an exactly similar process we see a variously-coloured expanse, and collect from it a space occupied by solid objects. In both cases the act of interpretation is become so habitual that we can hardly stop short at the mere impression of sense.

4. But yet there are various ways in which we may satisfy ourselves that these two parts of the process of seeing objects are distinct. To separate these operations is precisely the task which the artist has to execute in making a drawing of what he sees. He has to recover the consciousness of his real and genuine sensations, and to discern the lines of objects as they appear. This at first he finds difficult; for he is tempted to draw what he knows of the forms of visible objects, and not what he sees: but as he improves in his art, he learns to put on paper what he sees only, separate from what he infers, in order that thus the inference, and with it a conception like that of the reality, may be left to the spectator. And thus the natural process of vision is the habit of seeing that which cannot be seen; and the difficulty of the art of drawing consists in not seeing more than is visible.

5. But again; even in the simplest drawing we exhibit something which we do not see. However slight is our representation of objects, it contains something which we create for ourselves. For we draw an *outline*. Now an outline has no existence in nature. There are no visible lines presented to the eye by a group of figures. We separate each figure from the rest, and the boundary by which we do this is the outline of the figure; and the like may be said of each member of every figure. A painter of our own times has made this remark in a work upon his art*. "The effect which natural objects produce upon our

* PHILLIPS *on Painting.*

sense of vision is that of a number of parts, or distinct masses of form and colour, and not of lines. But when we endeavour to represent by painting the objects which are before us, or which invention supplies to our minds, the first and the simplest means we resort to is this picture, by which we separate the form of each object from those that surround it, marking its boundary, the extreme extent of its dimensions in every direction, as impressed on our vision: and this is termed drawing its outline."

5. Again, there are other ways in which we see clear manifestations of the act of thought by which we assign to the parts of objects their relations in space, the impressions of sense being merely subservient to this act. If we look at a medal through a glass which inverts it, we see the figures upon it become concave depressions instead of projecting convexities; for the light which illuminates the nearer side of the convexity, will be transferred to the opposite side by the apparent inversion of the medal, and will thus imply a hollow in which the side nearest the light gathers the shade. Here our decision as to which part is nearest to us, has reference to the side from which the light comes. In other cases it is more spontaneous. If we draw black outlines, such as represent the edges of a cube seen in perspective, certain of the lines will cross each other ; and we may make this cube appear to assume two different positions, by determining that the lines which belong to one end of the cube shall be understood to be before or to be behind those which they cross. Here an act of the will, operating upon the same sensible image, gives us two cubes, occupying two entirely different positions. Again, many persons may have observed that when a windmill in motion at a distance from us, (so that the outline of the sails only is seen,) stands obliquely to the eye, we may, by an effort of thought, make the

obliquity assume one or the other of two positions; and as we do this, the sails, which in one instance appear to turn from right to left, in the other case turn from left to right. A person a little familiar with this mental effort can invert the motion as often as he pleases, so long as the conditions of form and light do not offer a manifest contradiction to either position.

Thus we have these abundant and various manifestations of the activity of the mind, in the process by which we collect from vision the relations of solid space of three dimensions. But we must further make some remarks on the process by which we perceive mere visible figure; and also on the mode in which we perceive the relations of space by the touch; and first of the latter subject.

6. The opinion above illustrated, that our sight does not give us a direct knowledge of the relations of solid space, and that this knowledge is acquired only by an inference of the mind, was first clearly taught by the celebrated Bishop Berkeley*, and is a doctrine now generally assented to by metaphysical speculators.

But does the sense of *touch* give us directly a knowledge of space? This is a question which has attracted considerable notice in recent times; and new light has been thrown upon it in a degree which is very remarkable, when we consider that the philosophy of perception has been a prominent subject of inquiry from the earliest times. Two philosophers, advancing to this inquiry from different sides, the one a metaphysician, the other a physiologist, have independently arrived at the conviction that the long current opinion, according to which we acquire a knowledge of space by the sense of touch, is erroneous. And the doctrine which they teach instead of the ancient error, has a very important bearing upon the principle which we are endeavouring to establish,—

* *Theory of Vision.*

that our knowledge of space and its properties is derived rather from the active operations than from the passive impressions of the percipient mind.

Undoubtedly the persuasion that we acquire a know-ledge of form by the touch is very obviously suggested by our common habits. If we wish to know the form of any body in the dark, or to correct the impressions conveyed by sight, when we suspect them to be false, we have only, it seems to us, at least at first, to stretch forth the hand and touch the object; and we learn its shape with no chance of error. In these cases, form appears to be as immediate a perception of the sense of touch, as colour is of the sense of sight.

7. But is this perception really the result of the passive sense of touch merely? Against such an opinion Dr. Brown, the metaphysician of whom I speak, urges* that the feeling of touch alone, when any object is applied to the hand, or any other part of the body, can no more convey the conception of form or extension, than the sensation of an odour or a taste can do, except we have already some knowledge of the relative position of the parts of our bodies; that is, except we are already in possession of an idea of space, and have in our minds referred our limbs to their positions ; which is to suppose the conception of form already acquired.

8. By what faculty then do we originally acquire onr conceptions of the relations of position? Brown answers by the *muscular sense;* that is, the conscious exertions of the various muscles by which we move our limbs. When we feel out the form and position of bodies by the hand, our knowledge is acquired, not by the mere touch of the body, but by perceiving the course the fingers must take in order to follow the surface of the body, or to pass from one body to another. We are

* *Lectures,* vol. i. p. 459, (1824).

conscious of the slightest of the volitions by which we thus feel out form and place; we know whether we move the finger to the right or left, up or down, to us or from us, through a large or a small space; and all these conscious acts are bound together and regulated in our minds by an idea of an extended space in which they are performed. That this idea of space is not borrowed from the sight, and transferred to the muscular feelings by habit, is evident. For a man born blind can feel out his way with his staff, and has his conceptions of position determined by the conditions of space, no less than one who has the use of his eyes. And the muscular consciousness which reveals to us the position of objects and parts of objects when we feel them out by means of the hand, shews itself in a thousand other ways, and in all our limbs: for our habits of standing, walking, and all other attitudes and motions, are regulated by our feeling of our position and that of surrounding objects. And thus we cannot touch any object without learning something respecting its position; not that the sense of touch directly conveys such knowledge; but we have already learnt, from the muscular sense, constantly exercised, the position of the limb which the object thus touches.

9. The justice of this distinction will, I think, be assented to by all persons who attend steadily to the process itself, and might be maintained by many forcible reasons. Perhaps one of the most striking evidences in its favour is that, as I have already intimated, it is the opinion to which another distinguished philosopher, Sir Charles Bell, has been led, reasoning entirely upon physiological principles. From his researches it resulted that besides the nerves which convey the impulse of the will from the brain to the muscle, by which every motion of our limbs is produced, there is another set of nerves which carry back to the brain a sense of the condition of the

muscle, and thus regulate its activity; and give us the
consciousness of our position and relation to surrounding
objects. The motion of the hand and fingers, or the consciousness of this motion, must be combined with the
sense of touch properly so called, in order to make an
inlet to the knowledge of such relations. This consciousness of muscular exertion, which he called a sixth sense*,
is our guide, Sir C. Bell shows, in the common practical
government of our motions; and he states that having
given this explanation of perception as a physiological
doctrine, he had with satisfaction seen it confirmed by
Dr. Brown's speculations.

10. Thus it appears that our consciousness of the relations of space is inseparably and fundamentally connected with our own actions in space. We perceive only
while we act; our sensations require to be interpreted by
our volitions. The apprehension of extension and figure
is far from being a process in which we are inert and
passive. We draw lines with our fingers; we construct
surfaces by curving our hands; we generate spaces by the
motion of our arms. When the geometer bids us form
lines, or surfaces, or solids by motion, he intends his in-
junction to be taken as hypothetical only; we need only
conceive such motions. But yet this hypothesis represents truly the origin of our knowledge; we perceive by
motion at first, as we conceive afterwards. Or if not
always by actual motion, at least by potential. If we
perceive the length of a staff by holding its two ends in
our two hands without running the finger along it, this is
because by habitual motion we have already acquired a
measure of the distance of our hands in any attitude of
which we are conscious. Even in the simplest case, our
perceptions are derived not from the touch, but from the

* *Bridgewater Treatise*, p. 195. *Phil. Trans.*, 1826, p. ii.,
p. 167.

sixth sense; and this sixth sense at least, whatever may
be the case with the other five, implies an active mind
along with the passive sense.

10. Upon attentive consideration, it will be clear that
a large portion of the perceptions respecting space which
appear at first to be obtained by sight alone, are, in fact;
acquired by means of this sixth sense. Thus we consider
the visible sky as a single surface surrounding us and returning into itself, and thus forming a hemisphere. But
such a mode of conceiving an object of vision could never
have occurred to us, if we had not been able to turn our
heads, to follow this surface, to pursue it till we find it returning into itself. And when we have done this, we
necessarily represent it to ourselves as a concave inclosure
within which we are. The sense of sight alone, without
the power of muscular motion, could not have led us to
view the sky as a vault or hemisphere. Under such circumstances, we should have perceived only what was presented to the eye in one position; and if different
appearances had been presented in succession, we could
not have connected them as parts of the same picture,
for want of any perception of their relative position.
They would have been so many detached and incoherent visual sensations. The muscular sense connects
their parts into a whole, making them to be only different
portions of one universal scene.

11. These considerations point out the fallacy of a very
curious representation made by Dr. Reid, of the convictions to which man would be led, if he possessed vision
without the sense of touch. To illustrate this subject,
Reid uses the fiction of a nation whom he terms the *Ido-
menians,* who have no sense except that of sight. He
describes their notions of the relations of space as being
entirely different from ours. The axioms of their geometry are quite contradictory to our axioms. For example,

it is held to be self-evident among them that two straight
lines which intersect each other once, must intersect a
second time; that the three angles of any triangle are
greater than two right angles; and the like.   These paradoxes are obtained by tracing the relations of lines on the
surface of a concave sphere, which surrounds the spectator, and on which all visible appearances may be supposed to be presented to him.   But from what is said
above it appears that the notion of such a sphere, and
such a connexion of visible objects which are seen in different directions, cannot be arrived at by sight alone.
When the spectator combines in his conception the relations of long-drawn lines and large figures, as he sees
them by turning his head to the right and to the left, upwards and downwards, he ceases to be an Idomenian.
And thus our conceptions of the properties of space derived through the exercise of one mode of perception are
not at variance with those obtained in another way; but
all such conceptions, however produced or suggested, are
in harmony with each other; being, as has already been
said, only different aspects of the same idea.

12. If our perceptions of the position of objects
around us do not depend on the sense of vision alone, but
on the muscular feeling brought into play when we turn
our head, it will obviously follow that the same is true
when we turn the eye instead of the head.   And thus
we may learn the form of objects, not by looking at
them with a fixed gaze, but by following the boundary of
them with the eye.   While the head is held perfectly
still, the eye can rove along the outlines of visible objects,
scrutinize each point in succession, and leap from one
point to another; each such act being accompanied by a
muscular consciousness which makes us aware of the
direction in which the look is travelling.   And we may
thus gather information concerning the figures and places

which we trace out with the visual ray, as the blind man learns the forms of things which he traces out with his staff, being conscious of the motions of his hand.

13. This view of the mode in which the eye perceives position, which is thus supported by the analogy of other members employed for the same purpose, is further confirmed by Sir Charles Bell by physiological reasons. He teaches us that* when an object is seen we employ two senses: there is an impression on the retina; but we receive also the idea of position or relation in space, which it is not the office of the retina to give, by our consciousness of the efforts of the voluntary muscles of the eye: and he has traced in detail the course of the nerves by which these muscles convey their information. The constant *searching* motion of the eye, as he terms it†, is the means by which we become aware of the position of objects about us.

14. It is not to our present purpose to follow the physiology of this subject; but we may notice that Sir C. Bell has examined the special circumstances which belong to this operation of the eye. We learn from him that the particular point of the eye which thus traces the forms of visible objects is a part of the retina which has been termed the *sensible spot;* being that part which is *most* sensible to the impressions of light and colour. This part, indeed, is not a spot of definite size and form, for it appears that proceeding from a certain point of the retina, the sensibility diminishes on every side by degrees. And the searching motion of the eye arises from the desire which we instinctively feel of receiving upon the sensible spot the image of the object to which the attention is directed. We are uneasy and impatient till the eye is turned so that this is effected. And as our attention is

---

* *Phil. Trans.*, 1823. On the Motions of the Eye.

† *Bridgewater Treatise*, p. 282.

transferred from point to point of the scene before us, the
eye, and this point of the eye in particular, travel along
with the thoughts; and the muscular sense which tells
us of these movements of the organ of vision, conveys
to us a knowledge of the forms and places which we thus
successively survey.

15. How much of activity there is in the process by
which we perceive the outlines of objects appears further
from the language by which we describe their forms.
We apply to them not merely adjectives of form, but
verbs of motion.   An abrupt hill *starts* out of the plain;
a beautiful figure has a *gliding* outline.   We have

> The windy summit, wild and high,
> Roughly *rushing* on the sky.

These terms express the course of the eye as it follows
the lines by which such forms are bounded and marked.
In like manner another modern poet* says of Soracte,
that it

> From out the plain
> *Heaves* like a long-swept wave about to break,
> And on the curl *hangs pausing.*

Thus the muscular sense, which is inseparably connected with an act originating in our own mind, not only
gives us all that portion of our perceptions of space in
which we use the sense of touch, but also, at least in a
great measure, another large portion of such perceptions,
in which we employ the sense of sight.   As we have
before seen that our *knowledge* of solid space and its
properties is not conceivable in any other way than as the
result of a mental act, governed by conditions depending
on its own nature; so it now appears that our *perceptions*
of visible figure are not obtained without an act performed
under the same conditions.   The sensations of touch and
sight are subordinated to an idea which is the basis of
our speculative knowledge concerning space and its rela-

* BYRON, *Ch. Har.* IV., St. 75,

tions; and this same idea is disclosed to our consciousness by its practically regulating our intercourse with the
external world.

By considerations such as have been adduced and
referred to, it is proved beyond doubt, that in a great
number of cases our knowledge of form and position is
acquired from the muscular sense, and not from sight
directly :—for instance, in all cases in which we have
before us large objects and extensive spaces. Whether
in any case the eye gives us a direct perception of form,
we shall not here further inquire. Another opportunity
of discussing this subject will occur hereafter.

We now quit the consideration of the properties of
Space, and consider the Idea of Time.

<hr>

CHAPTER VI.

OF THE IDEA OF TIME.

1. RESPECTING the Idea of Time, we may make several
of the same remarks which we made concerning the idea
of space, in order to shew that it is not borrowed from
experience; but is a bond of connexion among the
impressions of sense, derived from a peculiar activity of
the mind, and forming a foundation both of our experience
and of our speculative knowledge.

Time is not a notion obtained by experience. Experience, that is, the impressions of sense and our consciousness of our thoughts, gives us various perceptions; and different successive perceptions considered
together exemplify the notion of change. But this very
connexion of different perceptions,—this successiveness,
—presupposes that the perceptions exist *in time*. That
things happen either together, or one after the other, is

intelligible only by assuming time as the condition under which they are presented to us.

Thus time is a necessary condition in the presentation of all occurrences to our minds. We cannot conceive this condition to be taken away. We can conceive time to go on while nothing happens in it ; but we cannot conceive anything to happen while time does not go on.

It is clear from this that time is not an impression derived from experience, in the same manner in which we derive from experience our information concerning the objects which exist, and the occurrences which take place in time. The objects of experience can easily be conceived to be, or not to be :—to be absent as well as present. Time always is, and always is present, and even in our thoughts we cannot form the contrary supposition.

2. Thus time is something distinct from the *matter* or substance of our experience, and may be considered as a necessary *form* which that matter (the experience of change) must assume, in order to be an object of contemplation to the mind. Time is one of the necessary conditions under which we apprehend the information which our senses and consciousness give us. By considering time as a form which belongs to our power of apprehending occurrences and changes, and under which alone all such experience can be accepted by the mind, we explain the necessity, which we find to exist, of conceiving all such changes as happening in time ; and we thus see that time is not a property perceived as existing in objects, or as conveyed to us by our senses ; but a condition impressed upon our knowledge by the constitution of the mind itself ; involving an act of thought as well as an impression of sense.

3. We showed that space is an idea of the mind, or

form of our perceiving power, independent of experience, by pointing out that we possess necessary and universal truths concerning the relations of space, which could never be given by means of experience; but of which the necessity is readily conceivable, if we suppose them to have for their basis the constitution of the mind. There exist also respecting number, many truths absolutely necessary, entirely independent of experience and anterior to it; and so far as the conception of number depends upon the idea of time, the same argument might be used to show that the idea of time is not derived from experience, but is a result of the native activity of the mind: but we shall defer all views of this kind till we come to the consideration of Number.

4. Some persons have supposed that we obtain the notion of time from the perception of motion. But it is clear that the perception of motion, that is, change of place, presupposes the conception of time, and is not capable of being presented to the mind in any other way. If we contemplate the same body as being in different places at different times, and connect these observations, we have the conception of motion, which thus presupposes the necessary conditions that existence in time implies. And thus we see that it is possible there should be necessary truths concerning all motion, and consequently concerning those motions which are the objects of experience: but that the source of this necessity is the Ideas of time and space, which, being universal conditions of knowledge residing in the mind, afford a foundation for necessary truths.

Chapter VII.

OF SOME PECULIARITIES OF THE IDEA OF TIME.

1. The Idea of Time, like the Idea of Space, offers to our notice some characters which do not belong to our fundamental ideas generally, but which are deserving of remark. These characters are, in some respects, closely similar with regard to time and to space, while, in other respects, the peculiarities of these two ideas are widely different. We shall point out some of these characters.

Time is not a general *abstract* notion collected from experience; as, for example, a certain general conception of the relations of things. For we do not consider particular *times* as examples of Time in general, (as we consider particular causes to be examples of Cause,) but we conceive all particular times to be parts of a single and endless Time. This continually-flowing and endless time is what offers itself to us when we contemplate any series of occurrences. All actual and possible times exist as parts, in this original and general time. And since all particular times are considered as derivable from time in general, it is manifest that the notion of time in general cannot be derived from the notions of particular times. The notion of time in general is therefore not a general conception gathered from experience.

2. Time is infinite. Since all actual and possible times exist in the general course of time, this general time must be infinite. All limitation merely divides, and does not terminate, the extent of absolute time. Time has no beginning and no end; but the beginning and the end of every other existence takes place in it.

3. Time, like space, is not only a form of perception

but of *intuition*. We contemplate events as taking place *in* time. We consider its parts as added to one another, and events as filling a larger or smaller extent of such parts. The time which any event takes up is the sum of all such parts, and the relation of the same to time is fully understood when we can clearly see what portions of time it occupies, and what it does not. Thus the relation of known occurrences to time is perceived by intuition; and time is a form of intuition of the external world.

5. Time is conceived as a quantity of one dimension; it has great analogy with a line, but none at all with a surface or solid. Time may be considered as consisting of a series of instants, which are before and after one another; and they have no other relation than this, of before and after. Just the same would be the case with a series of points taken along a line; each would be after those on one side of it, and before those on another. Indeed the analogy between time and space of one dimension is so close, that the same terms are applied to both ideas, and we hardly know to which they originally belong. Times and lines are alike called *long* and *short;* we speak of the *beginning* and *end* of a line; of a *point* of time, and of the *limits* of a portion of duration.

6. But as has been said, there is nothing in time which corresponds to more than one dimension in space, and hence nothing which has any obvious analogy with figure. Time resembles a line indefinitely extended both ways; all partial times are portions of this line; and no mode of conceiving time suggests to us a line making any angle with the original line, or any other combination which might give rise to figures of any kind. The analogy between time and space, which in many circumstances is so clear, here disappears altogether. Spaces of two and of three dimensions, planes and solids, have

nothing to which we can compare them in the conceptions arising out of time.

7. As figure is a conception solely appropriate to space, there is also a conception which peculiarly belongs to time, namely, the conception of recurrence of times similarly marked ; or, as it may be termed, *rhythm*, using this word in a general sense. The term rhythm is most commonly used to designate the recurrence of times marked by the syllables of a verse, or the notes of a melody : but it is easy to see that the general conception of such a recurrence does not depend on the mode in which it is impressed upon the sense. The forms of such recurrence are innumerable. Thus in such a line as

Quádrupedánte putrém sonitú quatit úngula cámpum,

we have alternately one long or forcible syllable, and two short or light ones, recurring over and over. In like manner in our own language, in the line

At the clóse of the dáy when the hámlet is stíll,

we have two light and one strong syllable repeated four times over. Such repetition is the essence of versification. The same kind of rhythm is one of the main elements of music, with this difference only, that in music the forcible syllables are made so for the purposes of rhythm by their length only ; for example, if either of the above lines were imitated by a melody in the most simple and obvious manner, each strong syllable would occupy exactly twice as much time as two of the weaker ones. Something very analogous to such rhythm may be traced in other parts of poetry and art, which we need not here dwell upon. But in reference to our present subject, we may remark that by the introduction of such rhythm, the flow of time, which appears otherwise so perfectly simple and homogeneous, admits of an infinite number of varied yet regular modes of progress. All the kinds of versification which occur in all languages, and the still

more varied forms of recurrence of notes of different
lengths, which are heard in all the varied strains of melodies, are only examples of such modifications, or configurations as we may call them, of time.   They involve relations of various portions of time, as figures involve relations of various portions of space.   But yet the analogy
between rhythm and figure is by no means very close;
for in rhythm we have relations of quantity alone in the
parts of time, whereas in figure we have relations not
only of quantity, but of a kind altogether different,—
namely, of position.   On the other hand, a *repetition* of
similar elements, which does not necessarily occur in
figures, is quite essential in order to impress upon us that
measured progress of time of which we here speak.
And thus the ideas of time and space have each its peculiar and exclusive relations; position and figure belonging only to space, while repetition and rhythm are appropriate to time.

8. One of the simplest forms of recurrence is *alternation*, as when we have alternate strong and slight syllables.   For instance,—

Awáke, aríse, or bé for éver fáll'n.

Or without any subordination, as when we reckon numbers, and call them in succession, odd, even, odd, even.

9. But the simplest of all forms of recurrence is that
which has no variety;—in which a series of units, each
considered as exactly similar to the rest, succeed each
other; as one, one, one, and so on.   In this case, however,
we are led to consider each unit with reference to all that
have preceded; and thus the series one, one, one, and so
forth, becomes one, two, three, four, five, and so on; a
series with which all are familiar, and which may be continued without limit.

We thus collect from that repetition of which time
admits, the conception of *Number*.

10. The relations of position and figure are the subject of the science of geometry; and are, as we have already said, traced into a very remarkable and extensive body of truths, which rests for its foundations on axioms involved in the Idea of Space.   There is, in like manner, a science of great complexity and extent, which has its foundation in the Idea of Time.   But this science, as it is usually pursued, applies only to the conception of Number, which is, as we have said, the simplest result of repetition.   This science is *Theoretical Arithmetic*, or the speculative doctrine of the properties and relations of numbers; and we must say a few words concerning the principles which it is requisite to assume as the basis of this science.

Chapter VIII.

OF THE AXIOMS WHICH RELATE TO NUMBER.

1. The foundations of our speculative knowledge of the relations and properties of Number, as of Space, are contained in the mode in which we represent to ourselves the magnitudes which are the subjects of our reasonings. To express these foundations in axioms in the case of number is a matter requiring some consideration, for the same reason as in the case of geometry; that is, because these axioms are principles which we assume as true, without being aware that we have made any assumption; and we cannot, without careful scrutiny, determine when we have stated in the form of axioms, all that is necessary for the formation of the science, and no more than is necessary. We will, however, attempt to detect the principles which really must form the basis of theoretical arithmetic.

2. Why is it that three and two are equal to four and

one? Because if we look at five things of any kind, we *see* that it is so. The five are four and one; they are also three and two. The truth of our assertion is involved in our being able to conceive the number five at all. We perceive this truth by *intuition*, for we cannot see, or imagine we see, five things, without perceiving also that the assertion above stated is true.

But how do we state in words this fundamental principle of the doctrine of numbers? Let us consider a very simple case. If we wish to show that seven and two are equal to four and five, we say that seven are four and three, *therefore* seven and two are four and three and two; and because three and two are five, this is four and five. The axioms by which mathematical reasoners justify the first inference (marked by the conjunctive word *therefore*), is by saying that " When equals are added to equals the wholes are equal," and that thus, since seven is equal to three and four, if we add two to both, seven and two are equal to four and three and two.

3. Such axioms as this, that when equals are added to equals the wholes are equal, are, in fact, expressions of the general condition of intuition, by which a whole is contemplated as made up of parts, and as identical with the aggregate of the parts. And a yet more general form in which we might more adequately express this condition of intuition would be this; that "Two magnitudes are equal when they can be divided into parts which are equal, each to each." Thus in the above example, seven and two are equal to four and five, because each of the two sums can be divided into the parts, four, three, and two.

4. In all these cases a person who had never seen such axioms enunciated in a verbal form would employ the same reasoning as a practised mathematician, in order to satisfy himself that the proposition was true. The

steps of the reasoning, being seen to be true by intuition, would carry an entire conviction, whether or not the argument were made verbally complete.   Hence the axioms may appear superfluous, and on this account such axioms have often been spoken contemptuously of as empty and barren assertions.   In fact, however, although they cannot supply the deficiency of the clear intuition of number and space in the reasoner himself, and although when he possesses such a faculty, he will reason rightly if he have never heard of such axioms, they still have their place properly at the beginning of our treatises on the science of quantity; since they express, as simply as words can express, those conditions of the intuition of magnitudes on which all reasoning concerning quantity must be based; and are necessary when we want, not only to see the truth of the elementary reasonings on these subjects, but to put such reasonings in a formal and logical shape.

5. We have considered the axioms which we have suggested above as the basis of all arithmetical operations of the nature of addition.   But it is easily seen that the same principle may be carried into other cases; as for instance, multiplication, which is merely a repeated addition, and admits of the same kind of evidence. Thus five times three are equal to three times five; why is this?   If we arrange fifteen things in five rows of three, it is seen by looking, or by imaginary looking, which is intuition, that they may also be taken as three rows of five.   And thus the principle that those wholes are equal which can be resolved into the same partial magnitudes, is immediately applicable in this as in the other case.

6. We may proceed to higher numbers, and may find ourselves obliged to use artificial nomenclature and notation in order to represent and reckon them; but the

reasoning in these cases also is still the same. And the usual artifice by which our reasoning in such instances is assisted is, that the number which is the root of our scale of notation (which is *ten* in our usual system), is alternately separated into parts and treated as a single thing. Thus 47 and 35 are 82; for 47 is four tens and seven; 35 is three tens and five; whence 47 and 35 are seven tens and twelve; that is, 7 tens, 1 ten, and 2; which is 8 tens and 2, or 82. The like reasoning is applicable in other cases. And since the most remote and complex properties of numbers are obtained by a prolongation of a course of reasoning exactly similar to that by which we thus establish the most elementary propositions, we have in the principles just noticed, the foundation of the whole of Theoretical Arithmetic.

CHAPTER IX.

OF THE PERCEPTION OF TIME AND NUMBER.

1. OUR perception of the passage of time involves a series of acts of memory. This is easily seen and assented to, when large intervals of time and a complex train of occurrences are concerned. But since memory is requisite in order to apprehend time in such cases, we cannot doubt that the same faculty must be concerned in the shortest and simplest cases of succession; for it will hardly be maintained that the process by which we contemplate the progress of time is different when small and when large intervals are concerned. If memory be absolutely requisite to connect two events which begin and end a day, and to perceive a tract of time between them, it must be equally indispensable to connect the beginning and end of a minute, or a second; though in

this case the effort may be smaller, and consequently more easily overlooked. In common cases, we are unconscious of the act of thought by which we recollect the preceding instant, though we perceive the effort when we recollect some distant event. And this is analogous to what happens in other instances. Thus, we walk without being conscious of the volitions by which we move our muscles; but, in order to leap, a distinct and manifest exertion of the same muscles is necessary. Yet no one will doubt that we walk as well as leap by an act of the will exerted through the muscles; and in like manner our consciousness of small as well as large intervals of time involves something of the nature of an act of memory.

2. But this constant and almost imperceptible kind of memory, by which we connect the beginning and end of each instant as it passes, may very fitly be distinguished in common cases from manifest acts of recollection, although it may be difficult or impossible to separate the two operations in general. This perpetual and latent kind of memory may be termed a *sense of successiveness*; and must be considered as an internal sense by which we perceive ourselves existing in time, much in the same way as by our external or muscular sense we perceive ourselves existing in space. And both our internal thoughts and feelings, and the events which take place around us, are apprehended as objects of this internal sense, and thus as taking place in time.

3. In the same manner in which our interpretation of the notices of the muscular sense implies the power of moving our limbs, and of touching at will this object or that; our apprehension of the relations of time by means of the internal sense of successiveness implies a power of recalling what has past, and of retaining what is passing. We are able to seize the occurrences which have

just taken place, and to hold them fast in our minds so as mentally to measure their distance in time from occurrences now present. And thus, this sense of successiveness, like the muscular sense with which we have compared it, implies activity of the mind itself, and is not a sense passively receiving impressions.

4. The conception of *Number* appears to require the exercise of the same sense of succession. At first sight, indeed, we seem to apprehend Number without any act of memory, or any reference to time: for example, we look at a horse, and see that his legs are four; and this we seem to do at once, without reckoning them. But it is not difficult to see that this seeming instantaneousness of the perception of small numbers is an illusion. This resembles the many other cases in which we perform short and easy acts so rapidly and familiarly that we are unconscious of them; as in the acts of seeing, and of articulating our words. And this is the more manifest, since we begin our acquaintance with number by counting even the smallest numbers. Children and very rude savages must use an effort to reckon even their five fingers, and find a difficulty in going further. And persons have been known who were able by habit, or by a peculiar natural aptitude, to count by dozens as rapidly as common persons can by units. We may conclude, therefore, that when we appear to catch a small number by a single glance of the eye, we do in fact count the units of it in a regular, though very brief succession. To count requires an act of memory. Of this we are sensible when we count very slowly, as when we reckon the strokes of a church clock; for in such a case we may forget in the intervals of the strokes, and *miscount*. Now it will not be doubted that the nature of the process in counting is the same whether we count fast or slow. There is no definite speed of reckoning at which the faculties which

it requires are changed; and therefore memory, which is requisite in some cases, must be so in all.

The act of counting, (one, two, three, and so on,) is the foundation of all our knowledge of number. The intuition of the relations of number involves this act of counting; for, as we have just seen, the conception of number cannot be obtained in any other way. And thus the whole of theoretical arithmetic depends upon an act of the mind, and upon the conditions which the exercise of that act implies. These have been already explained in the last chapter.

5. But if the apprehension of number be accompanied by an act of the mind, the apprehension of *rhythm* is so still more clearly. All the forms of versification and the *measures* of melodies are the creations of man, who thus realises in words and sounds the forms of recurrence which rise within his own mind. When we hear in a quiet scene any rapidly-repeated sound, as those made by the hammer of the smith or the saw of the carpenter, every one knows how insensibly we throw these noises into a rhythmical form in our own apprehension. We do this even without any suggestion from the sounds themselves. For instance, if the beats of a clock or watch be ever so exactly alike, we still reckon them alternately tick-*tack*, tick-*tack*. That this is the case, may be proved by taking a watch or clock of such a construction that the returning swing of the pendulum is silent, and in which therefore all the beats are rigorously alike: we shall find ourselves still reckoning its sounds as tick-*tack*. In this instance it is manifest that the rhythm is entirely of our own making. In melodies, also, and in verses in which the rhythm is complex, obscure, and difficult, we perceive something is required on our part; for we are often incapable of contributing our share, and thus lose the sense of the measure alto-

gether. And when we consider such cases, and attend to what passes within us when we catch the measure, even of the simplest and best-known air, we shall no longer doubt that an act of our own thoughts is requisite in such cases, as well as impressions on the sense. And thus the conception of this peculiar modification of time, which we have called rhythm, like all the other views which we have taken of the subject, shows that we must, in order to form such conceptions, supply a certain idea by our own thoughts, as well as merely receive by senses, whether external or internal, the impressions of appearances and collections of appearances.

CHAPTER X.

OF MATHEMATICAL REASONING.

1. *Discursive Reasoning.*—We have thus seen that our notions of space, time, and their modifications, necessarily involve a certain activity of the mind; and that the conditions of this activity form the foundations of those sciences which have the relations of space, time, and number for their object. Upon the fundamental principles thus established, the various sciences which are included in the term Pure Mathematics, (Geometry, Algebra, Trigonometry, Conic Sections and the rest of the Higher Geometry, the Differential Calculus, and the like,) are built up by a series of reasonings. These reasonings are subject to the rules of logic, as we have already remarked; nor is it necessary here to dwell long on the nature and rules of such processes. But we may here notice that such processes are termed *discursive*, in opposition to the operations by which we acquire our fundamental principles, which are, as we have seen, *intui-*

tive. This opposition was formerly very familiar to our writers, as Milton :—

> . . . . Thus the soul reason receives,
> Discursive or intuitive.——*Paradise Lost*, v. 438.

For in such reasonings we obtain our conclusions, not by looking at our conceptions steadily in one view, which is intuition, but by passing from one view to another, like those who run from place to place (*discursus*). Thus a straight line may be at the same time a side of a triangle and a radius of a circle: and in the first proposition of Euclid a line is considered, first in one of these relations, and then in the other, and thus the sides of a certain triangle are proved to be equal. And by this " discourse of reason," as by our older writers it was termed, we set forth from those axioms which we perceive by intuition, travel securely over a vast and varied region, and become possessed of a copious store of mathematical truths.

2. *Technical Terms of Reasoning.*—The reasoning of mathematics, thus proceeding from a few simple principles to many truths, is conducted according to the rules of Logic. If it be necessary, mathematical proofs may be reduced to logical forms, and expressed in Syllogisms, consisting of major, minor, and conclusion. But in most cases the syllogism is of that kind which is called by logical writers an *enthymeme;* a word which implies something existing in the thoughts only, and which designates a syllogism in which one of the premises is understood, and not expressed. Thus we say in a mathematical proof, " because the point c is the centre of the circle a b, a c is equal to b c;" not stating the *major,*—that all lines drawn from the centre of a circle to the circumference are equal ; or introducing it only by a transient reference to the definition of a circle. But the enthymeme is so constantly used in all habitual forms of reasoning, that it does not occur to us as being anything peculiar in mathematical works.

The propositions which are proved to be generally true are termed *theorems:* but when anything is required to be done, as to draw a line or a circle under given conditions, this proposition is a *problem.* A theorem requires demonstration; a problem, solution. And for both purposes the mathematician usually makes a *construction.* He directs us to draw certain lines, circles, or other curves, on which is to be founded his demonstration that his theorem is true, or that his problem is solved. Sometimes, too, he establishes some *lemma,* or preparatory proposition, before he proceeds to his main task; and often he deduces from his demonstration some conclusion in addition to that which was the professed object of his proposition; and this is termed a *corollary.*

These technical terms are noted here, not as being very important, but in order that they may not sound strange and unintelligible if we should have occasion to use some of them. There is, however, one technical distinction more peculiar, and more important.

3. *Geometrical Analysis and Synthesis.*—In geometrical reasoning such as we have described, we introduce at every step some new consideration; and it is by combining all these considerations, that we arrive at the conclusion, that is, the demonstration of the proposition. Each step tends to the final result, by exhibiting some part of the figure under a new relation. To what we have already proved is added something more; and hence this process is called *Synthesis,* or putting together. The proof flows on, receiving at every turn new contributions from different quarters; like a river fed and augmented by many tributary streams. And each of these tributaries flows from some definition or axiom as its fountain, or is itself formed by the union of smaller rivulets which have sources of this kind. In descending along its course, the synthetical proof gathers all these accessions into one common trunk, the proposition finally proved.

But we may proceed in a different manner. We may begin from the formed river, and ascend to its sources. We may take the proposition of which we require a proof, and may examine what the supposition of its truth implies. If this be true, then something else may be seen to be true; and from this, something else, and so on. We may often in this way discover of what simpler propositions our theorem or solution is compounded, and may resolve these in succession, till we come to some proposition which is obvious. This is geometrical *Analysis.* Having succeeded in this analytical process, we may invert it; and may descend again from the simple and known propositions, to the proof of a theorem, or the solution of a problem, which was our starting-place.

This process resembles, as we have said, tracing a river to its sources. As we ascend the stream, we perpetually meet with bifurcations; and some sagacity is needed to enable us to see which, in each case, is the main stream: but if we proceed in our research, we exhaust the unexplored valleys, and finally obtain a clear knowledge whence the waters flow. Analytical is sometimes confounded with symbolical reasoning, on which subject we shall make a remark in the next chapter. The object of that chapter is to notice certain other fundamental principles and ideas, not included in those hitherto spoken of, which we find thrown in our way as we proceed in our mathematical speculations. It would detain us too long, and involve us in subtle and technical disquisitions, to examine fully the grounds of these principles; but Mathematics hold so important a place in relation to the inductive sciences, that I shall briefly notice the leading ideas which the ulterior progress of the subject involves.

CHAPTER XI.

OF THE FOUNDATIONS OF THE HIGHER MATHEMATICS.

1. *The Idea of a Limit.*—The general truths concerning relations of space which depend upon the axioms and definitions contained in Euclid's *Elements,* and which involve only properties of straight lines and circles, are termed Elementary Geometry: all beyond this belongs to the Higher Geometry. To this latter province appertain, for example, all propositions respecting the lengths of any portions of curve lines; for these cannot be obtained by means of the principles of the Elements alone. Here then we must ask to what other principles the geometer has recourse, and from what source these are drawn. Is there any origin of geometrical truth which we have not yet explored?

The *Idea of a Limit* supplies a new mode of establishing mathematical truths. Thus with regard to the length of any portion of a curve, a problem which we have just mentioned; a curve is not made up of straight lines, and therefore we cannot by means of any of the doctrines of elementary geometry measure the length of any curve. But we may make up a figure nearly resembling any curve by putting together many short straight lines, just as a polygonal building of very many sides may nearly resemble a circular room. And in order to approach nearer and nearer to the curve, we may make the sides more and more small, more and more numerous. We may then possibly find some mode of measurement, some relation of these small lines to other lines, which is not disturbed by the multiplication of the sides however far it be carried. And thus we may do what is equivalent to measuring the curve itself; for by multiplying the sides we may approach more and more closely to the curve till no appreciable difference remains. The curve line is the *Limit* of the polygon; and in this process we proceed on the *Axiom*, that "What is true up to the limit is true at the limit."

This mode of conceiving mathematical magnitudes is of wide extent and use; for every curve may be considered as the limit of some polygon; every varied magnitude, as the limit of some aggregate of simpler forms; and thus the relations of the elementary figures enable us to advance to the properties of the most complex cases.

A Limit is a peculiar and fundamental conception, the use of which in proving the propositions of the Higher Geometry cannot be superseded by any combination of other hypotheses and definitions*. The axiom just noticed, that what is true up to the limit is true at the limit, is involved in the very conception of a limit: and this principle, with its consequences, leads to all the results which form the subject of the higher mathematics, whether proved by the consideration of evanescent triangles,

---

* This assertion cannot be fully proved and illustrated without a reference to mathematical reasonings which would not be generally intelligible. I have shown the truth of the assertion in my *Thoughts on the Study of Mathematics*, annexed to the *Principles of English University Education*. The proof is of this kind :—The ultimate equality of an arc of a curve and the corresponding periphery of a polygon, when the sides of the polygon are indefinitely increased in number, is *evident*. But this truth cannot be proved from any other axiom. For if we take the supposed axiom, that a curve is always less than the including broken line, this is not true, except with a condition; and in tracing the import of this condition, we find its necessity becomes evident only when we introduce a reference to a Limit. And the same is the case if we attempt to supersede the notion of a Limit in proving any other simple and evident proposition in which that notion is involved. Therefore these evident truths are *self*-evident, *in virtue of the Idea of a Limit.*

by the processes of the Differential Calculus, or in any other way.

The ancients did not expressly introduce this conception of a Limit into their mathematical reasonings; although in the application of what is termed the Method of Exhaustions, (in which they show how to exhaust the difference between a polygon and a curve, or the like,) they were in fact proceeding upon an obscure apprehension of principles equivalent to those of the Method of Limits. Yet the necessary fundamental principle not having, in their time, been clearly developed, their reasonings were both needlessly intricate and imperfectly satisfactory. Moreover they were led to put in the place of axioms, assumptions which were by no means self-evident; as when Archimedes assumed, for the basis of his measure of the circumference of the circle, the proposition that a circular arch is necessarily less than two lines which inclose it, joining its extremities. The reasonings of the older mathematicians, which professed to proceed upon such assumptions, led to true results in reality, only because they were guided by a latent reference to the limiting case of such assumptions. And this latent employment of the conception of a Limit, reappeared in various forms during the early period of modern mathematics; as for example, in the *Method* of *Indivisibles* of Cavalleri, and the *Characteristic Triangle* of Barrow; till at last Newton distinctly referred such reasonings to the conception of a Limit, and established the fundamental principles and processes which that conception introduces, with a distinctness and exactness which required little improvement to make it as unimpeachable as the demonstrations of geometry. And when such processes as Newton thus deduced from the conception of a Limit are represented by means of general algebraical symbols instead of geometrical diagrams, we have then before us the Method of Fluxions, or the Differential Calculus; a mode of treating mathematical problems justly considered as the principal weapon by which the splendid triumphs of modern mathematics have been achieved.

2. *The Use of General Symbols.*—The employment of algebraical symbols, of which we have just spoken, has been another of the main instruments to which the successes of modern mathematics are owing. And here again the processes by which we obtain our results depend for their evidence upon a fundamental conception, —the conception of *arbitrary symbols* as the *Signs* of quantity and its relations; and upon a corresponding axiom, that "The interpretation of such symbols must be perfectly general." In this case, as in the last, it was only by degrees that mathematicians were led to a just apprehension of the grounds of their reasoning. For symbols were at first used only to represent numbers considered with regard to their numerical properties; and thus the science of algebra was formed. But it was found, even in cases belonging to common algebra, that the symbols often admitted of an interpretation which went beyond the limits of the problem, and which yet was not unmeaning, since it pointed out a question closely analogous to the question proposed. This was the case, for example, when the answer was a *negative quantity;* for when Descartes had introduced the mode of representing curves by means of algebraical relations among the symbols of the *co-ordinates,* or distances of each of their points from fixed lines, it was found that negative quantities must be dealt with as not less truly significative than positive ones. And as the researches of mathematicians proceeded, other cases also were found, in which the symbols, although destitute of meaning according to the original conventions of their institution, still pointed

out truths which could be verified in other ways; as in the cases in which what are called *impossible quantities* occur. Such processes may usually be confirmed upon other principles, and the truth in question may be established by means of a demonstration in which no such seeming fallacies defeat the reasoning. But it has also been shown in many such cases, that the process in which some of the steps appear to be without real meaning, does in fact involve a valid proof of the proposition. And what we have here to remark is, that this is not true accidentally or partially only, but that the results of systematic symbolical reasoning must always express general truths, by their nature, and do not, for their justification, require each of the steps of the process to represent some definite operation upon quantity. *The absolute universality of the interpretation of symbols* is the fundamental principle of their use. This has been shown very ably by Professor Peacock in his *Algebra.* He has there illustrated, in a variety of ways, this principle: that "If general symbols express an identity when they are supposed to be of any special nature, they must also express an identity when they are general in their nature." And thus this universality of symbols is a principle in addition to those we have already noticed; and is a principle of the greatest importance in the formation of mathematical science, according to the wide generality which such science has in modern times assumed.

3. *Connexion of Symbols and Analysis.*—Since in our symbolical reasoning our symbols thus reason for us, we do not necessarily here, as in geometrical reasoning, go on adding carefully one known truth to another, till we reach the desired result. On the contrary, if we have a theorem to prove or a problem to solve which can be brought under the domain of our symbols, we may at once state the given but unproved truth, or the given

combination of unknown quantities, in its symbolical form. After this first process we may then proceed to trace, by means of our symbols, what other truth is involved in the one thus stated, or what the unknown symbols must signify; resolving step by step the symbolical assertion with which we began, into others more fitted for our purpose. The former process is a kind of *synthesis*, the latter is termed *analysis*. And although symbolical reasoning does not necessarily imply such analysis; yet the connexion is so familiar, that the term analysis is frequently used to designate symbolical reasoning.

CHAPTER XII.

THE DOCTRINE OF MOTION.

1. *Pure Mechanism.*—THE doctrine of Motion, of which we have here to speak, is that in which motion is considered quite independently of its cause, force; for all consideration of force belongs to a class of ideas entirely different from those with which we are here concerned. In this view it may be termed the *pure* doctrine of motion, since it has to do solely with space and time, which are the subjects of pure mathematics. Although the doctrine of motion in connexion with force, which is the subject of mechanics, is by far the most important form in which the consideration of motion enters into the formation of our sciences, the pure doctrine of motion, which treats of space, time, and velocity, might be followed out so as to give rise to a very considerable and curious body of science. Such a science is the science of Mechanism, independent of Force, and considered as the solution of a problem which may be thus enunciated: " To communicate any given motion from a first mover to

a given body." The science which should have for its object to solve all the various cases into which this problem would ramify, might be termed *Pure Mechanism* in contradistinction to *Mechanics Proper*, or *Machinery*, in which Force is taken into consideration. The greater part of the machines which have been constructed for use in manufactures have been practical solutions of some of the cases of this problem. We have also important contributions to such a science in the works of mathematicians; for example, the various investigations and demonstrations which have been published respecting the form of the Teeth of Wheels, and Mr. Babbage's memoir* on the Language of Machinery. There are also several works which contain collections of the mechanical contrivances which have been invented for the purpose of transmitting and modifying motion, and these works may be considered as treatises on the science of Pure Mechanism. But this science has not yet been reduced to the systematic simplicity which is desirable, nor indeed generally recognised as a separate science. It has been confounded, under the common name of Mechanics, with the other science, Mechanics Proper, or Machinery, which considers the effect of *force* transmitted by mechanism from one part of a material combination to another. For example, the *Mechanical Powers*, as they are usually termed, (the Lever, the Wheel and Axle, the Inclined Plane, the Wedge, and the Screw,) have almost always been treated with reference to the relation between the *Power* and the *Weight*, and not primarily as a mode of changing the velocity and kind of the motion. The science of pure motion has not generally been separated from the science of motion viewed with reference to its causes.

* *On a Method of expressing by Signs the Action of Machinery.* Phil. Trans., 1826, p. 250.

Recently, indeed, the necessity of such a separation has been seen by those who have taken a philosophical view of science. Thus this necessity has been urged by M. Ampere, in his *Essai sur la Philosophie des Sciences* (1834): "Long," he says, (p. 50), "before I employed myself upon the present work, I had remarked that it is usual to omit, in the beginning of all books treating of sciences which regard motion and force, certain considerations which, duly developed, must constitute a special science: of which science certain parts have been treated of, either in memoirs or in special works; such, for example, as that of Carnot upon Motion considered geometrically, and the essay of Lanz and Betancourt upon the Composition of Machines." He then proceeds to describe this science nearly as we have done, and proposes to term it *Kinematics* (*Cinématique*), from κίνημα, motion.

2. *Formal Astronomy.*—I shall not attempt here further to develop the form which such a science must assume. But I may notice one very large province which belongs to it. When men had ascertained the apparent motions of the sun, moon, and stars, to a moderate degree of regularity and accuracy, they tried to conceive in their minds some mechanism by which these motions might be produced; and thus they in fact proposed to themselves a very extensive problem in *Kinematics*. This, indeed, was the view originally entertained of the nature of the science of astronomy. Thus Plato in the seventh Book of his *Republic**, speaks of astronomy as the doctrine of the motion of solids, meaning thereby, spheres. And the same was a proper description of the science till the time of Kepler, and even later: for Kepler endeavoured, though in vain, to conjoin with the knowledge of the motions of the heavenly bodies, those true mechanical conceptions which converted formal into physical astronomy†.

* P. 528.   † *Hist. Ind. Sc.*, ii. 130.

The astronomy of the ancients admitted none but uniform circular motions, and could therefore be completely cultivated by the aid of their elementary geometry. But the pure science of motion might be extended to all motions, however varied as to the speed or the path of the moving body. In this form it must depend upon the doctrine of limits; and the fundamental principle of its reasonings would be this: That velocity is measured by the Limit of the *space* described, considered with reference to the *times* in which it is described. I shall not further pursue this subject; and in order to complete what I have to say respecting the Pure Sciences, I have only a few words to add respecting their bearing on Inductive Science in general.

CHAPTER XIII.

OF THE APPLICATION OF MATHEMATICS TO THE INDUCTIVE SCIENCES.

1. ALL objects in the world which can be made the subjects of our contemplation are subordinate to the conditions of Space, Time, and Number; and on this account the doctrines of pure mathematics have most numerous and extensive applications in every department of our investigations of nature. And there is a peculiarity in these Ideas, which has caused the mathematical sciences to be, in all cases, the first successful efforts of the awakening speculative powers of nations at the commencement of their intellectual progress. Conceptions derived from these Ideas are from the very first perfectly precise and clear, so as to be fit elements of scientific truths. This is not the case with the other conceptions which form the subjects of scientific inquiries. The conception

of *statical force,* for instance, was never presented in a distinct form till the works of Archimedes appeared the conception of *accelerating force* was confused, in the mind of Kepler and his contemporaries, and only became clear enough for purposes of sound scientific reasoning in the succeeding century: the just conception of chemical *composition* of elements gradually, in modern times, emerged from the erroneous and vague notions of the ancients. If we take works published on such subjects before the epoch when the foundations of the true science were laid, we find the knowledge not only small, but worthless. The writers did not see any evidence in what we now consider as the axioms of the science; nor any inconsistency where we now see self-contradiction. But this was never the case with speculations concerning space and number. From their first rise, these were true as far as they went. The Geometry and Arithmetic of the Greeks and Indians, even in their first and most scanty form, contained none but true propositions. Men's intuitions upon these subjects never allowed them to slide into error and confusion; and the truths to which they were led by the first efforts of their faculties, so employed, form part of the present stock of our mathematical knowledge.

2. But we are here not so much concerned with mathematics in their pure form, as with their application to the phenomena and laws of nature. And here also the very earliest history of civilization presents to us some of the most remarkable examples of man's success in his attempts to attain to science. Space and time, position and motion, govern all visible objects; but by far the most conspicuous examples of the relations which arise out of such elements, are displayed by the ever-moving luminaries of the sky, which measure days, and months, and years, by their motions, and man's place on the earth by their position.

Hence the sciences of space and number were from the first cultivated with peculiar reference to Astronomy. I have elsewhere* quoted Plato's remark,—that it is absurd to call the science of the relations of space *geometry*, the measure of the earth, since its most important office is to be found in its application to the heavens. And on other occasions also it appears how strongly he, who may be considered as the representative of the scientific and speculative tendencies of his time and country, had been impressed with the conviction, that the formation of a science of the celestial motions must depend entirely upon the progress of mathematics. In the Epilogue to the *Dialogue on the Laws*†, he declares mathematical knowledge to be the first and main requisite for the astronomer, and describes the portions of it which he holds necessary for astronomical speculators to cultivate. These seem to be, Plane Geometry, Theoretical Arithmetic, the Application of Arithmetic to planes and to solids, and finally the doctrine of Harmonics. Indeed the bias of Plato appears to be rather to consider mathematics as the essence of the science of astronomy, than as its instrument; and he seems disposed, in this as in other things, to disparage observation, and to aspire after a science founded upon demonstration alone. "An astronomer," he says in the same place, "must not be like Hesiod and persons of that kind, whose astronomy consists in noting the settings and risings of the stars; but he must be one who understands the revolutions of the celestial spheres, each performing its proper cycle."

A large portion of the mathematics of the Greeks, so long as their scientific activity continued, was directed towards astronomy. Besides many curious propositions of plane and solid Geometry, to which their astronomers

* *Hist. Ind. Sc.*, i. 161.   † *Epinomis*, p. 990.

were led, their Arithmetic, though very inconvenient in
its fundamental assumptions, was cultivated to a great
extent; and the science of Trigonometry, in which problems concerning the relations of space were resolved by
means of tables of numerical results previously obtained,
was created. Menelaus of Alexandria wrote six Books
on Chords, probably containing methods of calculating
Tables of these quantities; such Tables were familiarly
used by the later Greek astronomers. The same author
also wrote three Books on Spherical Trigonometry, which
are still extant.

3. The Greeks, however, in the first vigour of their pursuit of mathematical truth, at the time of Plato and soon
after, had by no means confined themselves to those
propositions which had a visible bearing on the phenomena of nature; but had followed out many beautiful trains of research, concerning various kinds of
figures, for the sake of their beauty alone; as for instance
in their doctrine of Conic Sections, of which curves they
had discovered all the principal properties. But it is
curious to remark, that these investigations, thus pursued
at first as mere matters of curiosity and intellectual
gratification, were destined, two thousand years later, to
play a very important part in establishing that system
of the celestial motions which succeeded the Platonic
scheme of cycles and epicycles. If the properties of the
conic sections had not been demonstrated by the Greeks,
and thus rendered familiar to the mathematicians of succeeding ages, Kepler would probably not have been able
to discover those laws respecting the orbits and motions
of the planets which were the occasion of the greatest
revolution that ever happened in the history of science.

4. The Arabians, who, as I have elsewhere said, added
little of their own to the stores of science which they
received from the Greeks, did however make some very

important contributions in those portions of pure mathematics which are subservient to astronomy. Their adoption of the Indian mode of computation by means of the Ten Digits, 1, 2, 3, 4, 5, 6, 7, 8, 9, 0, and by the method of Local Values, instead of the cumbrous sexagesimal arithmetic of the Greeks, was an improvement by which the convenience and facility of numerical calculations were immeasurably augmented. The Arabians also rendered several of the processes of trigonometry much more commodious, by using the Sine of an arc instead of the Chord; an improvement which Albategnius appears to claim for himself*; and by employing also the Tangents of arcs, or, as they called them†, *upright shadows.*

5. The constant application of mathematical knowledge to the researches of Astronomy, and the mutual influence of each science on the progress of the other, has been still more conspicuous in modern times. Newton's Method of Prime and Ultimate Ratios, which we have already noticed as the first correct exposition of the doctrine of a Limit, is stated in a series of Lemmas, or preparatory theorems, prefixed to his *Treatise on the System of the World.* Both the properties of curve lines and the doctrines concerning force and motion, which he had to establish, required that the common mathematical methods should be methodized and extended. If Newton had not been a most expert and inventive mathematician, as well as a profound and philosophical thinker, he could never have made any one of those vast strides in discovery of which the rapid succession in his work strikes us with wonder‡. And if we see that the great task begun by him, goes on more slowly in the hands of his immediate successors, and lingers a little before its full completion, we perceive that this arises, in a great measure, from

---

* DELAMBRE, Ast., *M. A.*, p. 12.　　　　† *Ibid.*, p. 17.
‡ *Hist. Ind. Sc.*, ii., 155. 167. 176.

the defect of the mathematical methods then used. Newton's synthetical modes of investigation, as we have elsewhere observed, were an instrument*, powerful indeed
in his mighty hand, but too ponderous for other persons
to employ with effect. The countrymen of Newton
clung to it the longest, out of veneration for their
master ; and English cultivators of physical astronomy
were, on that very account, left behind the progress of
mathematical science in France and Germany, by a wide
interval, which they have only recently recovered. On
the Continent, the advantages offered by a familiar use of
symbols, and by attention to their symmetry and other
relations, were accepted without reserve. In this manner
the Differential Calculus of Leibnitz, which was in its
origin and signification identical with the Method of
Fluxions of Newton, soon surpassed its rival in the
extent and generality of its application to problems.
This Calculus was applied to the science of mechanics, to
which it, along with the symmetrical use of co-ordinates,
gave a new form ; for it was soon seen that the most
difficult problems might in general be reduced to finding
integrals, which is the reciprocal process of that by which
differentials are found ; so that all difficulties of physical
astronomy were reduced to difficulties of symbolical calculation, these, indeed, being often sufficiently stubborn.
Clairaut, Euler, and D'Alembert employed the increased
resources of mathematical science upon the Theory of
the Moon, and other questions relative to the system of
the world ; and thus began to pursue such inquiries in
the course in which mathematicians are still labouring up
to the present day. This course was not without its checks
and perplexities. We have elsewhere quoted† Clairaut's
expression when he had obtained the very complex
differential equations which contain the solution of the

---

* *Hist. Ind. Sc.*, ii., 167.   † *Ib.*, ii., 103.

problem of the moon's motion: " Now integrate them who can !" But in no very long time they were integrated, at least approximately; and the methods of approximation have since then been improved; so that now, with a due expenditure of labour, they may be carried to any extent which is thought desirable. If the methods of astronomical observation should hereafter reach a higher degree of exactness than they now profess, so that irregularities in the motions of the sun, moon, and planets, shall be detected which at present escape us, the mathematical part of the theory of universal gravitation is in such a condition that it can soon be brought into comparison with the newly-observed facts. Indeed at present the mathematical theory is in advance of such observations. It can venture to suggest what may afterwards be detected, as well as to explain what has already been observed. This has happened recently; for Professor Airy has calculàted the law and amount of an inequality depending upon the mutual attraction of the Earth and Venus; of which inequality (so small is it,) it remains to be determined whether its effect can be traced in the series of astronomical observations.

6. As the influence of mathematics upon the progress of astronomy is thus seen in the cases in which theory and observation confirm each other, so this influence appears in another way, in the very few cases in which the facts have not been fully reduced to an agreement with theory. The most conspicuous case of this kind is the state of our knowledge of the Tides. This is a portion of astronomy: for the Newtonian theory asserts these curious phenomena to be the result of the attraction of the sun and moon. Nor can there be any doubt that this is true, as a general statement; yet the subject is up to the present time a blot on the perfection of the theory of universal gravitation; for we are very far from being able in this, as in the

other parts of astronomy, to show that theory will exactly
account for the time, and magnitude, and all other circumstances of the phenomenon at every place on the
earth's surface.   And what is the portion of our mathematics which is connected with this solitary signal defect
in astronomy?   It is the mathematics of the Motion of
Fluids; a portion in which extremely little progress has
been made, and in which all the more general problems
of the subject have hitherto remained entirely insoluble.
The attempts of the greatest mathematicians, Newton,
Maclaurin, Bernoulli, Clairaut, Laplace, to master such
questions, all involve some gratuitous assumption, which
is introduced because the problem cannot otherwise be
mathematically dealt with: these assumptions confessedly
render the result defective, and how defective it is hard to
say.   And it was probably precisely the absence of a theory
which could be reasonably expected to agree with the
observations, which made Observations of this very curious
phenomenon, the Tides, to be so much neglected as till
very recently they were.   Of late years such observations
have been pursued, and their results have been resolved
into empirical laws, so that the rules of the phenomena
have been ascertained, although the dependence of these
rules upon the lunar and solar forces has not been shown.
Here then we have a portion of our knowledge relating to
facts undoubtedly dependent upon universal gravitation,
in which Observation has outstripped Theory in her progress, and is compelled to wait till her usual companion
overtakes her.   This is a position of which Theory has
usually been very impatient, and we may expect that she
will be no less so in the present instance.

7. It would be easy to show from the history of other
sciences, for example, Mechanics and Optics, how essential
the cultivation of pure mathematics has been to their
progress.   The parabola was already familiar among

mathematicians when Galileo discovered that it was the theoretical path of a Projectile; and the extension and generalization of the Laws of Motion could never have been effected, unless the differential and integral calculus had been at hand, ready to trace the results of every hypothesis which could be made. D'Alembert's mode of expressing the Third Law of Motion in its most general form*, if it did not prove the law, at least reduced the application of it to analytical processes which could be performed in most of those cases in which they were needed. In many instances the demands of mechanical science suggested the extension of the methods of pure analysis. The problem of Vibrating Strings gave rise to the Calculus of Partial Differences, which was still further stimulated by its application to the motions of fluids and other mechanical problems. And we have in the writings of Lagrange and Laplace other instances equally remarkable of new analytical methods, to which mechanical problems, and especially cosmical problems, have given occasion.

8. The progress of Optics as a science has, in like manner, been throughout dependent upon the progress of pure mathematics. The first rise of geometry was followed by some advances, slight ones no doubt, in the doctrine of Reflection and in Perspective. The law of Refraction was traced to its consequences by means of trigonometry, which indeed was requisite to express the law in a simple form. The steps made in optical science by Descartes, Newton, Euler, and Huyghens, required the geometrical skill which those philosophers possessed. And if Young and Fresnel had not been, each in his peculiar way, persons of eminent mathematical endowments, they would not have been able to bring the Theory of Undulations and Interferences into a condition in which it could be tested by experiments. We may see how unexpectedly

* *Hist. Ind. Sc.*, ii. 89.

recondite parts of pure mathematics may bear upon physical science, by calling to mind a circumstance already noticed in the History of Science*;—that Fresnel obtained one of the most curious confirmations of the theory (the laws of Circular Polarization by reflection) through an interpretation of an algebraical expression, which, according to the original conventional meaning of the symbols, involved an impossible quantity. We have already remarked, that in virtue of the principle of the generality of symbolical language, such an interpretation may often point out some real and important analogy.

8. From this rapid sketch it may be seen how important an office in promoting the progress of the physical sciences belongs to mathematics. Indeed in the progress of many sciences every step has been so intimately connected with some advance in mathematics, that we can hardly be surprised if some persons have considered mathematical reasoning to be the most essential part of such sciences; and have overlooked the other elements which enter into their formation. How erroneous this view is we shall best see by turning our attention to the other Ideas besides those of space, number, and motion, which enter into some of the most conspicuous and admired portions of what is termed exact science; and by showing that the clear and distinct developement of such Ideas is quite as necessary to the progress of exact and real knowledge as an acquaintance with arithmetic and geometry.

* Vol. ii. 445.

BOOK III.

THE PHILOSOPHY OF THE MECHANICAL SCIENCES.

CHAPTER I.

OF THE MECHANICAL SCIENCES.

In the History of the Sciences, that class of which we here speak occupies a conspicuous and important place; coming into notice immediately after those parts of astronomy which require for their cultivation merely the ideas of space, time, motion, and number. It appears from our History that certain truths concerning the *equilibrium* of bodies were established by Archimedes; that, after a long interval of inactivity, his principles were extended and pursued further in modern times: and that to these doctrines concerning equilibrium and the forces which produce it, (which constitute the science *Statics*,) were added many other doctrines concerning the *motions* of bodies, considered also as produced by forces, and thus the science of *Dynamics* was produced. The assemblage of these sciences composes the province of *Mechanics*. Moreover, philosophers have laboured to make out the laws of the equilibrium of *fluid* as well as solid bodies; and hence has arisen the science of *Hydrostatics*. And the doctrines of Mechanics have been found to have a most remarkable bearing upon the motions of the heavenly bodies; with reference to which, indeed, they were at first principally studied. The explanation of

those cosmical facts by means of mechanical principles
and their consequences, forms the science of *Physical
Astronomy*. These are the principal examples of mechanical science; although some other portions of Physics,
as Magnetism and Electrodynamics, introduce mechanical doctrines very largely into their speculations.

Now in all these sciences we have to consider *Forces*.
In all mechanical reasonings forces enter, either as
producing motion, or as prevented from doing so by other
forces. Thus force, in its most general sense, is the *cause*
of motion, or of tendency to motion; and in order to
discover the principles on which the mechanical sciences
truly rest, we must examine the nature and origin of our
knowledge of Causes.

In these sciences, however, we have not to deal with
Cause in its more general acceptation, in which it applies
to all kinds of agency, material or immaterial;—to the
influence of thought and will, as well as of bodily pressure
and attractive force. Our business at present is only
with such causes as immediately operate upon matter.
We shall nevertheless, in the first place, consider the
nature of Cause in its most general form; and afterwards
narrow our speculations so as to direct them specially
to the mechanical sciences.

CHAPTER II.

OF THE IDEA OF CAUSE.

1. WE see in the world around us a constant succession of causes and effects connected with each other.
The laws of this connexion we learn in a great measure
from experience, by observation of the occurrences which
present themselves to our notice, succeeding one another.

But in doing this, and in attending to this succession of appearances, of which we are aware by means of our senses, we supply from our own minds the Idea of Cause. This Idea, as we have already shown with respect to other Ideas, is not derived from experience, but has its origin in the mind itself;—is introduced into our experience by the active, and not by the passive part of our nature.

By Cause we mean some quality, power, or efficacy, by which a state of things produces a succeeding state. Thus the motion of bodies from rest is produced by a cause which we call Force: and in the particular case in which bodies fall to the earth, this force is termed Gravity. In these cases, the Conceptions of Force and Gravity receive their meaning from the Idea of Cause which they involve: for Force is conceived as the Cause of Motion. That this Idea of Cause is not derived from experience, we prove (as in former cases) by this consideration: that we can make assertions, involving this idea, which are rigorously necessary and universal; whereas knowledge derived from experience can only be true as far as experience goes, and can never contain in itself any evidence whatever of its necessity. We assert that "Every event must have a cause:" and this proposition we know to be true, not only probably, and generally, and as far as we can see: but we cannot suppose it to be false in any single instance. We are as certain of it as of the truths of arithmetic or geometry. We cannot doubt that it must apply to all events past and future, in every part of the universe, just as truly as to those occurrences which we have ourselves observed. *What* causes produce what effects;—what is the cause of any particular event; what will be the effect of any peculiar process; these are points on which experience may enlighten us. Observation and experience may be requisite, to enable us to judge respecting such matters. But

that every event has *some* cause, Experience cannot prove
any more than she can disprove. She can add nothing
to the evidence of the truth, however often she may
exemplify it. This doctrine, then, cannot have been
acquired by her teaching : and the Idea of Cause, which
the doctrine involves, and on which it depends, cannot
have come into our minds from the region of observation.

2. That we do, in fact, apply the Idea of Cause in a
more extensive manner than could be justified, if it were
derived from experience only, is easily shown. For from
the principle that everything must have a cause, we not
only reason concerning the succession of events which
occur in the progress of the world, and which form the
course of experience ; but we infer that .the world itself
must have a cause ; that the chain of events connected
by common causation, must have a First Cause of a
nature different from the events themselves. This we
are entitled to do, if our Idea of Cause be independent of,
and superior to, experience : but if we have no Idea of
Cause except such as we gather from experience, this
reasoning is altogether baseless and unmeaning.

3. Again ; by the use of our powers of observation,
we are aware of a succession of appearances and events.
But none of our senses or powers of external observation
can detect in these appearances the power or quality
which we call Cause. Cause is that which connects one
event with another ; but no sense or perception discloses
to us, or can disclose, any connexion among the events
which we observe. We see that one occurrence follows
another, but we can never see anything which shows that
one occurrence *must* follow another. We have already
noticed*, that this truth has been urged by metaphysicians in modern times, and generally assented to by those

* Book i., chap. 13.

who examine carefully the connexion of their own thoughts. The arguments are, indeed, obvious enough. One ball strikes another and causes it to move forwards. But by what compulsion? Where is the necessity? If the mind can see any circumstance in this case which makes the result inevitable, let this circumstance be pointed out. But, in fact, there is no such discoverable necessity; for we can conceive this event not to take place at all. The struck ball may stand still, for aught we can see. " But the laws of motion will not allow it to do so." Doubtless they will not. But the laws of motion are learnt from experience, and therefore can prove no necessity. Why should not the laws of motion be other than they are? Are they necessarily true? That they are necessarily such as do actually regulate the impact of bodies, is at least no obvious truth; and therefore this necessity cannot be, in common minds, the ground of connecting the impact of one ball with the motion of another. And assuredly, if this fail, no other ground of such necessary connexion can be shown. In this case, then, the events are not seen to be necessarily connected. But if this case, where one ball moves another by impulse, be not an instance of events exhibiting a necessary connexion, we shall look in vain for any example of such a connexion. There is, then, no case in which events can be observed to be necessarily connected: our idea of causation, which implies that the event is necessarily connected with the cause, cannot be derived from observation.

4. But it may be said, we have not any such idea of cause, implying necessary connexion with effect, and a quality by which this connexion is produced. We see nothing but the succession of events; and by cause we mean nothing but a certain succession of events;—namely, a constant, unvarying succession. Cause and effect are

only two events of which the second invariably follows
the first. We delude ourselves when we imagine that
our idea of causation involves anything more than this.

To this I reply by asking, what then is the meaning
of the maxim above quoted, and allowed by all to be
universally and necessarily true, that every event must
have a cause? Let us put this maxim into the language
of the explanation just noticed; and it becomes this:—
"Every event must have a certain other event invariably
preceding it." But why must it? Where is the necessity?
Why must like events always be preceded by like, except
so far as other events interfere? That there is such a
necessity, no one can doubt. All will allow that if a stone
ascend because it is thrown upwards in one case, a stone
which ascends in another case has also been thrown upwards, or has undergone some equivalent operation. All
will allow that in this sense, every kind of event must
have some other specific kind of event preceding it. But
this turn of men's thoughts shows that they see in events
a connexion which is not mere succession. They see in
cause and effect, not merely what does, often or always,
precede and follow, but what *must* precede and follow.
The events are not only conjoined, they are connected.
The cause is more than the prelude, the effect is more
than the sequel, of the fact. The cause is conceived not
as a mere occasion; it is a power, an efficacy, which has
a real operation.

5. Thus we have drawn from the maxim, that every
effect must have a cause, arguments to show that we
have an idea of cause which is not borrowed from experience, and which involves more than mere succession.
Similar arguments might be derived from any other
maxims of universal and necessary validity, which we
can obtain concerning cause: as, for example, the maxims
that causes are measured by their effects, and that reac-

tion is equal and opposite to action. These maxims we shall soon have to examine; but we may observe here, that the necessary truth which belongs to them, shows that they, and the ideas they involve, are not the mere fruits of observation; while their meaning, containing, as it does, something quite different from the mere conception of succession of events, proves that such a conception is far from containing the whole import and signification of our idea of cause.

The progress of the opinions of philosophers on the points discussed in this chapter, has been one of the most remarkable parts of the history of Metaphysics in modern times: and I shall therefore briefly notice some of its features.

CHAPTER III.

MODERN OPINIONS RESPECTING THE IDEA OF CAUSE.

1. Towards the end of the seventeenth century there existed in the minds of many of the most vigorous and active speculators of the European literary world, a strong tendency to ascribe the whole of our knowledge to the teaching of experience. This tendency, with its consequences, including among them the reaction which was produced when the tenet had been pushed to a length manifestly absurd, has exercised a very powerful influence upon the progress of metaphysical doctrines up to the present time. I proceed to notice some of the most prominent of the opinions which have thus obtained prevalence among philosophers, so far as the Idea of Cause is concerned.

Locke was one of the metaphysicians who produced the greatest effect in diffusing this opinion, of the exclusive dependence of our knowledge upon experience. Agreeably to this general system, he taught* that our ideas of Cause and Effect are got from observation of the things about us. Yet notwithstanding this tenet of his, he endeavoured still to employ these ideas in reasoning on subjects which are far beyond all limits of experience : for he professed to prove, from our idea of Causation, the existence of the Deity†.

Hume noticed this obvious inconsistency; but declared himself unable to discover any remedy for a defect so fatal to the most important parts of our knowledge. He could see, in our belief of the succession of canse and effect, nothing but the habit of associating in our minds what had often been associated in our experience. He therefore maintained that we could not, with logical propriety, extend our belief of such a succession to cases entirely distinct from all those of which our experience consisted. We see, he said, an actual *conjunction* of two events; but we can in no way detect a necessary *connexion*; and therefore we have no means of inferring cause from effect, or effect from cause‡. The only way in which we recognise cause and effect in the field of our experience, is as an unfailing sequence : we look in vain for anything which can assure us of an infallible consequence. And since experience is the only source of our knowledge, we cannot with any justice assert that the world in which we live must necessarily have had a cause.

2. This doctrine, taken in conjunction with the known scepticism of its author on religious points, produced a considerable fermentation in the speculative world. The solution of the difficulty thus thrown before philosophers, was by no means obvious. It was vain to endeavour to find in experience any other property of a cause than a

---

* *Essay on the Human Understanding*, b. ii., c. 26.   † B. iv., c. 10.
‡ HUME's *Phil. of the Human Mind*, vol. i., p. 94.

constant sequence of the effect. Yet it was equally vain to try to persuade men that they had no idea of cause; or even to shake their belief in the cogency of the familiar arguments concerning the necessity of an original cause of all that is and happens. Accordingly these hostile and apparently irreconcilable doctrines,—the indispensable necessity of a cause of every event, and the impossibility of our knowing such a necessity,—were at last allowed to encamp side by side. Reid, Beattie, and others, formed one party, who showed how widely and constantly the idea of a cause pervades all the processes of the human mind: while another sect, including Brown, and apparently Stewart, maintained that this idea is always capable of being resolved into a constant sequence; and these latter reasoners tried to obviate the dangerous and shocking inferences which some persons might try to draw from their opinion, by declaring the maxim that " Every event must have a cause," to be an instinctive law of belief, or a fundamental principle of the human mind *

3. While this series of discussions was going on in Britain, a great metaphysical genius in Germany was unravelling the perplexity in another way. Kant's speculations originated, as he informs us, in the trains of thought to which Hume's writings gave rise; and the *Kritik der Reinen Vernunft,* or *Examination of the Pure Reason,* was published in 1787, with the view of showing the true nature of our knowledge.

Kant's solution of the difficulties just mentioned differs materially from that above stated. According to Brown†, succession observed and cause inferred,—the memory of past conjunctions of events and the belief of similar future conjunctions,—are facts, independent, so far as we can discover, but inseparably combined by a

---

* STEWART's *Active Powers,* vol. i., p. 347. BROWN's *Lectures,* vol. i., p. 115.   † *Lect.,* vol. i., p. 114.

law of our mental nature. According to Kant, causality is an inseparable condition of our experience : a connexion in events is requisite to our apprehending them *as* events. Future occurrences must be connected by causation as the past have been, because we cannot think of past, present, and future, without such connexion. We cannot fix the mind upon occurrences, without including these occurrences in a series of causes and effects. The relation of causation is a condition under which we think of events, as the relations of space are a condition under which we see objects.

4. On a subject so abstruse, it is not easy to make our distinctions very clear. Some of Brown's illustrations appear to approach very near to the doctrine of Kant. Thus he says*, "The *form* of bodies is the relation of their elements to each other in space,—the *power* of bodies is their relation to each other in time." Yet notwithstanding such approximations in expression, the Kantian doctrine appears to be different from the views of Stewart and Brown, as commonly understood. According to the Scotch philosophers, the cause and the effect are two things, connected in our minds by a law of our nature. But this view requires us to suppose that we can conceive the law to be absent, and the course of events to be unconnected. If we can understand what is the special force of this law, we must be able to imagine what the case would be if the law were non-existing. We must be able to conceive a mind which does not connect effects with causes. The Kantian doctrine, on the other hand, teaches that we cannot imagine events liberated from the connexion of cause and effect : this connexion is a condition of our conceiving any real occurrences : we cannot think of a real sequence of things, except as involving the operation of causes. In the Scotch system,

* *Lect.*, i., p. 127.

the past and the future are in their nature independent,
but bound together by a rule; in the German system,
they share in a common nature and mutual relation, by
the act of thought which makes them past and future.
In the former doctrine cause is a tie which binds; in the
latter it is a character which pervades and shapes events.
The Scotch metaphysicians only assert the *universality* of
the relation; the German attempts further to explain its
necessity.

This being the state of the case, such illustrations as
that of Dr. Brown quoted above, in which he represents
cause as a relation of the same kind with form, do not
appear exactly to fit his opinions. Can the relations of
figure be properly said to be connected with each other
by a law of our nature, or a tendency of our mental constitution? Can we ascribe it to a law of our thoughts,
that we believe the three angles of a triangle to be equal
to two right angles? If so, we must give the same
reason for our belief that two straight lines cannot
inclose a space; or that three and two are five. But will
any one refer us to an ultimate law of our constitution
for the belief that three and two are five? Do we not
see that they are so, as plainly as we see that they are
three and two? Can we imagine laws of our constitution abolished, so that three and two shall make something different from five;—so that an inclosed space shall
lie between two straight lines;—so that the three angles
of a plane triangle shall be greater than two right angles?
We cannot conceive this. If the numbers *are* three and
two; if the lines *are* straight; if the triangle *is* a rectilinear triangle, the consequences are inevitable. We
cannot even imagine the contrary. We do not want a
law to direct that things should be what they are. The
relation, then, of cause and effect, being of the same kind
as the necessary relations of figure and number, is not

properly spoken of as established in our minds by a special law of our constitution : for we reject that loose and inappropriate phraseology which speaks of the relations of figure and number as determined by laws of belief.

5. In the present work, we accept and adopt, as the basis of our inquiry concerning our knowledge, the existence of necessary truths concerning causes, as there exist necessary truths concerning figure and number. We find such truths universally established and assented to among the cultivators of science, and among speculative men in general. All mechanicians agree that reaction is equal and opposite to action, both when one body presses another, and when one body communicates motion to another. All reasoners join in the assertion not only that every observed change of motion has had a cause, but that every change of motion must have a cause. Here we have certain portions of substantial and undoubted knowledge. Now the essential point in the view which we must take of the idea of cause is this,—that our view must be such as to form a solid basis for our knowledge. We have, in the Mechanical Sciences, certain universal and necessary truths on the subject of causes. Now any view which refers our belief in causation to mere experience or habit, cannot explain the possibility of such necessary truths, since experience and habit can never lead to a perception of necessary connexion. But a view which teaches us to acknowledge axioms concerning cause, as we acknowledge axioms concerning space, will lead us to look upon the science of mechanics as equally certain and universal with the science of geometry; and will thus materially affect our judgment concerning the nature and claims of our scientific knowledge.

Axioms concerning cause, or concerning force, which as we shall see, is a modification of cause, will flow from an idea of cause, just as axioms concerning space and number flow from the ideas of space and time. And thus the propositions which constitute the science of mechanics prove that we possess an idea of cause, in the same sense in which the propositions of geometry and arithmetic prove our possession of the ideas of space and of time or number.

6. The idea of cause, like the ideas of space and time, is a part of the *active* powers of the mind. The relation of cause and effect is a relation or condition under which events are apprehended, which relation is not given by observation, but supplied by the mind itself. According to the views which explain our apprehension of cause by reference to habit, or to a supposed law of our mental nature, causal connexion is a consequence of agencies which the mind passively obeys; but according to the view to which we are led, this connexion is a result of faculties which the mind actively exercises. And thus the relation of cause and effect is a condition of our apprehending successive events, a part of the mind's constant and universal activity, a source of necessary truths; or to sum all this in one phrase, a Fundamental Idea.

Chapter IV.

OF THE AXIOMS WHICH RELATE TO THE IDEA OF CAUSE.

1. *Cause is an abstract Term.*—We have now to express, as well as we can, the fundamental character of that Idea of Cause, of which we have just proved the existence. This may be done, at least for purposes of reasoning, in this as in former instances, by means of axioms. I shall state the principal axioms which belong to this subject, referring the reader to his own thoughts for the axiomatic evidence which belongs to them.

But I must first observe that in order to express general and abstract truths concerning cause and effect, these terms, *cause* and *effect*, must be understood in a general and abstract manner. When one event gives rise to another, the first *event* is, in common language, often called the cause, and the second the effect. Thus the meeting of two billiard balls may be said to be the cause of one of them turning aside out of the path in which it was moving. For our present purposes, however, we must not apply the term cause to such occurrences as this meeting and turning, but to a certain conception, *force*, abstracted from all such special events, and considered as a quality or property by which one body affects the motion of the other. And in like manner in other cases, cause is to be conceived as some abstract quality, power, or efficacy, by which change is produced; a quality not identical with the events, but disclosed by means of them. Not only is this abstract mode of conceiving force and cause useful in expressing the fundamental principles of science; but it supplies us with the only mode by which such principles can be stated in a general manner, and made to lead to substantial truth and real knowledge.

Understanding cause, therefore, in this sense, we proceed to our Axioms.

2. First Axiom. *Nothing can take place without a Cause.*

Every event, of whatever kind, must have a Cause in the sense of the term which we have just indicated; and that it must, is a universal and necessary proposition to which we irresistibly assent as soon as it is understood. We believe each appearance to come into existence, we conceive every change to take place, not only with something preceding it, but something by which it is made to be what it is. An effect without a cause;—an event without a preceding condition involving the efficacy by which the event is produced ;—are suppositions which we cannot for a moment admit. That the connexion of effect with cause is universal and necessary, is a universal and constant conviction of mankind. It persists in the minds of all men, undisturbed by all the assaults of sophistry and scepticism ; and, as we have seen in the last chapter, remains unshaken, even when its foundations seem to be ruined. This axiom expresses, to a certain extent, our Idea of cause; and when that idea is clearly apprehended, the axiom requires no proof, and indeed admits of none which makes it more evident. That notwithstanding its simplicity, it is of use in our speculations, we shall hereafter see; but in the first place, we must consider the other axioms belonging to this subject.

3. Second Axiom. *Effects are proportional to their Causes, and Causes are measured by their Effects.*

We have already said that cause is that quality or power in the circumstances of each case by which the effect is produced; and this power, an abstract property of the condition of things to which it belongs, can in no way fall directly under the cognisance of the senses. Cause, of whatever kind, is not apprehended as including objects and events which share its nature by being co-extensive with certain portions of it, as space and time are. It cannot therefore, like them, be measured by repetition of its own parts, as space is measured by repetition of inches, and time by repetition of minutes. Causes may be greater or less; as, for instance, the force of a man is greater than the force of a child. But how much is the one greater than the other? How are we to compare the abstract conception, force, in such cases as these?

To this the obvious and only answer is, that we must compare causes by means of their effects; that we must compare force by something which force can do. The child can lift one fagot; the man can lift ten such fagots: we have here a means of comparison. And whether or not the rule is to be applied in this manner, that is, by the number of the things operated on, (a question which we shall have to consider hereafter,) it is clear that this form of rule, namely, a reference to some effect or other as our measure, is the right, because the only possible form. The cause determines the effect. The cause being the same, the effect must be same. The connexion of the two is governed by a fixed and inviolable rule. It admits of no ambiguity. Every degree of intensity in the cause has some peculiar modification of the effect corresponding to it. Hence the effect is an unfailing index of the amount of the cause; and if it be a measurable effect, gives a measure of the cause. We can have no other measure; but we need no other, for this is exact, sufficient, and complete.

It may be said, that various effects are produced by the same cause. The sun's heat melts wax and expands quicksilver. The force of gravity causes bodies to move downwards if they are free, and to press down upon their supports if they are supported. Which of the effects is to be taken as the measure of heat or of gravity in these cases? To this we reply, that if we had merely different states of the same cause to compare, any of the effects might be taken. The sun's heat on different days might be measured by the expansion of quicksilver, or by the quantity of wax melted. The force of gravity, if it were different at different places, might be measured by the spaces through which a given weight would bend an elastic support, or by the spaces through which a body would fall in a given time. All these measures are consistent with the general character of our idea of cause.

4. *Limitation of the Second Axiom.*—But there may be circumstances in the nature of the case which may

further determine the kind of effect which we must take for the measure of the cause. For example, if causes are conceived to be of such a nature as to be capable of addition, the effects taken as their measure must conform to this condition. This is the case with mechanical causes. The weights of two bodies are the causes of the pressure which they exert downwards; and these weights are capable of addition. The weight of the two is the sum of the weight of each. We are therefore not at liberty to say that weights shall be measured by the spaces through which they bend a certain elastic support, except we have first ascertained that the whole weight bends it through a space equal to the sum of the inflections produced by the separate weights. Without this precaution, we might obtain inconsistent results. Two weights, each of the magnitude 3 as measured by their effects, might, if we took the inflections for the effects, be together equal to 5 or to 7 by the same kind of measurement. For the inflection produced by two weights of 3 might, for aught we can see beforehand, be more or less than twice as great as the inflection produced by one weight of 3. That forces are capable of addition, is a condition which limits, and, as we shall see, rigorously fixes, the kind of effects which are to be taken as their measures.

Causes which are thus capable of addition are to be measured by the repeated addition of equal quantities. Two such causes are *equal* to each other when they produce exactly the same effect. So far our axiom is applied directly. But these two causes can be *added* together; and being thus added, they are *double* of one of them; and the cause composed by addition of *three* such, is *three* times as great as the first; and so on for any measure whatever. By this means, and by this means only, we have a complete and consistent measure of those causes which are so conceived as to be subject to this condition of being added and multiplied.

Causes are, in the present chapter, to be understood in the widest sense of the term; and the axiom now under our consideration applies to them, whenever they are of such a nature as to admit of any measure at all. But the cases which we have more particularly in view are mechanical causes, the causes of the motion and of the equilibrium of bodies. In these cases, forces are conceived as capable of addition; and what has been said of the measure of causes in such cases, applies peculiarly to mechanical forces. Two weights, placed together, may be considered as a single weight, equal to the *sum* of the two. Two pressures, pushing a body in the same direction at the same point, are identical in all respects with some single pressure, their *sum*, pushing in like manner; and this is true whether or not they put the body in motion. In the cases of mechanical forces, therefore, we take some certain effect, velocity generated or weight supported, which may fix the *unit* of force; and we then measure all other forces by the successive repetition of this unit, as we measure all spaces by the successive repetition of our unit of lineal measure.

But these steps in the formation of the science of Mechanics will be further explained, when we come to follow our axioms concerning cause into their application in that science. At present we have, perhaps, sufficiently explained the axiom that causes are measured by their effects, and we now proceed to a third axiom, also of great importance.

5. Third Axiom. *Reaction is equal and opposite to Action.*

In the case of mechanical forces, the action of a cause often takes place by an operation of one body upon another; and in this case, the action is always and inevitably accompanied by an *opposite* action. If I press a stone with my hand, the stone presses my hand in return. If one ball strike another and put it in motion, the second ball diminishes the motion of the first. In these cases the operation is mutual; the Action is accompanied by a Reaction. And in all such cases the Reaction is a force of exactly of the same nature as the Action, exerted in an opposite direction. A pressure exerted upon a body at rest is resisted and balanced by another pressure: when the pressure of one body puts another in motion, the body, though it yields to the force, nevertheless exerts upon the pressing body a force like that which it suffers.

Now the axiom asserts further, that this Reaction is *equal*, as well as opposite, to the Action. For the Reaction is an effect of the Action, and is determined by it. And since the two, Action and Reaction, are forces of the same nature, each may be considered as cause and as effect; and they must, therefore, determine each other by a common rule. But this consideration leads necessarily to their equality: for since the rule is mutual, if we could for an instant suppose the Reaction to be less than the Action, we must, by the same rule, suppose the Action to be less than the Reaction. And thus Action and Reaction, in every such case, are rigorously equal to each other.

It is easily seen that this axiom is not a proposition which is, or can be, proved by experience; but that its truth is anterior to special observation, and depends on our conception of Action and Reaction. Like our other axioms, this has its source in an Idea; namely, the Idea of Cause, under that particular condition in which cause and effect are mutual. The necessary and universal truth which we cannot help ascribing to the axiom, shows that it is not derived from the stores of experience, which can never contain truths of this character.

Accordingly, it was asserted with equal confidence and generality by those who did not refer to experience for their principles, and by those who did. Leonicus Tomæus, a commentator of Aristotle, whose work was published in 1552, and therefore at a period when no right opinions concerning mechanical reaction were current, at least in his school, says, in his remarks on the Author's Questions concerning the communication of motion, that "Reaction is equal and contrary to Action." The same principle was taken for granted by all parties, in all the controversies concerning the proper measure of force, of which we shall have to speak: and would be rigorously true, as a law of motion, whichever of the rival interpretations of the measure of the term "Action" we were to take.

6. *Extent of the Third Axiom.*—It may naturally be asked whether this third axiom respecting causation extends to any other cases than those of mechanical action, since the notion of cause in general has certainly a much wider extent. For instance, when a hot body heats a cold one, is there necessarily an equal reaction of the second body upon the first? Does the snowball cool the boy's hand exactly as much as the hand heats the snow? To this we reply, that, in every case in which one body acts upon another by its physical qualities, there must be some reaction. No body can affect another without being itself also affected. But in any physical change the *action* exerted is an abstract term which may be variously understood. The hot hand may *melt* a cold body, or may *warm* it: which kind of effect is to be taken as action? This remains to be determined by other considerations.

In all cases of physical change produced by one body in another, it is generally possible to assume such a meaning of action, that the reaction shall be of the same nature as the action; and when this is done, the third axiom of causation, that reaction is equal to action, is universally true. Thus if a hot body heat a cold one, the change may be conceived as the transfer of a certain substance, *heat* or *caloric*, from the first body to the second. On this supposition, the first body loses just as much heat as the other gains; action and reaction are equal. But if the reaction be of a different kind to the action we can no longer apply the axiom. If a hot body *melt* a cold one, the latter *cools* the former: here, then, is reaction; but so long as the action and reaction are stated in this form, we cannot assert any equality between them.

In treating of the secondary mechanical sciences, we shall see further in what way we may conceive the physical action of one body upon another, so that the same axioms which are the basis of the science of Mechanics shall apply to changes not at first sight manifestly mechanical.

The three axioms of causation which we have now stated are the fundamental maxims of all reasoning concerning causes as to their quantities; and it will be shown in the sequel that these axioms form the basis of the science of Mechanics, determining its form, extent, and certainty. We must, however, in the first place, consider how we acquire those conceptions upon which the axioms now established are to be employed.

CHAPTER V.

OF THE ORIGIN OF OUR CONCEPTIONS OF FORCE AND MATTER.

1. *Force.*—When the faculties of observation and thought are developed in man, the idea of causation is applied to those changes which we see and feel in the

state of rest and motion of bodies around us. And when our abstract conceptions are thus formed and named, we become possessed of the term *Force,* to denote that property which is the cause of motion produced, changed, or prevented. This conception is, it would seem, mainly and primarily suggested by our consciousness of the exertions by which we put bodies in motion. The Latin and Greek words for force, *vis,* F*ìs,* were probably, like all abstract terms, derived at first from some sensible object. The original meaning of the Greek word was a *muscle* or *tendon.* Its first application as an abstract term is accordingly to muscular force.

> Δεύτερος αὖτ' Αἴας πολὺ μείζονα λᾶαν ἀείρας
> ἧκ' ἐπιδινήσας, ἐπέρεισε δὲ Ϝ͂ΙΝ' ἀπελεθρον.
>
> Then Ajax a far heavier stone upheaved,
> He whirled it, and impressing Force intense
> Upon the mass, dismist it.

The property by which bodies affect each other's motions, was naturally likened to that energy which we exert upon them with similar effect: and thus the labouring horse, the rushing torrent, the descending weight, the elastic bow, were said to exert force. Homer* speaks of the *force* of the river, F*ìs* ποταμοῖο; and Hesiod† of the *force* of the north wind, F*ìs* ἀνέμου βορέαο.

Thus man's general notion of force was probably first suggested by his muscular exertions, that is, by an act depending upon that muscular sense, to which, as we have already seen, the perception of space is mainly due, And this being the case, it will be easily understood that the *Direction* of the force thus exerted is perceived by the muscular sense, at the same time that the force itself is perceived; and that the direction of any other force is understood by comparison with force which man must exert to produce the same effect, in the same manner as force itself is so understood.

* *Il.* xxi.   † *Op. et D.*

This abstract notion of Force long remained in a very vague and obscure condition, as may be seen by referring to the History for the failures of attempts at a science of force and motion, made by the ancients and their commentators in the middle ages. By degrees, in modern times, we see the scientific faculty revive. The conception of force becomes so far distinct and precise that it can be reasoned upon in a consistent manner, with demonstrated consequences; and a genuine science of Mechanics comes into existence. The foundations of this science are to be found in the Axioms concerning causation which we have already stated; these axioms being interpreted and fixed in their application by a constant reference to observed facts, as we shall show. But we must, in the first place, consider further those primary processes of observation by which we acquire the first materials of thought on such subjects.

2. *Matter.*—The conception of Force, as we have said, arises with our consciousness of our own muscular exertions. But we cannot imagine such exertions without also imagining some bodily substance against which they are exercised. If we press, we press something : if we thrust or throw, there must be something to resist the thrust or to receive the impulse. Without body, muscular force cannot be exerted, and force in general is not conceivable.

Thus Force cannot exist without *Body* on which it acts. The two conceptions, Force and Matter, are coexistent and correlative. Force implies resistance; and the force is effective only when the resistance is called into play. If we grasp a stone, we have no hold of it till the closing of the hand is resisted by the solid texture of the stone. If we push open a gate, we must surmount the opposition which it exerts while turning on its hinges. However slight the resistance be, there

must be some resistance, or there would be no force.
If we imagine a state of things in which objects do not
resist our touch, they must also cease to be influenced by
our strength.   Such a state of things we sometimes
imagine in our dreams; and such are the poetical pictures
of the regions inhabited by disembodied spirits.   In
these, the figures which appear are conspicuous to the
eye, but impalpable like shadow or smoke; and as they
do not resist the corporeal impressions, so neither do
they obey them.   The spectator tries in vain to strike
or to grasp them.

> Et ni cana vates tenues sine corpore vitas
> Admoneat volitare cavâ sub imagine formæ,
> Irruat ac frustra ferro diverberet umbras.
>
> The Sibyl warns him that there round him fly
> Bodiless things, but substance to the eye;
> Else had he pierced those shapes with life-like face,
> And smitten, fierce, the unresisting space.
>
>   Neque illum
> Prensantem nequicquam umbras et multa volentem
> Dicere, preterea vidit.
>
> He grasps her form, and clutches but the shade.

Such may be the circumstances of the unreal world of
dreams, or of poetical fancies approaching to dreams:
for in these worlds our imaginary perceptions are bound
by no rigid conditions of force and reaction.   In such
cases, the mind casts off the empire of the idea of cause,
as it casts off even the still more familiar sway of the
ideas of space and time.   But the character of the
material world in which we live when awake is, that we
have at every instant and at every place, force operating
on matter and matter resisting force.

3. *Solidity.*—From our consciousness of muscular
exertion, we derive, as we have seen, the conception of
force, and with that also the conception of matter.   We
have already shown, in a former chapter, that the same

part of our frame, the muscular system, is the organ by which we perceive extension and the relations of space. Thus the same organ gives us the perception of body as resisting force, and as occupying space ; and by combining these conditions we have the conception of *solid* extended bodies. In reality, this resistance is inevitably presented to our notice in the very facts from which we collect the notion of extension. For the action of the hand and arm by which we follow the forms of objects, implies that we apply our fingers to their surface; and we are stopped there by the resistance which the body offers. This resistance is precisely that which is requisite in order to make us conscious of our muscular effort*. Neither touch, nor any other mere passive sensation, could produce the perception of extent, as we have already urged : nor could the muscular sense lead to such a perception, except the extension of the muscles were felt to be resisted. And thus the perception of resistance enters the mind along with the perception of extended bodies. All the objects with which we have to do are not only extended but solid.

This sense of the term solidity, (the general property of all matter,) is different to that in which we oppose solidity to *fluidity.* We may avoid ambiguity by opposing *rigid* to fluid bodies. By solid bodies, as we now speak of them, we mean only such as resist the pressure which we exert, so long as their parts continue in their places. By fluid bodies, we mean those whose parts are, by a slight pressure, removed out of their places. A drop of water ceases to prevent the contact of our two hands, not by ceasing to have solidity in this sense, but by being thrust out of the way. If it could remain in its place, it could not cease to exercise its resistance to our pressure, except by ceasing to be matter altogether.

* Brown's *Lectures,* i., 466.

The perception of solidity, like the perception of extension, implies an act of the mind, as well as an impression of the senses: as the perception of extension implics the idea of space, so the perception of solidity implies the idea of action and reaction. That an idea is involved in our knowledge on this subject appears, as in other instances, from this consideration, that the convictions of persons, even of those who allow of no ground of knowledge but experience, do in fact go far beyond the possible limits of experience. Thus Locke says*, that "the bodies which we daily handle hinder by an *insurmountable* force the approach of the parts of our hands that press them." Now it is manifest that our observation can never go to this length. By our senses we can only perceive that bodies resist the greatest actual forces that we exert upon them. But our conception of force carries us further: and since, so long as the body is there to receive the action of the force, it must suffer the whole of that action, and must react as much as it suffers: it is therefore true, that so long as the body remains there, the force which is exerted upon it can never surmount the resistance which the body exercises. And thus this doctrine, that bodies resist the intrusion of other bodies by an insurmountable force, is in fact a consequence of the axiom that the reaction is always equal to the action.

4. *Inertia.*—But this principle of the equality of action and reaction appears also in another way. Not only when we exert force upon bodies at rest, but when, by our exertions, we put them in motion, they react. If we set a large stone in motion, the stone resists; for the operation requires an effort. By increasing the effort, we can increase the effect, that is, the motion produced; but the resistance still remains. And the greater the stone moved, the greater is the effort requisite to move it.

* *Essay*, b. ii., c. 4.

There is, in every case, a resistance to motion, which shows itself, not in preventing the motion, but in a reciprocal force, exerted backwards upon the agent by which the motion is produced. And this resistance resides in each portion of matter, for it is increased as we add one portion of matter to another. We can push a light boat rapidly through the water; but we may go on increasing its freight, till we are barely able to stir it. This property of matter, then, by which it resists the reception of motion, or rather by which it reacts and requires an adequate force in order that any motion may result, is called its inertness, or *inertia*. That matter has such a property, is a conviction flowing from that idea of a reaction equal and opposite to the action, which the conception of all force involves. By what laws this inertia depends on the magnitude, form, and material of the body, must be the subject of our consideration hereafter. But that matter has this inertia, in virtue of which, as the matter is greater, the velocity which the same effort can communicate to it is less, is a principle inseparable from the notion of matter itself.

Hermann says that Kepler first introduced this "most significant word" *inertia*. Whether it is to be found in earlier writers I know not; Kepler certainly does use it familiarly in those attempts to assign physical reasons for the motions of the planets which were among the main occasions of the discovery of the true laws of mechanics. He assumes the slowness of the motions of the planets to increase, (other causes remaining the same,) as the inertia increases; and though, even in this assumption, there is an error involved, (if we adopt that interpretation of the term *inertia* to which subsequent researches led,) the introduction of such a word was one step in determining and expressing those laws of motion which depend on the fundamental principle of the equality of action and reaction.

5. We have thus seen, I trust in a satisfactory manner, the origin of our conceptions of Force, Matter, Solidity, and Inertness. It has appeared that the organ by which we obtain such conceptions is that very muscular frame, which is the main instrument of our perceptions of space; but that, besides bodily sensations, these ideal conceptions, like all the others which we have hitherto considered, involve also an habitual activity of the mind, giving to our sensations a meaning which they could not otherwise possess. And among the ideas thus brought into play, is an idea of action with an equal and opposite reaction, which forms a foundation for universal truths to be hereafter established respecting the conceptions thus obtained.

We must now endeavour to trace in what manner these fundamental principles and conceptions are unfolded by means of observation and reasoning, till they become an extensive yet indisputable science.

CHAPTER VI.

OF THE ESTABLISHMENT OF THE PRINCIPLES
OF STATICS.

1. *Object of the Chapter.*—In the present and the succeeding chapters we have to show how the general axioms of Causation enable us to construct the science of Mechanics. We have to consider these axioms as moulding themselves, in the first place, into certain fundamental mechanical principles, which are of evident and necessary truth in virtue of their dependence upon the general axioms of Causation; and thus as forming a foundation for the whole structure of the science; a system of truths no less necessary than the fundamental principles, because derived from these by rigorous demonstration.

This account of the construction of the science of
Mechanics, however generally treated, cannot be otherwise than technical in its details, and will probably be
imperfectly understood by any one not acquainted with
Mechanics as a mathematical science.

I cannot omit this portion of my survey without
rendering my work incomplete; but I may remark that
the main purpose of it is to prove, in a more particular
manner, what I have already declared in general, that
there are in Mechanics no less than in Geometry, fundamental principles of axiomatic evidence and necessity;
—that these principles derive their axiomatic character
from the Idea which they involve, namely the Idea of
Cause;—and that through the combination of principles
of this kind, the whole science of Mechanics, including its
most complex and remote results, exists as a body of solid
and universal truths.

2. *Statics and Dynamics.*—We must first turn our
attention to a technical distinction of Mechanics into two
portions, according as the forces about which we reason
produce rest, or motion; the former portion is termed
Statics, the latter *Dynamics.* If a stone fall, or a weight
put a machine in motion, the problem belongs to Dynamics; but if the stone rest upon the ground, or a
weight be merely supported by a machine, without being
raised higher, the question is one of Statics.

3. *Equilibrium.*—In Statics, forces *balance* each other,
or keep each other *in equilibrium.* And forces which
directly balance each other, or keep each other in equilibrium, are necessarily and manifestly equal. If we see
two boys pull at two ends of a rope so that neither of
them in the smallest degree prevails over the other, we
have a case in which two forces are in equilibrium. The
two forces are evidently equal, and are a statical exemplification of action and reaction, such as are spoken of

in the third axiom concerning causes. Now the same exemplification occurs in every case of equilibrium. No point or body can be kept at rest except in virtue of opposing forces acting upon it; and these forces must always be equal in their opposite effect. When a stone lies on the floor, the weight of the stone downwards is opposed and balanced by an equal pressure of the floor upwards. If the stone rests on a slope, its tendency to slide is counteracted by some equal and opposite force, arising, it may be, from the resistance which the sloping ground opposes to any motion along its surface. Every case of rest is a case of equilibrium : every case of equilibrium is a case of equal and opposite forces.

The most complex frame-work on which weights are supported, as the roof of a building, or the cordage of a machine, are still examples of equilibrium. In such cases we may have many forces all combining to balance each other; and the equilibrium will depend on various conditions of direction and magnitude among the forces. And in order to understand what are these conditions, we must ask, in the first place, what we understand by the magnitude of such forces;—what is the measure of statical forces.

4. *Measure of Statical Forces.*—At first we might expect, perhaps, that since statical forces come under the general notion of Cause, the mode of measuring them would be derived from the second axiom of Causation, that causes are measured by their effects. But we find that the application of this axiom is controlled by the limitation which we noticed, after stating that axiom; namely, the condition that the causes shall be capable of addition. Further, as we have seen, a statical force produces no other effect than this, that it balances some other statical force; and hence the measure of statical forces is necessarily dependent upon their balancing, that is, upon the equality of action and reaction.

That *statical forces are capable of addition* is involved
in our conception of such forces. When two men pull
at a rope in the same direction, the forces which they
exert are added together. When two heavy bodies are
put into a basket suspended by a string, their weights are
added, and the sum is supported by the string.

Combining these considerations, it will appear that
the measure of statical forces is necessarily given at once
by the fundamental principle of the equality of action
and reaction. Since two opposite forces which balance
each other are equal, each force is measured by that
which it balances; and since forces are capable of addition, a force of any magnitude is measured by adding together a proper number of such equal forces. Thus a heavy
body which, appended to some certain elastic branch of a
tree, would bend it down through one inch, may be taken
as a unit of weight. Then if we remove this first body,
and find a second heavy body which will also bend the
branch through the same space, this is also a unit of
weight; and in like manner we might go on to a third
and a fourth equal body; and adding together the two,
or the three, or the four heavy bodies, we have a force
twice, or three times, or four times the unit of weight.
And with such a collection of heavy bodies, or *weights*, we
can readily measure all other forces; for the same principle of the equality of action and reaction leads at once
to this maxim, that any statical force is measured by the
weight which it would support.

As has been said, it might at first have been supposed
that we should have to apply, in this case, the axiom that
causes are measured by their effects in another manner;
that thus, if that body were a unit of weight which bent
the bough of a tree through one inch, *that* body would be
two units which bent it through *two* inches, and so on.
But, as we have already stated, the measures of weight

must be subject to this condition, that they are susceptible of being added: and therefore we cannot take the deflexion of the bough for our measure, till we have ascertained, that which experience alone can teach us, that under the burden of two equal weights, the deflexion will be twice as great as it is with one weight, which is not true, or at least is neither obviously nor necessarily true. In this, as in all other cases, although causes must be measured by their effects, we learn from experience only how the effects are to be interpreted, so as to give a true and consistent measure.

With regard, however, to the measure of statical force, and of weight, no difficulty really occurred to philosophers from the time when they first began to speculate on such subjects; for it was easily seen that if we take any uniform material, as wood, or stone, or iron, portions of this which are geometrically equal, must also be equal in statical effect; for this was implied in the very hypothesis of a uniform material. And a body ten times as large as another of the same substance, will be of ten times the weight. But before men could establish by reasoning the conditions under which weights would be in equilibrium, some other principles were needed in addition to the mere measure of forces. The principles introduced for this purpose still resulted from the conception of equal action and reaction; but it required no small clearness of thought to select them rightly, and to employ them successfully. This, however, was done, to a certain extent, by the Greeks; and the treatise of Archimedes *On the Centre of Gravity*, is founded on principles which may still be considered as the genuine basis of statical reasoning. I shall make a few remarks on the most important principle among those which Archimedes thus employs.

5. *The Centre of Gravity.*—The most important of

the principles which enter into the demonstration of
Archimedes is this: that "Every body has a centre of
gravity;" meaning by the centre of gravity, a point at
which the whole matter of the body may be supposed to
be collected, to all intents and purposes of statical
reasoning. This principle has been put in various forms
by succeeding writers: for instance, it has been thought
sufficient to assume a case much simpler than the general
one; and to assert that two *equal* bodies have their
centre of gravity in the point midway between them. It
is to be observed, that this assertion not only implies
that the two bodies will *balance* upon a support placed
at that midway point, but also, that they will exercise,
upon such a support, a *pressure equal to their sum;*
for this point being the centre of gravity, the whole
matter of the two bodies may be conceived to be collected there, and therefore the whole weight will press
there. And thus the principle in question amounts to
this, that *when two equal heavy bodies are supported on the
middle point between them, the pressure upon the support is
equal to the sum of the weights of the bodies.*

A clear understanding of the nature and grounds of
this principle is of great consequence: for in it we have
the foundation of a large portion of the science of
Mechanics. And if this principle can be shown to be
necessarily true, in virtue of our Fundamental Ideas, we
can hardly doubt that there exist many other truths of
the same kind, and that no sound view of the evidence
and extent of human knowledge can be obtained, so long
as we mistake the nature of these, its first principles.

The above principle, that the pressure on the support
is equal to the sum of the bodies supported, is often
stated as an axiom in the outset of books on Mechanics.
And this appears to be the true place and character of
this principle, in accordance with the reasonings which

we have already urged. The axiom depends upon our conception of action and reaction. That the two weights are supported, implies that the supporting force must be equal to the force or weight supported.

In order further to show the foundation of this principle, we may ask the question: if it be not an axiom, deriving its truth from the fundamental conception of equal action and reaction, which equilibrium always implies, what is the origin of its certainty? The principle is never for an instant denied or questioned: it is taken for granted, even before it is stated. No one will doubt that it is not only true, but true with the same rigour and universality as the axioms of Geometry. Will it be said, that it is borrowed from experience? Experience could never prove a principle to be universally and rigorously true. Moreover, when from experience we prove a proposition to possess great exactness and generality, we approach by degrees to this proof: the conviction becomes stronger, the truth more secure, as we accumulate trials. But nothing of this kind is the case in the instance before us. There is no gradation from less to greater certainty;—no hesitation which precedes confidence. From the first, we know that the axiom is exactly and certainly true. In order to be convinced of it, we do not require many trials, but merely a clear understanding of the assertion itself.

But, in fact, not only are trials not necessary to the proof, but they do not strengthen it. Probably no one ever made a trial for the purpose of showing that the pressure upon the support is equal to the sum of the two weights. Certainly no person with clear mechanical conceptions ever wanted such a trial to convince him of the truth; or thought the truth clearer after the trial had been made. If to such a person, an experiment were shown which seemed to contradict the principle, his conclusion would be, not that the principle was doubtful, but that the apparatus was out of order. Nothing can be less like collecting truth from experience.

We maintain, then, that this equality of mechanical action and reaction, is one of the principles which do not flow from, but regulate our experience. To this principle, the facts which we observe must conform; and we cannot help interpreting them in such a manner that they shall be exemplifications of the principle. A mechanical pressure not accompanied by an equal and opposite pressure, can no more be given by experience, than two unequal right angles. With the supposition of such inequalities, space ceases to be space, force ceases to be force, matter ceases to be matter. And this equality of action and reaction, considered in the case in which two bodies are connected so as to act on a single support, leads to the axiom which we have stated above, and which is one of the main foundations of the science of Mechanics.

6. *Oblique Forces.*—By the aid of this axiom and a few others, the Greeks made some progress in the science of Statics. But after a short advance, they arrived at another difficulty, that of Oblique Forces, which they never overcame; and which no mathematician mastered till modern times. The unpublished manuscripts of Leonardo da Vinci, written in the fifteenth century, and the works of Stevinus and Galileo, in the sixteenth, are the places in which we find the first solid grounds of reasoning on the subject of forces acting obliquely to each other. And mathematicians, having thus become possessed of all the mechanical principles which are requisite in problems respecting equilibrium, soon framed a complete science of Statics. Succeeding writers presented this science in forms variously modified; for it was found, in Mechanics as in Geometry, that various

propositions might be taken as the starting points ; and
that the collection of truths which it was the mechanician's business to include in his course, might be
traversed by various routes, each path offering a series
of satisfactory demonstrations. The fundamental conceptions of force and resistance, like those of space and
number, could be contemplated under different aspects,
each of which might be made the basis of axioms,
or of principles employed as axioms. Hence the
grounds of the truth of Statics may be stated in various
ways ; and it would be a task of some length to examine
all these completely, and to trace them to their Fundamental Ideas. This I shall not undertake here to do ;
but the philosophical importance of the subject makes it
proper to offer a few remarks on some of the main
principles involved in the different modes of presenting
Statics as a rigorously demonstrated science.

7. *A Force may be supposed to act at any Point of its
Direction.*—It has been stated in the history of Mechanics[*],
that Leonardo da Vinci and Galileo obtained the true
measure of the effect of oblique forces, by reasonings
which were, in substance, the same. The principle of
these reasonings is that expressed at the head of this
paragraph ; and when we have a little accustomed ourselves to contemplate our conceptions of force, and its
action on matter, in an abstract manner, we shall have
no difficulty in assenting to the principle in this general
form. But it may, perhaps, be more obvious at first in a
special case.

If we suppose a wheel, moveable about its axis, and
carrying with it in its motion a weight, (as, for example,
one of the wheels by means of which the large bells of a
church are rung,) this weight may be supported by means
of a rope (not passing along the circumference of the wheel,

[*] *Hist. Ind. Sc.*, ii. pp. 17 and 122.

as is usual in the case of bells,) but fastened to one of the spokes of the wheel. Now the principle which is enunciated above asserts, that if the rope pass in a straight line across several of the spokes of the wheel, it makes no difference in the mechanical effect of the force applied, for the purpose of putting the bell in motion, to *which* of these spokes the rope is *fastened*. In each case, fastening the rope to the wheel merely serves to enable the force to produce motion about the centre; and so long as the force acts in the same line, the effect is the same, at whatever point of the rope the line of action finishes.

This axiom very readily aids us in estimating the effect of oblique forces. For when a force acts on one of the arms of a lever at any oblique angle, we suppose another arm projecting from the centre of motion, like another spoke of the same wheel, so situated that it is perpendicular to the force. This arm we may, with Leonardo, call the *virtual lever ;* for, by the axiom, we may suppose the force to act where the line of its direction meets this arm ; and thus we reduce the case to that in which the force acts perpendicularly on the arm.

The ground of this axiom is, that matter, in Statics, is necessarily conceived as *transmitting* force. That force can be transmitted from one place to another, by means of matter;—that we can push with a rod, pull with a rope,—are suppositions implied in our conceptions of force and matter. Matter is, as we have said, that which receives the impression of force, and the modes just mentioned, are the simplest ways in which that impression operates. And since, in any of these cases, the force might be resisted by a reaction equal to the force itself, the reaction in each case would be equal, and, therefore, the action in each case is necessarily equal ; and thus the forces must be transmitted, from one point to another, without increase or diminution.

This property of matter, of transmitting the action of force, is of various kinds. We have the coherence of a rope which enables us to pull, and the rigidity of a staff, which enables us to push with it in the direction of its length; and again, the same staff has a rigidity of another kind, in virtue of which we can use it as a lever; that is, a rigidity to resist flexure, and to transmit the force which turns a body round a fulcrum. There is, further, the rigidity by which a solid body resists *twisting*. Of these kinds of rigidity, the first is that to which our axiom refers; but in order to complete the list of the elementary principles of Statics, we ought also to lay down axioms respecting the other kinds of rigidity*. These, however, I shall not here state, as they do not involve any new principle. Like the one just considered, they form part of our fundamental conception of matter; they are not the results of any experience, but are the hypotheses to which we are irresistibly led, when we would liberate our reasonings concerning force and matter from a dependence on the special results of experience. We cannot even conceive (that is, if we have any clear mechanical conceptions at all) the force exerted by the point of a staff and resisting the force which we steadily impress on the head of it, to be different from the impressed force.

8. *Forces may have equivalent Forces substituted for them. The Parallelogram of Forces.*—It has already been observed, that in order to prove the doctrines of Statics, we may take various principles as our starting points, and may still find a course of demonstration by which the leading propositions belonging to the subject may be established. Thus, instead of beginning our reasonings, as in the last section we supposed them to

---

* Such axioms are given in a little work (*The Mechanical Euclid* which I published on the Elements of Mechanics.

commence, with the case in which forces act upon different points of the same body in the same line of force, and counteract each other in virtue of the intervening matter by which the effect of force is transferred from one point to another, we may suppose different forces to act at the same point, and may thus commence our reasonings with a case in which we have to contemplate force, without having to take into our account the resistance or rigidity of matter. Two statical forces, thus acting at a mathematical point, are equivalent, in all respects, to some single force acting at the same point; and would be kept in equilibrium by a force equal and opposite to that single force. And the rule by which the single force is derived from the two, is commonly termed *the parallelogram of forces;* the proposition being this,—That if the two forces be represented in magnitude and direction by the two sides of a parallelogram, the resulting force will be represented in the same manner by the diagonal of the parallelogram. This proposition has very frequently been made, by modern writers, the commencement of the science of Mechanics: a position for which, by its simplicity, it is well suited; although, in order to deduce from it the other elementary propositions of the science, as, for instance, those respecting the lever, we require the axiom stated in the last section.

9. *The Parallelogram of Forces is a necessary Truth.*
—In the series of discussions in which we are here engaged, our main business is to ascertain the nature and grounds of the certainty of scientific truths. We have, therefore, to ask whether this proposition, the parallelogram of forces, be a necessary truth; and if so, on what grounds its necessity ultimately rests. We shall find that this, like the other fundamental doctrines of Statics, justly claims a demonstrative certainty. Daniel Bernoulli, in 1726, gave the first proof of this important

proposition on pure statical principles; and thus, as he says*, "proved that statical theorems are not less necessarily true than geometrical are." If we examine this proof of Bernoulli, in order to discover what are the principles on which it rests, we shall find that the reasoning employs in its progress such axioms as this;— That if from forces which are in equilibrium at a point be taken away other forces which are in equilibrium at the same point, the remainder will be in equilibrium; and generally ;—That if forces can be resolved into other equivalent forces, these may be separated, grouped, and recombined, in any new manner, and the result will still be identical with what it was at first. Thus in Bernoulli's proof, the two forces to be compounded are represented by P and Q; P is resolved into two other forces, X and U; and Q into two others, Y and V, under certain conditions. It is then assumed that these forces may be grouped into the pairs X, Y, and U, V: and when it has been shown that X and Y are in equilibrium, they may, by what has been said, be removed, and the forces P, Q, are equivalent to U, V; which, being in the same direction by the course of the construction, have a result equal to their sum.

It is clear that the principles here assumed are genuine axioms, depending upon our conception of the nature of equivalence of forces, and upon their being capable of addition and composition. If the forces P, Q, be *equivalent* to forces X, U, Y, V, they are equivalent to these forces added and compounded in any order; just as a geometrical figure is, by our conception of space, equivalent to its parts added together in any order. The apprehension of forces as having magnitude, as made up of parts, as capable of composition, leads to such axioms in Statics, in the same manner as the like

* *Comm. Petrop.* vol. i.

apprehension of space leads to the axioms of Geometry. And thus the truths of Statics, resting upon such foundations, are independent of experience in the same manner in which geometrical truths are so. The proof of the parallelogram of forces thus given by Daniel Bernoulli, as it was the first, is also one of the most simple proofs of that proposition which have been devised up to the present day. Many other demonstrations, however, have been given of the same proposition. Jacobi, a German mathematician, has collected and examined eighteen of these*. They all depend either upon such principles as have just been stated; That forces may in every way be replaced by those which are equivalent to them;—or else upon those previously stated, the doctrine of the lever, and the transfer of a force from one point to another of its direction. In either case, they are necessary results of our statical conceptions, independent of any observed laws of motion, and indeed, of the conception of actual motion altogether.

There is another class of alleged proofs of the parallelogram of forces, which involve the consideration of the motion produced by the forces. But such reasonings are, in fact, altogether irrelevant to the subject of Statics. In that science, forces are not measured by the motion which they produce, but by the forces which they will balance, as we have already seen. The combination of two forces employed in producing motion in the same body, either simultaneously or successively, belongs to that part of Mechanics which has motion for its subject, and is to be considered in treating of the laws of motion. The composition of motion, (as when a man moves in a

* These are by the following mathematicians; D. Bernoulli (1726); Lambert (1771); Scarella (1756); Venini (1764); Araldi (1806); Wachter (1815); Kæstner; Marini; Eytelwein; Salimbeni; Duchayla; two different proofs by Foncenex (1760); three by D'Alembert; and those of Laplace and M. Poisson.

ship while the ship moves through the water,) has constantly been confounded with the composition of force. But though this has been done by very eminent mathematicians, it is quite necessary for us to keep the two subjects distinct, in order to see the real nature of the evidence of truth in either case. The conditions of equilibrium of two forces on a lever, or of three forces at a point, can be established without any reference whatever to any motions which the forces might, under *other* circumstances, produce. And because this can be done, to do so is the only scientific procedure. To prove such propositions by any other course, would be to support truth by extraneous and inconclusive reasons; which would be foreign to our purpose, since we seek not only knowledge, but the grounds of our knowledge.

10. *The Centre of Gravity seeks the lowest place.*—The principles which we have already mentioned afford a sufficient basis for the science of Statics in its most extensive and varied applications; and the conditions of equilibrium of the most complex combinations of machinery may be deduced from these principles with a rigour not inferior to that of geometry. But in some of the more complex cases, the results of long trains of reasoning may be foreseen, in virtue of certain maxims which appear to us self-evident, although it may not be easy to trace the exact dependence of these maxims upon our fundamental conceptions of force and matter. Of this nature is the maxim now stated ;—That in any combination of matter any how supported, the Centre of Gravity will descend into the lowest position which the connexion of the parts allows it to assume by descending. It is easily seen that this maxim carries to a much greater extent the principle which the Greek mathematicians assumed, that every body has a Centre of Gravity, that is, a point in which, if the whole matter of the body be collected,

the effect will remain unchanged. For the Greeks asserted this of a single rigid mass only; whereas, in the maxim now under our notice, it is asserted of any masses, connected by strings, rods, joints, or in any manner. We have already seen that more modern writers on mechanics, desirous of assuming as fundamental no wider principles than are absolutely necessary, have not adopted the Greek axiom in all its generality, but have only asserted that two *equal* weights have a centre of gravity midway between them. Yet the principle that every body, however irregular, has a centre of gravity, and will be supported if that centre is supported, and not otherwise, is so far evident, that it might be employed as a fundamental truth, if we could not resolve it into any simpler truths: and, historically speaking, it was assumed as evident by the Greeks. In like manner the still wider principle, that a collection of bodies, as, for instance, a flexible chain hanging upon one or more supports, has a centre of gravity; and that this point will descend to the lowest possible situation, as a single body would do, has been adopted at various periods in the history of mechanics; and especially at conjunctures when mathematical philosophers have had new and difficult problems to contend with. For in almost every instance it has only been by repeated struggles that philosophers have reduced the solution of such problems to a clear dependence upon the most simple axioms.

11. *Stevinus's Proof for Oblique Forces.*—We have an example of this mode of dealing with problems, in Stevinus's mode of reasoning concerning the Inclined Plane; which, as we have stated in the History of Mechanics, was the first correct published solution of that problem. Stevinus supposes a loop of chain, or a loop of string loaded with a series of equal balls at equal distances, to hang over the Inclined Plane; and his reasoning proceeds upon this assumption,—That

such a loop so hanging will find a certain position in which it will rest: for otherwise, says he *, its motion must go on for ever, which is absurd.   It may be asked how this absurdity of a perpetual motion appears; and it will perhaps be added, that although the impossibility of a machine with such a condition may be proved as a remote result of mechanical principles, this impossibility can hardly be itself recognised as a self-evident truth. But to this we may reply, that the impossibility is really evident in the case contemplated by Stevinus; for we cannot conceive a loop of chain to go on through all eternity, sliding round and round upon its support, by the effect of its own weight.   And the ground of our conviction that this cannot be, seems to be this consideration; that when the chain moves by the effect of its weight, we consider its motion as the result of an effort to reach some certain position, in which it can rest; just as a single ball in a bowl moves till it comes to rest at the lowest point of the bowl.   Such an effect of weight in the chain, we may represent to ourselves by conceiving all the matter of the chain to be collected in one single point, and this single heavy point to hang from the support in some way or other, so as fitly to represent the mode of support of the chain.   In whatever manner this heavy point (the centre of gravity of the chain) be supported and controlled in its movements, there will still be some position of rest which it will seek and find.   And thus there will be some corresponding position of rest for the chain; and the interminable shifting from one position to another, with no disposition to rest in any position, cannot exist.

Thus the demonstration of the property of the Inclined Plane by Stevinus, depends upon a principle which, though far from being the simplest of those to which the case can be reduced, is still both true and evident: and the evidence of this principle, depending

* STEVIN. *Statique*, livre i., prop. 19.

upon the assumption of a centre of gravity, is of the same
nature as the evidence of the Greek statical demonstrations, the earliest real advances in the science.

12. *Principle of Virtual Velocities.* — We have
referred above to an assertion often made, that we
may, from the simple principles of Mechanics, demonstrate the impossibility of a perpetual motion. In reality,
however, the simplest proof of that impossibility, in
a machine acted upon by weight only, arises from the
very maxim above stated, that the centre of gravity seeks
and finds the lowest place; or from some similar proposition. For if, as is done by many writers, we profess
to prove the impossibility of a perpetual motion by means
of that proposition which includes the conditions of equilibrium, and is called the *Principle of Virtual Velocities**,
we are under the necessity of first proving in a general
manner that principle. And if this be done by a mere
enumeration of cases, (as by taking those five cases which
are called the *mechanical powers*,) there may remain some
doubts whether the enumeration of possible mechanical
combinations be complete. Accordingly, some writers
have attempted independent and general proofs of the
Principle of Virtual Velocities; and these proofs rest
upon assumptions of the same nature as that now under
notice. This is, for example, the case with Lagrange's
proof, which depends upon what he calls the *Principle
of Pulleys*. For this principle is,—That a weight any how
supported, as by a string passing round any number of
pulleys any how placed, will be at rest then only, when
it cannot get lower by any small motion of the pulleys.
And thus the maxim that a weight will descend if it can,
is assumed as the basis of this proof.

There is, as we have said, no need to assume such
principles as these for the foundation of our mechanical
science. But it is, on various accounts, useful to direct

* See *Hist. Ind. Sci.*, ii. 41,

our attention to those cases in which truths, apprehended
at first in a complex and derivative form, have afterwards
been reduced to their simpler elements; in which, also,
sagacious and inventive men have fixed upon those
truths as self-evident, which now appear to us only certain in virtue of demonstration. In these cases we can
hardly doubt that such men were led to assert the doctrines which they discovered, not by any capricious conjecture or arbitrary selection, but by having a keener
and deeper insight than other persons into the relations
which were the object of their contemplation; and in the
science now spoken of, they were led to their assumptions
by possessing clearly and distinctly the conceptions of
mechanical cause and effect,—action and reaction,—force,
and the nature of its operation.

13. *Fluids press Equally in all Directions.*—The doctrines which concern the equilibrium of fluids depend on
principles no less certain and simple than those which
refer to the equilibrium of solid bodies; and the Greeks,
who, as we have seen, obtained a clear view of some of
the principles of Statics, also made a beginning in the
kindred subject of Hydrostatics. We still possess a treatise of Archimedes *On Floating Bodies*, which contains
correct solutions of several problems belonging to this
subject, and of some which are by no means easy. In
this treatise, the fundamental assumption is of this kind:
"Let it be assumed that the nature of a fluid is such,
that the parts which are less pressed yield to those which
are more pressed." In this assumption or axiom it is
implied that a pressure exerted upon a fluid in one direction produces a pressure in another direction; thus, the
weight of the fluid which arises from a downward force
produces a lateral pressure against the sides of the containing vessel. Not only does the pressure thus diverge
from its original direction into all other directions, but it
is in all directions exactly equal, an equal extent of the

fluid being taken. This principle, which was involved in the reasoning of Archimedes, is still to the present day the basis of all hydrostatical treatises, and is expressed, as above, by saying that *fluids press equally in all directions.* Concerning this, as concerning previously-noticed principles, we have to ask whether it can rightly be said to be derived from experience. And to this the answer must still be, as in the former cases, that the proposition is not one borrowed from experience in any usual or exact sense of the phrase. I will endeavour to illustrate this. There are many elementary propositions in physics, our knowledge of which indisputably depends upon experience; and in these cases there is no difficulty in seeing the evidence of this dependence. In such cases, the experiments which prove the law are prominently stated in treatises upon the subject: they are given with exact measures, and with an account of the means by which errors were avoided: the experiments of more recent times have either rendered more certain the law originally asserted, or have pointed out some correction of it as requisite : and the names, both of the discoverers of the law and of its subsequent reformers, are well known. For instance, the proposition that "The elastic force of air varies as the density," was first proved by Boyle, by means of operations of which the detail is given in his *Defence* of his *Pneumatical Experiments**; and by Marriotte in his *Traité de l'Equilibre des Liquides*, from whom it has generally been termed Marriotte's law. After being confirmed by many other experimenters, this law was suspected to be slightly inaccurate, and a commission of the French Academy of Sciences was appointed, consisting of several distinguished philosophers†, to ascertain the truth or false-

---

* SHAW's *Boyle*, vol. ii., p. 671.

† The members were Prony, Arago, Ampère, Girard, and Dulong.

hood of this suspicion. The result of their investigations appeared to be, that the law is exact, as nearly as the inevitable inaccuracies of machinery and measures will allow us to judge. Here we have an example of a law which is of the simplest kind and form; and which yet is not allowed to rest upon its simplicity or apparent probability, but is rigorously tested by experience. In this case, the assertion, that the law depends upon experience, contains a reference to plain and notorious passages in the history of science.

Now with regard to the principle that fluids press equally in all directions, the case is altogether different. It is, indeed, often asserted in works on hydrostatics, that the principle is collected from experience, and sometimes a few experiments are described as exhibiting its effect; but these are such as to illustrate and explain, rather than to prove, the truth of the principle: they are never related to have been made with that exactness of precaution and measurement, or that frequency of repetition, which are necessary to establish a purely experimental truth. Nor did such experiments occur as important steps in the history of science. It does not appear that Archimedes thought experiment necessary to confirm the truth of the law as he employed it: on the contrary, he states it in exactly the same shape as the axioms which he employs in statics, and even in geometry; namely, as an assumption. Nor does any intelligent student of the subject find any difficulty in assenting to this fundamental principle of hydrostatics as soon as it is propounded to him. Experiment was not requisite for its discovery; experiment is not necessary for its proof at present; and

The experiments were extended to a pressure of twenty-seven atmospheres; and in no instance did the difference between the observed and calculated elasticity amount to one-hundredth of the whole; nor did the difference appear to increase with the increase of pressure.— FECHNER. *Repertorium*, i. 110.

we may add, that experiment, though it may make the proposition more readily intelligible, can add nothing to our conviction of its truth when it is once understood.

14. *Foundation of the above Axiom.*—But it will naturally be asked, What then is the ground of our conviction of this doctrine of the equal pressure of a fluid in all directions? And to this I reply, that the reasons of this conviction are involved in our idea of a fluid, which is considered as matter, and therefore as capable of receiving, resisting, and transmitting force according to the general conception of matter; and which is also considered as matter which has its parts perfectly moveable among one another. For it follows from these suppositions, that if the fluid be confined, a pressure which thrusts in one side of the containing vessel, may cause any other side to bulge outwards, if there be a part of the surface which has not strength to resist this pressure from within. And that this pressure when thus transferred into a direction different from the original one, is not altered in intensity, depends upon this consideration; that any difference in the two pressures would be considered as a defect of *perfect* fluidity, since the fluidity would be still more complete, if this entire and undiminished transmission of pressure in all directions were supposed. If, for instance, the lateral pressure were less than the vertical, this could be conceived no other way than as indicating some rigidity or adhesion of the parts of the fluid. When the fluidity is perfect, the two pressures which act in the two different parts of the fluid exactly balance each other: they are the action and the reaction, and must hence be equal by the same necessity as two directly opposite forces in statics.

But it may be urged, that even if we grant that this conception of a perfect fluid, as a body which has its parts perfectly moveable among each other, leads us necessarily to the principle of the equality of hydrostatic pressure in all directions, still this conception itself is obtained from experience, or suggested by observation. And to this we may reply, that the conception of a fluid, as contemplated in mechanical theory, cannot be said to be derived from experience, except in the same manner as the conception of a solid and rigid body may be said to be acquired by experience. For if we imagine a vessel full of small, smooth spherical balls, such a collection of balls would approach to the nature of a fluid, in having its parts moveable among each other; and would approach to perfect fluidity, as the balls became smoother and smaller. And such a collection of balls would also possess the statical properties of a fluid; for it would transmit pressure out of a vertical into a lateral (or any other) direction, in the same manner as a fluid would do. And thus a collection of solid bodies has the same property which a fluid has; and the science of Hydrostatics borrows from experience no principles beyond those which are involved in the science of Statics respecting solids. And since in this latter portion of science, as we have already seen, none of the principles depend for their evidence upon any special experience, the doctrines of Hydrostatics also are not proved by experience, but have a necessary truth borrowed from the relations of our ideas.

It is hardly to be expected that the above reasoning will, at first sight, produce conviction in the mind of the reader, except he have, to a certain extent, acquainted himself with the elementary doctrines of the science of Hydrostatics as usually delivered; and have followed, with clear and steady apprehension, some of the trains of reasoning by which the pressures of fluids are determined; as, for instance, the explanation of what is called the Hydrostatic Paradox. The necessity of such a discipline in order that the reader may enter fully into this part of our speculations, naturally renders them less popular; but this disadvantage is inevitable in our plan. We cannot expect to throw light upon philosophy by means of the advances which have been made in the mathematical and physical sciences, except we really understand the doctrines which have been firmly established in those sciences. This preparation for philosophizing may be somewhat laborious; but such labour is necessary if we would pursue speculative truth with all the advantages which the present condition of human knowledge places within our reach.

We may add, that the consequences to which we are directed by the preceding opinions, are of very great importance in their bearing upon our general views respecting human knowledge. I trust to be able to show, that some important distinctions are illustrated, some perplexing paradoxes solved, and some large anticipations of the future extension of our knowledge suggested, by means of the conclusions to which the preceding discussions have conducted us. But before I proceed to these general topics, I must consider the foundations of some of the remaining portions of Mechanics.

CHAPTER VII.

OF THE ESTABLISHMENT OF THE PRINCIPLES OF DYNAMICS.

1. In the History of Mechanics, I have traced the steps by which the three Laws of Motion and the other principles of mechanics were discovered, established, and extended to the widest generality of form and application. We have, in these laws, examples of principles which were, historically speaking, obtained by reference to experience. Bearing in mind the object and the result of the preceding discussions, we cannot but turn

with much interest to examine these portions of science;
to inquire whether there be any real difference in the
grounds and nature between the knowledge thus obtained,
and those truths which we have already contemplated;
and which, as we have seen, contain their own evidence,
and do not require proof from experiment.

2. *The First Law of Motion.*—The first law of motion
is, that *When a body moves not acted upon by any
force, it will go on perpetually in a straight line, and
with a uniform velocity.* Now what is the real ground
of our assent to this proposition? That it is not at first
sight a self-evident truth, appears to be clear; since from
the time of Aristotle to that of Galileo the opposite
assertion was held to be true; and it was believed that
all bodies in motion had, by their own nature, a constant
tendency to move more and more slowly, so as to stop at
last. This belief, indeed, is probably even now entertained by most persons, till their attention is fixed upon
the arguments by which the first law of motion is established. It is, however, not difficult to lead any person
of a speculative habit of thought to see that the retardation which constantly takes place in the motion of all
bodies when left to themselves, is, in reality, the effect of
extraneous forces which destroy the velocity. A top
ceases to spin because the friction against the ground and
the resistance of the air gradually diminish its motion,
and not because its motion has any internal principle of
decay or fatigue. This may be shown, and was, in fact,
shown by Hooke before the Royal Society, at the time
when the laws of motion were still under discussion, by
means of experiments in which the weight of the top is
increased, and the resistance to motion offered by its support, is diminished; for by such contrivances its motion is
made to continue much longer than it would otherwise
do. And by experiments of this nature, although we can
never remove the whole of the external impediments to

continued motion, and although, consequently, there will always be some retardation ; and an end of the motion of a body left to itself, however long it may be delayed, must at last come ; yet we can establish a conviction that if all resistance could be removed, there would be no diminution of velocity, and thus the motion would go on for ever.

If we call to mind the axioms which we formerly stated, as containing the most important conditions involved in the idea of Cause, it will be seen that our conviction in this case depends upon the first axiom of Causation, that nothing can happen without a cause. Every change in the velocity of the moving body must have a cause ; and if the change can, in any manner, be referred to the presence of other bodies, these are said to exert *force* upon the, moving body : and the conception of force is thus evolved from the general idea of cause. *Force is any cause which has motion, or change of motion, for its effect ;* and thus, all the change of velocity of a body which can be referred to extraneous bodies, as the air which surrounds it, or the support on which it rests, is considered as the effect of forces ; and this consideration looked upon as explaining the difference between the motion which really takes place in the experiment, and that which, as the law asserts, would take place if the body were not acted on by any forces.

Thus the truth of the first law of motion depends upon the axiom that no change can take place without a cause ; and follows from the definition of force, if we suppose that there can be none but an *external* cause of change. But in order to establish the law, it was necessary further to be assured that there is no *internal* cause of change of velocity belonging to all matter whatever, and operating in such a manner that the mere progress of time is sufficient to produce a diminution of velocity in all moving bodies. It appears from the history of mechanical science, that this latter step required a reference to observation and experiment; and that the first law of motion is so far, historically at least, dependent upon our experience.

But notwithstanding this historical evidence of the need which we have of a reference to observed facts, in order to place this first law of motion out of doubt, it has been maintained by very eminent mathematicians and philosophers, that the law is, in truth, evident of itself, and does not really rest upon experimental proof. Such, for example, is the opinion of D'Alembert*, who offers what is called an *à priori* proof of this law; that is, a demonstration derived from our ideas alone. When a body is put in motion, either, he says, the cause which puts it in motion at first, suffices to make it move one foot, or the continued action of the cause during this foot is requisite for the motion. In the first case, the same reason which made the body proceed to the end of the first foot will hold for its going on through a second, a third, a fourth foot, and so on for any number. In the second case, the same reason which made the force continue to act during the first foot, will hold for its acting, and therefore for the body moving during each succeeding foot. And thus the body, once beginning to move, must go on moving for ever.

It is obvious that we might reply to this argument, that the reasons for the body proceeding during each succeeding foot may not necessarily be all the same; for among these reasons may be the time which has elapsed; and thus the velocity may undergo a change as the time proceeds: and we require observation to inform us that it does not do so.

Professor Playfair has presented nearly the same argument, although in a different and more mathematical form†. If the velocity change, says he, it must change

---

* *Dynamique.*   † *Outlines,* &c., p. 26.·

according to some expression of calculation depending upon the time, or, in mathematical language, must be a *function* of the time. If the velocity diminish as the time increases, this may be expressed by stating the velocity in each case as a certain number, from which another quantity, or *term*, increasing as the time increases, is subtracted. But, Playfair adds, there is no condition involved in the nature of the case, by which the *coefficients*, or numbers which are to be employed, along with the number representing the time, in calculating this second term, can be determined to be of one magnitude rather than of any other. Therefore he infers there can be no such coefficients, and that the velocity is in each case equal to some constant number, independent of the time; and is therefore the same for all times.

In reply to this we may observe, that the circumstance of *our not seeing* in the nature of the case anything which determines for us the coefficients above spoken off, cannot prove that they have not some certain value *in nature*. We do not see in the nature of the case anything which should determine a body to fall sixteen feet in a second of time, rather than one foot or one hundred feet: yet in fact the space thus run through by falling bodies is determined to a certain magnitude. It would be easy to assign a mathematical expression for the velocity of a body, implying that one-hundredth of the velocity, or any other fraction, is lost in each second *: and where is the absurdity of supposing such an expression really to represent the velocity?

Most modern writers on mechanics have embraced the opposite opinion, and have ascribed our knowledge of

---

* This would be the case, if, t being the number of seconds elapsed, and C some constant quantity, the velocity were expressed by this mathematical formula,

$$C \left(\frac{99}{100} \right)^t$$

this first law of motion to experience.   Thus M. Poisson, one of the most eminent of the mathematicians who have written on this subject, says*, "We cannot affirm *à priori* that the velocity communicated to a body will not become slower and slower of itself, and end by being entirely extinguished.   It is only by experience and induction that this question can be decided."

Yet it cannot be denied that there is much force in those arguments by which it is attempted to shew that the First Law of Motion, such as we find it, is more consonant to our conceptions than any other would be.   The Law, as it exists, is the most simple that we can conceive. Instead of having to determine by experiments what is the law of the natural change of velocity, we find the Law to be that it does not change at all.   To a certain extent, the Law depends upon the evident axiom, that no change can take place without a cause.   But the question further occurs, whether the mere lapse of time may not be a cause of change of velocity.   In order to ensure this, we have recourse to experiment; and the result is that time alone does not produce any such change.   In addition to the conditions of change which we collect from our own ideas, we ask of experience what other conditions and circumstances she has to offer; and the answer is, that she can point out none.   When we have removed the alterations which external causes, in our very conception of them, occasion, there are no longer any alterations.   Instead of having to guide ourselves by experience, we learn that on this subject she has nothing to tell us.   Instead of having to take into account a number of circumstances, we find that we have only to reject all circumstances.   The velocity of a body remains unaltered by time alone, of whatever kind the body itself be.

But the doctrine that time alone is not a cause of

* Poisson. *Dynamique.* Ed. 2, Art. 113.

change of velocity in any body is further recommended to us by this consideration;—that time is conceived by us not as a cause, but only as a condition of other causes producing their effects. Causes operate in time; but it is only when the cause exists that the lapse of time can give rise to alterations. When therefore all external causes of change of velocity are supposed to be removed, the velocity must continue identical with itself, whatever the time which elapses. An eternity of negation can produce no positive result.

Thus, though the discovery of the First Law of Motion was made, historically speaking, by means of experiment, we have now attained a point of view in which we see that it might have been certainly known to be true independently of experience. This law in its ultimate form, when completely simplified and steadily contemplated, assumes the character of a self-evident truth. We shall find the same process to take place in other instances. And this feature in the progress of science will hereafter be found to suggest very important views with regard both to the nature and prospects of our knowledge.

2. *Gravity is a Uniform Force.*—We shall find observations of the same kind offering themselves in a manner more or less obvious, with regard to the other principles of Dynamics. The determination of the laws according to which bodies fall downwards by the common action of gravity, has already been noticed in the History of Mechanics*, as one of the earliest positive advances in the doctrine of motion. These laws were first rightly stated by Galileo, and established by reasoning and by experiment, not without dissent and controversy. The amount of these doctrines is this: That gravity is a uniform accelerating force; such a *uniform force* having this for its character, that it makes *the velocity increase in*

---

* *Hist. Ind. Sci.,* ii, 26.

exact proportion to the time of motion. The relation which the spaces described by the body bear to the times in which they are described, is obtained by mathematical deduction from this definition of the force.

The clear Definition of a uniform accelerating force, and the Proposition that gravity is such a force, were co-ordinate and contemporary steps in this discovery. In defining accelerating force, reference, tacit or express, was necessarily made to the second of the general axioms respecting causation,—That causes are measured by their effects. Force, in the cases now under our notice, is conceived to be, as we have already stated, (p. 209,) any cause which, acting from without, changes the motion of a body. It must, therefore, in this acceptation, be measured by the magnitude of the changes which are produced. But *in what manner* the changes of motion are to be employed as the measures of force, is learnt from observation of the facts which we see taking place in the world. Experience *interprets* the axiom of causation, from which otherwise we could not deduce any real knowledge. We may assume, in virtue of our general conceptions of force, that under the same circumstances, a greater change of motion implies a greater force producing it; but what are we to expect when the circumstances change? The weight of a body makes it fall from rest at first, and causes it to move more quickly as it descends lower. We may express this by saying, that gravity, the universal force which makes all terrestrial bodies fall when not supported, by its continuous action first *gives* velocity to the body when it has none, and afterwards *adds* velocity to that which the body already has. But how is the velocity added proportioned to the velocity which already exists? Force acting on a body at rest, and on a body in motion, appears under very different conditions;—how are the effects related? Let the force be conceived to be in both cases the *same,* since force is conceived to depend upon the extraneous bodies, and not upon the condition of the moving mass itself. But the force being the same, the effects may still be different. It is at first sight conceivable that the body, acted upon by the same gravity, may receive a less addition of velocity when it is already moving in the direction in which this gravity impels it; for if we ourselves push a body forwards, we can produce little additional effect upon it when it is already moving rapidly away from us. May it not be true, in like manner, that although gravity be always the same force, its effect depends upon the velocity which the body under its influence already possesses?

Observation and reasoning combined, as we have said, enabled Galileo to answer these questions. He asserted and proved that we may consistently and properly measure a force by the velocity which is by it generated in a body, in some certain time, as one second; and further, that if we adopt this measure, gravity will be a force of the same value under all circumstances of the body which it affects; since it appeared that, in fact, a falling body does receive equal increments of velocity in equal times from first to last.

If it be asked whether we could have known, anterior to, or independent of, experiment, that gravity is a uniform force in the sense thus imposed upon the term; it appears clear that we must reply, that we could not have attained to such knowledge, since other laws of the motion of bodies downwards are easily conceivable, and nothing but observation could inform us that one of these laws does not prevail in fact. Indeed, we may add, that the assertion that the force of gravity is uniform, is so far from being self-evident, that it is not even true; for gravity varies according to the distance from the centre of the earth; and although this variation is so small as to be, in the case of falling bodies, imperceptible, it negatives the rigorous uniformity of the force as completely, though not to the same extent, as if the weight of a body diminished in a marked degree, when it was carried from the lower to the upper room of a house. It cannot, then, be a truth independent of experience, that gravity is uniform.

Yet, in fact, the assertion that gravity is uniform was assented to, not only before it was proved, but even before it was clearly understood. It was readily granted by all, that bodies which fall freely are *uniformly* accelerated; but while some held the opinion just stated, that uniformly accelerated motion is that in which the velocity increases in proportion to the time, others maintained, that *that* is uniformly accelerated motion, in which the velocity increases in proportion to the space; so that, for example, a body in falling vertically through twenty feet should acquire twice as great a velocity as one which falls through ten feet.

These two opinions are both put forward by the interlocutors of Galileo's dialogue on this subject*. And the latter supposition is rejected, the author showing, not that it is inconsistent with experience, but that it is impossible in itself: inasmuch as it would inevitably lead to the conclusion, that the fall though a large and a small vertical space would occupy exactly the same time.

Indeed, Galileo assumes his definition of uniformly accelerated motion as one which is sufficiently recommended by its own simplicity. " If we attend carefully," he says, " we shall find that no mode of increase of velocity is more simple than that which adds equal increments in equal times. Which we may easily understand if we consider the close affinity of time and motion: for as the uniformity of motion is defined by the equality of spaces

* *Dialogo*, iii. p. 95.   † *Ibid.* p. 91.

described in equal times, so we may conceive the uniformity of acceleration to exist when equal velocities are added in equal times."

Galileo's mode of supporting his opinion, that bodies falling by the action of gravity are thus uniformly accelerated, consists, in the first place, in adducing the maxim that nature always employs the most simple means*. But he is far from considering this a decisive argument. " I," says one of his speakers, "as it would be very unreasonable in me to gainsay this or any other definition which any author may please to make, since they are all arbitrary, may still, without offence, doubt whether such a definition, conceived and admitted in the abstract, fits, agrees, and is verified in that kind of accelerated motion which bodies have when they descend naturally."

The experimental proof that bodies, when they fall downwards, are uniformly accelerated, is (by Galileo) derived from the inclined plane; and therefore assumes the proposition, that if such uniform acceleration prevail in vertical motion, it will also hold when a body is compelled to describe an oblique rectilinear path. This proposition may be shown to be true, if (assuming by anticipation the Third Law of Motion, of which we shall shortly have to speak,) we introduce the conception of a uniform statical force as the cause of uniform acceleration. For the force on the inclined plane bears a constant proportion to the vertical force, and this proportion is known from statical considerations. But in the work of which we are speaking, Galileo does not introduce this abstract conception of force as the foundation of his doctrines. Instead of this, he proposes, as a postulate sufficiently evident to be made the basis of his reasonings, That bodies which descend

* *Dialogo*, iii. p. 91.

down inclined planes of different inclinations, but of
the same vertical height, all acquire the same velocity*.
But when this postulate has been propounded by one
of the persons of the dialogue, another interlocutor says,
"You discourse very probably; but besides this like-
lihood, I wish to augment the probability so far, that
it shall be almost as complete as a necessary demonstration." He then proceeds to describe a very ingenious and simple experiment, which shows that when a
body is made to swing upwards at the end of a string,
it attains to the same height, whatever is the path it
follows, so long as it starts from the lowest point with
the same velocity. And thus Galileo's postulate is experimentally confirmed, so far as the force of gravity can
be taken as an example of the forces which the postulate
contemplates: and conversely, gravity is proved to be a
uniform force, so far as it can be considered clear that
the postulate is true of uniform forces.

When we have introduced the conception and definition of accelerating force, Galileo's postulate, that
bodies descending down inclined planes of the same
vertical height, acquire the same velocity, may, by a
few steps of reasoning, be demonstrated to be true of
uniform forces: and thus the proof that gravity, either in
vertical or oblique motion, is a uniform force, is confirmed
by the experiment above mentioned; as it also is, on
like grounds, by many other experiments, made upon
inclined planes and pendulums.

Thus the propriety of Galileo's conception of a uniform force, and the doctrine that gravity is a uniform
force, were confirmed by the same reasonings and experiments. We may make here two remarks; *First*, that the
conception, when established and rightly stated, appears
so simple as hardly to require experimental proof; a
remark which we have already made with regard to the

* *Dialogo*, iii. p. 36.

First Law of Motion: and *Second*, that the discovery of the real law of nature was made by assuming propositions which, without further proof, we should consider as very precarious, and as far less obvious, as well as less evident, than the law of nature in its simple form.

3. *The Second Law of Motion.*—When a body, instead of falling downwards from rest, is thrown in any direction, it describes a curve line, till its motion is stopped. In this, and in all other cases in which a body describes a curved path in free space, its motion is determined by the Second Law of Motion. The law, in its general form, is as follows:—When a body is thus cast forth and acted upon by a force in a direction transverse to its motion, the result is, That *there is combined with the motion with which the body is thrown, another motion, exactly the same as that which the same force would have communicated to a body at rest.*

It will readily be understood that the basis of this law is the axiom already stated, that effects are measured by their causes. In virtue of this axiom, the effect of gravity acting upon a body in a direction transverse to its motion, must measure the accelerative or deflective force of gravity under those circumstances. If this effect vary with the varying velocity and direction of the body thus acted upon, the deflective force of gravity also will vary with those circumstances. The more simple supposition is, that the deflective force of gravity is the same, whatever be the velocity and direction of the body which is subjected to its influence: and this is the supposition which we find to be verified by facts. For example, a ball let fall from the top of a ship's upright mast, when she is sailing steadily forward, will fall at the foot of the mast, just as if it were let fall while the ship were at rest; thus showing that the motion which gravity gives to the ball is compounded with the horizontal motion which the ball

shares with the ship from the first. This general and
simple conception of motions as *compounded* with one
another, represents, it is proved, the manner in which the
motion produced by gravity modifies any other motion
which the body may previously have had.

The discussions which terminated in the general
reception of this Second Law of Motion among mechanical writers, were much mixed up with the arguments for
and against the Copernican system, which system represented the earth as revolving upon its axis. For the
obvious argument against this system was, that if the
earth were thus in motion from west to east, a stone
dropt from the top of a tower would be left behind, the
tower moving away from it: and the answer was, that by
this law of motion, the stone would have the earth's
motion impressed upon it, as well as that motion which
would arise from its gravity to the earth; and that the
motion of the stone relative to the tower would thus be
the same as if both earth and tower were at rest. Galileo further urged, as a presumption in favour of the
opinion that the two motions,—the circular motion arising
from the rotation of the earth, and the downward motion
arising from the gravity of the stone, would be compounded in the way we have described, (neither of them
disturbing or diminishing the other,) — that the first
motion was in its own nature not liable to any change or
diminution*, as we learn from the First Law of Motion.
Nor was the subject lightly dismissed. The experiment
of the stone let fall from the top of the mast was made
in various forms by Gassendi; and in his Epistle, *De
Motu impresso a Motore translato*, the rule now in question
is supported by reference to these experiments. In this
manner, the general truth, the Second Law of Motion,
was established completely and beyond dispute.

* *Dialogo*, ii. p. 114.

But when this law had been proved to be true in a general sense, with such accuracy as rude experiments, like those of Galileo and Gassendi, would admit, it still remained to be ascertained (supposing our knowledge of the law to be the result of experience alone,) whether it were true with that precise and rigorous exactness which more refined modes of experimenting could test. We so willingly believe in the simplicity of laws of nature, that the rigorous accuracy of such a law, known to be at least approximately true, was taken for granted, till some ground for suspecting the contrary should appear. Yet calculations have not been wanting which might confirm the law as true to the last degree of accuracy. Laplace relates (*Syst. du Monde*, livre iv., chap. 16,) that at one time he had conceived it possible that the effect of gravity upon the moon might be slightly modified by the moon's direction and velocity; and that in this way an explanation might be found for the moon's *acceleration* (a deviation of her observed from her calculated place, which long perplexed mathematicians). But it was after some time discovered that this feature in the moon's motion arose from another cause; and the second law of motion was confirmed as true in the most rigorous sense.

Thus we see that although there were arguments which might be urged in favour of this law, founded upon the necessary relations of ideas, men became convinced of its truth only when it was verified and confirmed by actual experiment. But yet in this case again, as in the former ones, when the law had been established beyond doubt or question, men were very ready to believe that it was not a mere result of observation,—that the truth which it contained was not derived from experience,—that it might have been assumed as true in virtue of reasonings anterior to experience,—and that experiments served only to make the law more plain and intelligible, as visible diagrams in geometry serve to illustrate geometrical truths; our knowledge not being (they deemed) in mechanics, any more than in geometry, borrowed from the senses. It was thought by many to be self-evident, that the effect of a force in any direction cannot be increased or diminished by any motion transverse to the direction of the force which the body may have at the same time: or, to express it otherwise, that if the motion of the body be compounded of a horizontal and vertical motion, the vertical motion alone will be affected by the vertical force. This principle, indeed, not only has appeared evident to many persons, but even at the present day is assumed as an axiom by many of the most eminent mathematicians. It is, for example, so employed in the *Mécanique Celeste* of Laplace, which may be looked upon as the standard of mathematical mechanics in our time; and in the *Mécanique Analytique* of Lagrange, the most consummate example which has appeared of subtilty of thought on such subjects, as well as of power of mathematical generalization*. And thus we have here

---

* I may observe that the rule that we may *compound* motions, as the Law supposes, is involved in the step of *resolving* them; which is done in the passage to which I refer (*Méc. Analyt.* ptie. i., sect. i., art. 3, p. 225). " Si on conçoit que la mouvement d'un corps et les forces qui le sollicitent soient *decomposées* suivant trois lignes droites perpendiculaires entre elles, on pourra considérer séparément les mouvemens et les forces relatives à chacun a de ces trois directions. Car à cause de la perpendicularité des directions il est visible que chacun de ces mouvemens partiels peut être regardé comme indépendant des deux autres, et qu'il ne peut recevoir d'alteration que de la part de la force qui agit dans la direction de ce mouvement; l'on peut conclure que ces trois mouvements doivent suivre, chacun en particulier, les lois des mouvemens rectilignes accélérés ou retardés par les forces données." Laplace makes the same assumption in effect, (*Méc. Cel.* p. i, liv. i., art. 7,) by resolving the forces which act upon a point in three rectangular directions, and reasoning separately concerning each direction. But in his mode of treating the subject is involved a principle which belongs to the Third Law of Motion, namely, the doctrine that the velocity is as the force, of which we shall have to speak elsewhere.

another example of that circumstance which we have already noticed in speaking of the First Law of Motion, (p. 213,) and of the Law that Gravity is a uniform Force, (p. 218); namely, that the law, though historically established by experiments, appears, when once discovered and reduced to its most simple and general form, to be self-evident. I am the more desirous of drawing attention to this feature in various portions of the history of science, inasmuch as it will be found to lead to some very extensive and important views, hereafter to be considered.

4. *The Third Law of Motion.*—We have, in the definition of Accelerating Force, a measure of Forces, so far as they are concerned in producing motion. We had before, in speaking of the principles of statics, defined the measure of Forces or Pressures, so far as they are employed in producing equilibrium. But these two aspects of Force are closely connected; and we require a law which shall lay down the rule of their connexion. By the same kind of muscular exertion by which we can support a heavy stone, we can also put it in motion. The question then occurs, how is the rate and manner of its motion determined? The answer to this question is contained in the Third Law of Motion, and it is to this effect: that the *Momentum* which any pressure produces in the mass in a given time is proportional to the pressure. By momentum is meant the product of the numbers which express the velocity and the mass of the body: and hence, if the mass of the body be the same in the instances which we compare, the rule is,—That *the velocity is as the force which produces it;* and this is one of the simplest ways of expressing the Third Law of Motion.

In agreement with our general plan, we have to ask, What is the ground of this rule? What is the simplest and most satisfactory form to which we can reduce the proof of it? Or, to take an instance; if a double pressure be exerted against a given mass, so disposed as to be capable of motion, why must it produce twice the velocity in the same time?

To answer this question, suppose the double pressure to be resolved into two single pressures: one of these will produce a certain velocity; and the question is, why an equal pressure, acting upon the same mass, will produce an equal velocity *in addition* to the former? Or, stating the matter otherwise, the question is, why each of the two forces will produce its separate effect, unaltered by the simultaneous action of the other force?

This statement of the case makes it seem to approach very near to such cases as are included in the Second Law of Motion, and therefore it might appear that this Third Law has no grounds distinct from the Second. But it must be recollected that the word *force* has a different meaning in this case and in that; in this place it signifies *pressure;* in the statement of the Second Law its import was *accelerative* or *deflective force*, measured by the velocity or deflexion generated. And thus the Third Law of Motion, so far as our reasonings yet go, appears to rest on a foundation different from the Second.

Accordingly, that part of the Third Law of Motion which we are now considering, that the velocity generated is as the force, was obtained, in fact, by a separate train of research. The first exemplification of this law which was studied by mathematicians, was the motion of bodies upon inclined planes: for the force which urges a body down an inclined plane is known by statics, and hence the velocity of its descent was to be determined. Galileo originally* in his attempts to solve this problem of the descent of a body down an inclined plane, did not proceed

* *Dial. della Sc. Nuov.* iii., p. 96. See *Hist. Ind. Sci.*, ii., p. 47.

from the principle which we have stated, (the determination of the force which acts down the inclined plane from statical considerations,) obvious as it may seem; but assumed, as we have already seen, a proposition apparently far more precarious ;—namely, that a body sliding down a smooth inclined plane acquires always the same velocity, so.long as the *vertical* height fallen through is the same. And this conjecture, (for at first it was nothing more than a conjecture,) he confirmed by an ingenious experiment; in which bodies acquired or lost the same velocity by descending or ascending through the same height, although their paths were different in other respects.

This was the form in which the doctrine of the motion of bodies down inclined planes was at first presented in Galileo's *Dialogues* on the Science of Motion. But his disciple Viviani was dissatisfied with the assumption thus introduced; and in succeeding editions of the *Dialogues*, the apparent chasm in the reasoning was much narrowed, by making the proof depend upon a principle nearly identical with the third law of motion as we have just stated it. In the proof thus added, " We are agreed," says the interlocutor *, " that in a moving body the impetus, energy, momentum, or propension to motion, is as great as is the force or least resistance which suffices to sustain it ;" and the impetus or momentum, in the course of the proof, being taken to be as the velocity produced in a given time, it is manifest that the principle so stated amounts to this ; that the velocity produced is as the statical force. And thus this law of motion appears, in the school of Galileo, to have been suggested and established at first by experiment, but afterwards confirmed and demonstrated by *à priori* considerations.

We see, in the above reasoning, a number of abstract

* *Dialogo*, p. 104.

terms introduced which are not, at first at least, very distinctly defined, as impetus, momentum, &c. Of these, *momentum* has been selected, to express that quantity which, in a moving body, measures the statical force impressed upon the body. This quantity is, as we have just seen, proportional to the velocity in a given body. It is also, in different bodies, proportional to the mass of the body. This part of the third law of motion follows from our conception of matter in general as consisting of parts capable of addition. A double pressure must be required to produce the same velocity in a double mass; for if the mass be halved, each half will require an equal pressure; and the addition, both of the pressures and of the masses, will take place without disturbing the effects.

The measure of the quantity of matter of a body considered as affecting the velocity which pressure produces in the body, is termed its *inertia*, as we have already stated. (p. 182.) Inertia is the property by which a large mass of matter requires a greater force than a small mass, to give it an equal velocity. It belongs to each portion of matter; and portions of inertia are added whenever portions of matter are added. Hence *inertia is as the quantity of matter;* which is only another way of expressing this third law of motion, so far as quantity of matter is concerned.

But how do we know the quantity of matter of a body? We may reply, that we take the weight as the measure of the quantity of matter: but we may then be again asked, how it appears that the weight is proportional to the inertia; which it must be, in order that the quantity of matter may be proportional to both one and the other. We answer, that this appears to be true experimentally, because all bodies fall with equal velocities by gravity, when the known causes of difference are removed. The observations of falling bodies, indeed, are not susceptible of much exactness: but experiments leading to the same result, and capable of great precision, were made upon pendulums by Newton; as he relates in *The Principia*, book iii., prop. 6. They all agreed, he says, with perfect accuracy: and thus the weight and the inertia are proportional in all cases, and therefore each proportional to the quantity of matter as measured by the other.

The conception of inertia, as we have already seen in Chapter V., involves the notion of action and reaction; and thus the laws which involve inertia depend upon the idea of mutual causation. The rule, that the velocity is as the force, depends upon the principle of causation, that the effect is proportional to the cause; the effect being here so estimated as to be consistent both with the other laws of motion and with experiment.

But here, as in other cases, the question occurs again; Is experiment really requisite for the proof of this law? If we look to authorities, we shall be not a little embarrassed to decide. D'Alembert is against the necessity of experimental proof. "Why," says he*, "should we have recourse to this principle employed, at the present day, by everybody, that the force is proportional to the velocity? . . . a principle resting solely upon this vague and obscure axiom, that the effect is proportional to the cause. We shall not examine here," he adds, "if this principle is necessarily true; we shall only avow that the proofs which have hitherto been adduced do not appear to us unexceptionable: nor shall we, with some geometers, adopt it as a purely contingent truth; which would be to ruin the certainty of mechanics, and to reduce it to be nothing more than an experimental science. We shall content ourselves with observing," he proceeds, "that

* *Dynamique*, Pref. p. x.

certain or doubtful, clear or obscure, it is useless in mechanics, and consequently ought to be banished from the science." Though D'Alembert rejects the third law of motion in this form, he accepts one of equivalent import, which appears to him to possess axiomatic certainty; and this procedure is in consistence with the course which he takes, of claiming for the science of mechanics more than mere experimental truth. On the contrary, Laplace considers this third law as established by experiment. "Is the force," he says *, "proportional to the velocity? This," he replies, " we cannot know *à priori*, seeing that we are in ignorance of the nature of moving force: we must therefore, for this purpose, recur to experience; for all which is not a necessary consequence of the few data we have respecting the nature of things, is, for us, only a result of observation." And again he says †, " Here, then, we have two laws of motion,—the law of inertia [the first law of motion], and the law of the force proportional to the velocity,—which are given by observation. They are the most natural and the most simple laws which we can imagine, and without doubt they flow from the very nature of matter; but this nature being unknown, they are, for us, only observed facts: the only ones, however, which mechanics borrows from experience."

It will appear, I think, from the views given in this and several other parts of the present work, that we cannot with justice say that we have very " few data respecting the nature of things," in speculating concerning the laws of the universe; since all the consequences which flow from the relations of our fundamental ideas, necessarily regulate our knowledge of things, so far as we have any such knowledge. Nor can we say that the nature of matter is unknown to us, in any sense in which we can conceive knowledge as possible. The nature of matter is

* *Méc. Cel.* p. 15.   † P. 13.

no more unknown than the nature of space or of number.
In our conception of matter, as of space and of number,
are involved certain relations, which are the necessary
groundwork of our knowledge; and anything which is
independent of these relations, is not unknown, but
inconceivable.

It must be already clear to the reader, from the
phraseology employed by these two eminent mathematicians, that the question respecting the formation of the
third law of motion can only be solved by a careful consideration of what we mean by observation and experience, nature and matter. But it will probably be generally allowed, that, taking into account the explanations
already offered of the necessary conditions of experience
and of the conception of inertia, this law of motion, that
the inertia is as the quantity of matter, is almost or altogether self-evident.

5. *Action and Reaction are Equal in Moving Bodies.*
—When we have to consider bodies as acting upon one
another, and influencing each other's motions, the third
law of motion is still applied; but along with this, we
also employ the general principle that action and reaction
are equal and opposite. Action and reaction are here to
be understood as momentum produced and destroyed,
according to the measure of action established by the
third law of motion: and the cases in which this principle is thus employed form so large a portion of those in
which the third law of motion is used, that some writers
(Newton at the head of them) have stated the equality of
action and reaction as the third law of motion.

The third law of motion being once established, the
equality of action and reaction, in the sense of momentum
gained and lost, necessarily follows. Thus, if a weight
hanging by a string over the edge of a smooth level table
draw another weight along the table, the hanging weight

moves more slowly than it would do if not so connected, and thus loses velocity by the connexion; while the other weight gains by the connexion all the velocity which it has, for if left to itself it would rest. And the pressures which restrain the descent of the first body and accelerate that of the second, are equal at all instants of time, for each of these pressures is the tension of the string: and hence, by the third law of motion, the momentum gained by the one body, and the momentum lost by the other in virtue of the action of this string, are equal. And similar reasoning may be employed in any other case where bodies are connected.

The case where one body does not push or draw, but *strikes* another, appeared at first to mechanical reasoners to be of a different nature from the others; but a little consideration was sufficient to show that a blow is, in fact, only a short and violent pressure; and that, there-fore, the general rule of the equality of momentum lost and gained applies to this as well as to the other cases.

Thus, in order to determine the case of the direct action of bodies upon one another, we require no new law of motion. The equality of action and reaction, which enters necessarily into every conception of mechanical operation, combined with the measure of action as given by the third law of motion, enables us to trace the consequences of every case, whether of pressure or of impact.

6. *D'Alembert's Principle.*—But what will be the result when bodies do not act directly upon each other, but are *indirectly* connected in any way by levers, strings, pulleys, or in any other manner, so that one part of the system has a mechanical advantage over another? The result must still be determined by the principle that action and reaction balance each other. The action and reaction, being pressures in one sense, must balance each

other by the laws of statics, for these laws determine the equilibrium of pressure. Now action and reaction, according to their measures in the Third Law of Motion, are momentum gained and lost, when the action is direct; and except the indirect action introduce some modification of the law, they must have the same measure still. But, in fact, we cannot well conceive any modification of the law to take place in this case; for direct action is only one (the ultimate) case of indirect action. Thus if two heavy bodies act at different points of a lever, the action of each on the other is indirect; but if the two points come together, the action becomes direct. Hence the rule must be that which we have already stated; for if the rule were false for indirect action, it would also be false for direct action, for which case we have shown it to be true. And thus we obtain the general principle, that in any system of bodies which act on each other, action and reaction, estimated by momentum gained and lost, balance each other according to the laws of equilibrium. This principle, which is so general as to supply a key to the solution of all possible mechanical problems, is commonly called *D'Alembert's Principle*. The experimental proofs which convinced men of the truth of the third law of motion were, many or most of them, proofs of the law in this extended sense. And thus the proof of D'Alembert's Principle, both from the idea of mechanical action and from experience, is included in the proof of the law already stated.

7. *Connexion of Dynamical and Statical Principles.*— The principle of equilibrium of D'Alembert just stated, is the law which he would substitute for the third law of motion; and he would thus remove the necessity for an independent proof of that law. In like manner, the second law of motion is by some writers derived from the principle of the composition of statical forces; and they

would thus supersede the necessity of a reference to experiment in that case. Laplace takes this course, and thus, as we have seen, rests only the first and third law of motion upon experience. Newton, on the other hand, recognises the same connexion of propositions, but for a different purpose; for he derives the composition of statical forces from the second law of motion.

The close connexion of these three principles, the composition of (statical) forces, the composition of (accelerating) forces with velocities, and the measure of (moving) forces by velocities, cannot be denied; yet it appears to be by no means easy to supersede the necessity of independent proofs of the two last of these principles. Both may be proved or illustrated by experiment: and the experiments which prove the one are different from those which establish the other. For example, it appears by easy calculations, that when we apply our principles to the oscillations of a pendulum, the second law is proved by the fact, that the oscillations take place at the same rate in an east and west, and in a north and south direction: under the same circumstances, the third law is proved by our finding that the time of a small oscillation is proportional to the square root of the length of a pendulum; and similar differences might be pointed out in other experiments, as to their bearing upon the one law or the other.

8. *Mechanical Principles become gradually more simple and more evident.*—I will again point out in general two circumstances which I have already noticed in particular cases of the laws of motion. Truths are often at first assumed in a form which is far from being the most obvious or simple; and truths. once discovered are gradually simplified, so as to assume the appearance of self-evident truths.

The former circumstance is exemplified in several of

the instances which we have had to consider. The assumption that a perpetual motion is impossible preceded the knowledge of the first law of motion. The assumed equality of the velocities acquired down two inclined planes of the same height, was afterwards reduced to the third law of motion by Galileo himself. In the History*, we have noted Huyghens's assumption of the equality of the actual descent and potential ascent of the centre of gravity: this was afterwards reduced by Herman and the Bernoullis, to the statical equivalence of the solicitations of gravity and the vicarious solicitations of the effective forces which act on each point; and finally to the principle of D'Alembert, which asserts that the motions gained and lost balance each other.

This assertion of principles which now appear neither obvious nor self-evident, is not to be considered as a groundless assertion on the part of the discoverers by whom it was made. On the contrary, it is evidence of the deep sagacity and clear thought which were requisite in order to make such discoveries. For these results are really rigorous consequences of the laws of motion in their simplest form: and the evidence of them was probably present, though undeveloped, in the minds of the discoverers. We are told of geometrical students, who, by a peculiar aptitude of mind, perceived the evidence of some of the more advanced propositions of geometry without going through the introductory steps. We must suppose a similar aptitude for mechanical reasonings, which led Stevinus, Galileo, Newton, and Huyghens, to make those assumptions which finally resolved themselves into the laws of motion.

We may observe further, that the simplicity and evidence which the laws of mechanics have at length assumed, are much favoured by the usage of words among

* Vol. ii. p. 82.

the best writers on such subjects. Terms which originally, and before the laws of motion were fully known,
were used in a very vague and fluctuating sense, were
afterwards limited and rendered precise, so that assertions
which at first appear identical propositions become distinct and important principles. Thus *force, motion,
momentum,* are terms which were employed, though in a
loose manner, from the very outset of mechanical speculation. And so long as these words retained the vagueness
of common language, it would have been a useless and
barren truism to say that "the momentum is proportional
to the force," or that "a body loses as much motion as
it communicates to another." But when "momentum"
and "quantity of motion" are defined to mean the product of mass and velocity, these two propositions immediately become distinct statements of the third law of
motion and its consequences. In like manner, the assertion that "gravity is a uniform force" was assented to,
before it was settled what a uniform force was; but this
assertion only became significant and useful when that
point had been properly determined. The statement
that "when different motions are communicated to the
same body their effects are compounded," becomes the
second law of motion, when we define what composition
of motions is. And the same process may be observed
in other cases.

And thus we see how well the form which science
ultimately assumes is adapted to simplify it. The definitions which are adopted, and the terms which become
current in precise senses, produce a complete harmony
between the matter and the form of our knowledge; so
that truths which were at first unexpected and recondite,
became familiar phrases, and after a few generations
sound, even to common ears, like identical propositions.

9. *Controversy of the Measure of Force.*—In the His-

tory of Mechanics*, we have given an account of the controversy which, for some time, occupied the mathematicians of Europe, whether the forces of bodies in motion should be reckoned proportional to the velocity, or to the square of the velocity. We need not here recall the events of this dispute; but we may remark, that its history, as a metaphysical controversy, is remarkable in this respect, that it has been finally and completely settled; for it is now agreed among mathematicians that both sides were right, and that the results of mechanical action may be expressed with equal correctness by means of *momentum* and of *vis viva*. It is, in one sense, as D'Alembert has said†, a dispute about words; but we are not to infer that, on that account, it was frivolous or useless; for such disputes are one principal means of reducing the principles of our knowledge to their utmost simplicity and clearness. The terms which are employed in the science of mechanics are now liberated for ever, in the minds of mathematicians, from that ambiguity which was the battle-ground in the war of the *vis viva*.

But we may observe that the real reason of this controversy was exactly that tendency which we have been noticing : the disposition of man to assume in his speculations certain general propositions as true, and to fix the sense of terms so that they shall fall in with this truth. It was agreed, on all hands, that in the mutual action of

---

* Vol. ii. p. 87.

† D'Alembert has also remarked (*Dynamique*, Pref. xxi.,) that this controversy "shows how little justice and precision there is in the pretended axiom that causes are proportional to their effects." But this reflection is by no means well founded. For since both measures are true, it appears that causes may be *justly* measured by their effects, even when very different kinds of effects are taken. That the axiom does not point out one *precise* measure till illustrated by experience or by other considerations, we grant : but the same thing occurs in the application of other axioms also.

bodies the same quantity of force is always preserved; and the question was, by which of the two measures this rule could best be verified. We see, therefore, that the dispute was not concerning a definition merely, but concerning a definition combined with a general proposition. Such a question may be readily conceived to have been by no means unimportant; and we may remark, in passing, that such controversies, although they are commonly afterwards stigmatised as quarrels about words and definitions, are, in reality, events of considerable consequence in the history of science; since they dissipate all ambiguity and vagueness in the use of terms, and bring into view the conditions under which the fundamental principles of our knowledge can be most clearly and simply presented.

It is worth our while to pause for a moment on the prospect that we have thus obtained of the advance of knowledge, as exemplified in the history of mechanics. The general transformation of our views from vague to definite, from complex to simple, from unexpected discoveries to self-evident truths, from seeming contradictions to identical propositions, is very remarkable, but it is by no means peculiar to our subject. The same circumstances, more or less prominently, more or less developed, appear in the history of other sciences, according to the point of advance which each has reached. They bear upon very important doctrines respecting the prospects, the limits, and the very nature of our knowledge. And though these doctrines require to be considered with reference to the whole body of science, yet the peculiar manner in which they are illustrated by the survey of the history of mechanics, on which we have just been engaged, appears to make this a convenient place for introducing them to the reader.

Chapter VIII.

OF THE PARADOX OF UNIVERSAL PROPOSITIONS OBTAINED FROM EXPERIENCE.

1. It was formerly stated* that experience cannot establish any universal or necessary truths. The number of trials of any proposition is necessarily limited, and observation alone cannot give us any ground of extending the inference to untried cases. Observed facts have no visible bond of necessary connexion, and no exercise of our senses can enable us to discover such connexion. We can never acquire from a mere observation of facts, the right to assert that a proposition is true in all cases, and that it could not be otherwise than we find it to be.

Yet, as we have just seen in the history of the laws of motion, we may go on collecting our knowledge from observation, and enlarging and simplifying it, till it approaches or attains to complete universality and seeming necessity. Whether the laws of motion, as we now know them, can be rigorously traced to an absolute necessity in the nature of things, we have not ventured absolutely to pronounce. But we have seen that some of the most acute and profound mathematicians have believed that for these laws of motion, or some of them, there was such a demonstrable necessity compelling them to be such as they are, and no other. Most of those who have carefully studied the principles of mechanics will allow that some at least of the primary laws of motion approach very near to this character of necessary truth; and will confess that it would be difficult to imagine any other consistent scheme of fundamental principles. And almost all mathematicians will allow to these laws an absolute universality; so that we may apply them without scruple

* B. i., c. 12. Of Experience.

or misgiving, in cases the most remote from those to which our experience has extended. What astronomer would fear to refer to the known laws of motion in reasoning concerning the double stars; although these objects are at an immeasurably remote distance from that solar system which has been the only field of our observation of mechanical facts? What philosopher, in speculating respecting a magnetic fluid, or a luminiferous ether, would hesitate to apply to it the mechanical principles which are applicable to fluids of known mechanical properties? When we assert that the quantity of motion in the world cannot be increased or diminished by the mutual actions of bodies, does not every mathematician feel convinced that it would be an unphilosophical restriction to limit this proposition to such modes of action as we have tried?

Yet no one can, doubt that, in historical fact, these laws were collected from experience. That such is the case, is no matter of conjecture. We know the time, the persons, the circumstances, belonging to each step of each discovery. I have, in the History, given an account of these discoveries; and in the previous chapters of the present work, I have further examined the nature and the import of the principles which were thus brought to light.

Here, then, is an apparent contradiction. Experience, it would seem, has done that which we had proved that she cannot do. She has led men to propositions, universal at least, and to principles which appear to some persons necessary. What is the explanation of this contradiction, the solution of this paradox? Is it true that Experience can reveal to us universal and necessary truths? Does she possess some secret virtue, some unsuspected power, by which she can detect connexions and consequences which we have declared to be out of her sphere? Can she see more than mere appearances, and observe

more than mere facts? Can she penetrate, in some way,
to the nature of things? descend below the surface of
phenomena to their causes and origins, so as to be able to
say what can and what can not be; what occurrences are
partial, and what universal? If this be so, we have indeed mistaken her character and powers; and the whole
course of our reasoning becomes precarious and obscure.
But, then, when we return upon our path we cannot find
the point at which we deviated, we cannot detect the
false step in our deduction. It still seems that by experience, strictly so called, we cannot discover necessary
and universal truths. Our senses can give us no evidence
of a necessary connexion in phenomena. Our observation must be limited, and cannot testify concerning anything which is beyond its limits. A general view of our
faculties appears to prove it to be impossible that men
should do what the history of the science of mechanics
shows that they have done.

2. But in order to try to solve this Paradox, let us
again refer to the History of Mechanics. In the cases
belonging to that science, in which propositions of the
most unquestionable universality, and most approaching
to the character of necessary truths, (as, for instance, the
laws of motion,) have been arrived at, what is the source
of the axiomatic character which the propositions thus
assume? The answer to this question will, we may hope,
throw some light on the perplexity in which we appear to
be involved.

Now the answer to this inquiry is, that the laws of
motion borrow their axiomatic character from their
being merely *interpretations* of the Axioms of Causation.
Those axioms, being exhibitions of the Idea of Cause
under various aspects, are of the most rigorous universality and necessity. And so far as the laws of motion
are exemplifications of those axioms, these laws must be

no less universal and necessary. How these axioms are to be understood;—in what sense cause and effect, action and reaction, are to be taken, experience and observation did, in fact, teach inquirers on this subject; and without this teaching, the laws of motion could never have been distinctly known. If two forces act together, each must produce its effect, by the axiom of causation; and, therefore, the effects of the separate forces must be *compounded*. But a long course of discussion and experiment must instruct men of what kind this *composition* of forces is. Again; action and reaction must be equal; but much thought and some trial were needed to show what *action* and *reaction* are. Those metaphysicians who enunciated Laws of motion without reference to experience, propounded only such laws as were vague and inapplicable. But yet these persons manifested the indestructible conviction, belonging to man's speculative nature, that there exist Laws of motion, that is, universal formulæ, connecting the causes and effects when motion takes place. Those mechanicians, again, who observed facts involving equilibrium and motion, and stated some narrow rules, without attempting to ascend to any universal and simple principle, obtained laws no less barren and useless than the metaphysicians; for they could not tell in what new cases, or whether in any, their laws would be verified;—they needed a more general rule, to show them the limits of the rule they had discovered. They went wrong in each attempt to solve a new problem, because their interpretation of the terms of the axioms, though true, perhaps, in certain cases, was not right in general.

Thus Pappus erred in attempting to interpret as a case of the lever, the problem of supporting a weight upon an inclined plane; thus Aristotle erred in interpreting the doctrine that the weight of bodies is the cause of their fall; thus Kepler erred in interpreting the rule that the velocity of bodies depends upon the force; thus Bernoulli* erred in interpreting the equality of action and reaction upon a lever in motion. In each of these instances, true doctrines, already established, (whether by experiment or otherwise,) were erroneously applied. And the error was corrected by further reflection, which pointed out that another mode of interpretation was requisite, in order that the axiom which was appealed to in each case might retain its force in the most general sense. And in the reasonings which avoided or corrected such errors, and which led to substantial general truths, the object of the speculator always was to give to the acknowledged maxims which the Idea of Cause suggested, such a signification as should be consistent with their universal validity. The rule was not accepted as particular at the outset, and afterwards generalized more and more widely; but from the very first, the universality of the rule was assumed, and the question was, how it should be understood so as to be universally true. At every stage of speculation, the law was regarded as a general law. This was not an aspect which it gradually acquired, by the accumulating contributions of experience, but a feature of its original and native character. *What* should happen universally, experience might be needed to show: but that what happened should happen *universally,* was implied in the nature of knowledge. The universality of the laws of motion was not gathered from experience, however much the laws themselves might be so.

3. Thus we obtain the solution of our Paradox, so far as the case before us is concerned. The laws of motion borrow their *form* from the Idea of Causation, though their *matter* may be given by experience: and hence they possess a universality which experience cannot

* *Hist. Ind. Sc.,* ii. p. 83.

give. They are certainly and universally valid; and the only question for observation to decide is, how they are to be understood. They are like general mathematical formulæ, which are known to be true even while we are ignorant what are the unknown quantities which they involve. It must be allowed, on the other hand, that so long as these formulæ are not interpreted by a real study of nature, they are not only useless but prejudicial, filling men's minds with vague general terms, empty maxims, and unintelligible abstractions, which they mistake for knowledge. Of such perversion of the speculative propensities of man's nature, the world has seen too much in all ages. Yet we must not, on that account, despise these forms of truth, since without them, no general knowledge is possible. Without general terms, and maxims, and abstractions, we can have no science, no speculation; hardly, indeed, consistent thought or the exercise of reason. The course of real knowledge is, to obtain from thought and experience the right interpretation of our general terms, the real import of our maxims, the true generalizations which our abstractions involve.

4. If it be asked, how experience is able to teach us to interpret aright the general terms which the Axioms of Causation involve;—whence she derives the light which she is to throw on these general notions; the answer is obvious;—namely, that the relations of causation are the conditions of experience;—that the general notions are *exemplified* in the particular cases of which she takes cognizance. The events which take place about us, and which are the objects of our observation, we cannot conceive otherwise than as subject to the laws of cause and effect. Every event must have a cause;—every effect must be determined by its cause;— these maxims are true of the phenomena which form the materials of our experience. It is precisely to them,

that these truths apply. It is in the world which we have before our eyes, that these propositions are universally verified; and it is therefore by the observation of what we see, that we must learn how these propositions are to be understood. Every fact, every experiment, is an example of these statements; and it is therefore by attention to and familiarity with facts and experiments, that we learn the signification of the expressions in which the statements are made; just as in any other case we learn the import of language by observing the manner in which it is applied in known cases. Experience is the interpreter of nature; it being understood that she is to make her interpretation in that comprehensive phraseology which is the genuine language of science.

5. We may return for an instant to the objection, that experience cannot give us general truths, since, after any number of trials confirming a rule, we may, for aught we can foresee, have one which violates the rule. When we have seen a thousand stones fall to the ground, we may see one which does not fall under the same apparent circumstances. How then, it is asked, can experience teach us that *all* stones, rigorously speaking, will fall if unsupported? And to this we reply, that it is not true that we can conceive one stone to be suspended in the air, while a thousand others fall, without believing some peculiar cause to support it; and that, therefore, such a supposition forms no exception to the law, that gravity is a force by which *all* bodies are urged downwards. Undoubtedly we can conceive a body, when dropt or thrown, to move in a line quite different from other bodies: thus a certain missile * used by the natives of Australia, and lately brought to this country, when thrown from the hand in a proper manner, describes a curve, and returns to the place from whence it was thrown. But did any

* Called the Bo-me-rang.

one, therefore, even for an instant suppose that the laws of motion are different for this and for other bodies? On the contrary, was not every person of a speculative turn immediately led to inquire how it was that the known causes which modify motion, the resistance of the air and the other causes, produced in this instance so peculiar an effect? And if the motion had been still more unaccountable, it would not have occasioned any uncertainty whether it were consistent with the agency of gravity and the laws of motion. If a body suddenly alter its direction, or move in any other unexpected manner, we never doubt that there is a cause of the change. We may continue quite ignorant of the nature of this cause, but this ignorance never occasions a moment's doubt that the cause exists and is exactly suited to the effect. And thus experience can prove or discover to us general rules, but she can never prove that general rules do not exist. Anomalies, exceptions, unexplained phenomena, may remind us that we have much still to learn, but they can never make us suppose that truths are not universal. We may observe facts that show us we have not fully understood the meaning of our general laws, but we can never find facts which show our laws to have no meaning. Our experience is bound in by the limits of cause and effect, and can give us no information concerning any region where that relation does not prevail. The whole series of external occurrences and objects, through all time and space, exists only, and is conceived only, as subject to this relation; and therefore we endeavour in vain to imagine to ourselves when and where and how exceptions to this relation may occur. The assumption of the connexion of cause and effect is essential to our experience, as the recognition of the maxims which express this connexion is essential to our knowledge.

6. I have thus endeavoured to explain in some

measure how, at least in the field of our mechanical knowledge, experience can discover universal truths, though
she cannot give them their universality; and how such
truths, though borrowing their form from our ideas, cannot
be understood except by the actual study of external
nature. And thus with regard to the laws of motion,
and other fundamental principles of Mechanics, the
analysis of our ideas and the history of the progress of
the science well illustrate each other.

If the paradox of the discovery of universal truths by
experience be thus solved in one instance, a much wider
question offers itself to us;—How far the difficulty, and
how far the solution, are applicable to other subjects. It
is easy to see that this question involves most grave and
extensive doctrines with regard to the whole compass of
human knowledge : and the views to which we have been
led in the present Book of this work are, we trust, fitted
to throw much light upon the general aspect of the subject. But after discussions so abstract, and perhaps
obscure, as those in which we have been engaged for
some chapters, I willingly postpone to a future occasion
an investigation which may perhaps appear to most
readers more recondite and difficult still. And we have,
in fact, many other special fields of knowledge to survey,
before we are led by the order of our subject, to those
general questions and doctrines, those antitheses brought
into view and again resolved, which a view of the whole
territory of human knowledge suggests, and by which
the nature and conditions of knowledge are exhibited.

Before we quit the subject of mechanical science we
shall make a few remarks on another doctrine which
forms part of the established truths of the science,
namely, the doctrine of universal gravitation.

CHAPTER IX.

OF THE ESTABLISHMENT OF THE LAW OF UNIVERSAL GRAVITATION.

THE doctrine of universal gravitation is a feature of so much importance in the history of science that we shall not pass it by without a few remarks on the nature and evidence of the doctrine.

1. To a certain extent the doctrine of the attraction of bodies according to the law of the inverse square of the distance, exhibits in its progress among men the same general features which we have noticed in the history of the laws of motion. This doctrine was maintained *à priori* on the ground of its simplicity, and asserted positively, even before it was clearly understood :—notwithstanding this anticipation, its establishment on the ground of facts was a task of vast labour and sagacity :—when it had been so established in a general way, there occurred at later periods, an occasional suspicion that it might be approximately true only :—these suspicions led to further researches, which showed the rule to be rigorously exact :—and at present there are mathematicians who maintain, not only that it is true, but that it is a necessary property of matter. A very few words on each of these points will suffice.

2. I have shown in the *History of Science**, that the attraction of the sun according to the inverse square of the distance, had been divined by Bullialdus, Hooke, Halley, and others, before it was proved by Newton. Probably the reason which suggested this conjecture was that gravity might be considered as a sort of emanation; and that thus, like light or any other effect diffused from a

---

* Vol. ii., 148.

centre, it must follow the law just stated, the efficacy of the force being weakened in receding from the centre, exactly in proportion to the space through which it is diffused. It cannot be denied that such a view appears to be strongly recommended by analogy.

When it had been proved by Newton that the planets were really retained in their elliptical orbits by a central force, his calculations also showed that the above-stated *law* of the force must be at least very approximately correct, since otherwise the aphelia of the orbits could not be so nearly at rest as they were. Yet when it seemed as if the motion of the moon's apogee could not be accounted for without some new supposition, the *à priori* argument in favour of the inverse square did not prevent Clairaut from trying the hypothesis of a small term added to that which expressed the ancient law : but when, in order to test the accuracy of this hypothesis, the calculation of the motion of the moon's apogee was pushed to a greater degree of exactness than had been obtained before, it was found that the new term vanished of itself ; and that the inverse square now accounted for tke whole of the motion. And thus, as in the case of the second law of motion, the most scrupulous examination terminated in showing the simplest rule to be rigorously true.

3. Similar events occurred in the history of another part of the law of gravitation : namely, that the attraction is proportional to the quantity of matter attracted. This part of the law may also be thus stated, That the weight of bodies arising from gravity is proportional to their inertia ; and thus, that the accelerating force on all bodies under the same circumstances is the same. Newton made experiments which proved this with regard to terrestrial bodies ; for he found that, at the end of equal strings, balls of all substances, gold, silver, lead, glass, wood, &c., oscillated

in equal times*. But a few years ago, doubts arose among the German astronomers whether this law was rigorously true with regard to the planetary bodies. Some calculations appeared to prove, that the attraction of Jupiter as shown by the perturbations which he produces in the small planets Juno, Vesta, and Pallas, was different from the attraction which he exerts on his own satellites. Nor did there appear to these philosophers anything inconceivable in the supposition that the attraction of a planet might be thus *elective*. But when Mr. Airy obtained a more exact determination of the mass of Jupiter, as indicated by his effect on his satellites, it was found that this suspicion was unfounded; and that there was, in this case, no exception to the universality of the rule, that this cosmical attraction is in the proportion of the attracted mass.

4. Again: when it had thus been shown that a mutual attraction of parts, according to the law above mentioned, prevailed throughout the extent of the solar system, it might still be doubted whether the same law extended to other regions of the universe. It might have been perhaps imagined that each fixed star had its peculiar law of force. But the examination of the motions of double stars about each other, by the two Herschels and others, appears to show that they describe ellipses as the planets do: and thus extends the law of the inverse squares to parts of the universe immeasurably distant from the whole solar system.

5. Since every doubt which has been raised with regard to the universality and accuracy of the law of gravitation, has thus ended in confirming the rule, it is not surprising that men's minds should have returned with additional force to those views which had at first represented the law as a necessary truth, capable of being

* *Princ.* l. iii., Prop. 6.

established by reason alone.   When it had been proved by Newton that gravity is really a *universal* attribute of matter as far as we can learn, his pupils were not content without maintaining it be an *essential* quality.   This is the doctrine held by Cotes in the preface to the second edition of the *Principia* (1712) : "Gravity," he says, "is a primary quality of bodies, as extension, mobility, and impenetrability are."   But Newton himself by no means went so far.   In his second Letter to Bentley (1693), he says: "You sometimes speak of gravity as essential and inherent to matter; pray do not ascribe that notion to me.   The cause of gravity," he adds, "I do not pretend to know, and would take more time to consider of it."

Cotes maintains his opinion by urging, that we learn by *experience* that all bodies possess gravity, and that we do not learn in any other way that they are extended, moveable, or solid.   But we have already seen, that the ideas of space, time, and reaction, on which depend extension, mobility, and solidity, are not results, but conditions, of experience.   We cannot conceive a body except as extended; we cannot conceive it to exert mechanical action except with some kind of solidity.   But so far as our conceptions of body have hitherto been developed, we find no difficulty in conceiving two bodies which do not attract each other.

6. Newton lays down, in the second edition of the *Principia*, this "Rule of Philosophizing" (Book iii.); that "The qualities of bodies which cannot be made more or less intense, and which belong to all bodies on which we are able to make experiments, are to be held to be qualities of all bodies in general."   And this Rule is cited in the sixth proposition of the Third Book of the *Principia*, (Cor. 2,) in order to prove that gravity, proportional to the quantity of matter, may be asserted to be a quality of all bodies universally.   But we may remark that a Rule

of Philosophizing, itself of precarious authority, cannot authorize us in ascribing universality to an empirical result. Geometrical and statical properties are seen to be necessary, and *therefore* universal : but Newton appears disposed to assert a like universality of gravity, quite unconnected with any necessity. It would be a very inadequate statement, indeed a false representation, of statical truth, if we were to say, that because every body which has hitherto been tried *has been found* to have a centre of gravity, we venture to assert that all bodies whatever have a centre of gravity. And if we are ever able to assert the absolute universality of the law of gravitation, we shall have to rest this truth upon the clearer development of our ideas of matter and force ; not upon a Rule of Philosophizing, which, till otherwise proved, must be a mere rule of prudence, and which the opponent may refuse to admit.

7. Other persons, instead of asserting gravity to be in its own nature essential to matter, have made hypotheses concerning some mechanism or other, by which this mutual attraction of bodies is produced*. Thus the Cartesians ascribed to a vortex the tendency of bodies to a centre; Newton himself seems to have been disposed to refer this tendency to the elasticity of an ether; Le Sage propounded a curious hypothesis, in which this attraction is accounted for by the impulse of infinite streams of particles flowing constantly through the universe in all directions. In these speculations, the force of gravity is resolved into the pressure or impulse of solids or fluids. On the other hand, hypotheses have been propounded, in which the solidity, and other physical qualities of bodies, have been explained by representing the bodies as a collection of points, from which

---

* See VINCE, *Observations on the Hypotheses respecting Gravitation*, and the Critique of that work, *Edinb. Rev.* vol. xiii.

points repulsive, as well as attractive, forces emanate. This view of the constitution of bodies was maintained and developed by Boscovich, and is hence termed "Boscovich's Theory:" and the discussion of it will more properly come under our review at a future period, when we speak of the question whether bodies are made up of atoms. But we may observe, that Newton himself appears to have inclined, as his followers certainly did, to this mode of contemplating the physical properties of bodies. In his Preface to the *Principia*, after speaking of the central forces which are exhibited in cosmical phenomena, he says: "Would that we could derive the other phenomena of Nature from mechanical principles by the same mode of reasoning. For many things move me, so that I suspect all these phenomena may depend upon certain forces, by which the particles of bodies, through causes not yet known, are either impelled to each other and cohere according to regular figures, or are repelled and recede from each other: which forces being unknown, philosophers have hitherto made their attempts upon nature in vain."

8. But both these hypotheses;—that by which cohesion and solidity are reduced to attractive and repulsive forces, and that by which attraction is reduced to the impulse and pressure of media;—are hitherto merely modes of representing mechanical laws of nature; and cannot, either of them, be asserted as possessing any evident truth or peremptory authority to the exclusion of the other. This consideration may enable us to estimate the real weight of the difficulty felt in assenting to the mutual attraction of bodies not in contact with each other; for it is often urged that this attraction of bodies at a distance is an absurd supposition.

The doctrine is often thus stigmatised, both by popular and by learned writers. It was long received as a maxim in philosophy (as Monboddo informs us*), that a body cannot act *where* it is not, any more than *when* it is not. But to this we reply, that time is a necessary condition of our conception of causation, in a different manner from space. The action of force can only be conceived as taking place in a succession of moments, in each of which cause and effect immediately succeed each other: and thus the interval of time between a cause and its remote effect is filled up by a continuous succession of events connected by the same chain of causation. But in space, there is no such visible necessity of continuity; the action and reaction may take place at a distance from each other; all that is necessary being that they be equal and opposite.

Undoubtedly the existence of attraction is rendered more acceptable to common apprehension by supposing some intermediate machinery,—a cord, or rod, or fluid,— by which the forces may be conveyed from one point to another. But such images are rather fitted to satisfy those prejudices which arise from the earlier application of our ideas of force, than the real nature of those ideas. If we suppose two bodies to pull each other by means of a rod or a cord, we only suppose, in addition to those equal and opposite forces acting upon the two bodies, which forces are alone essential to mutual attraction, a certain power of resisting transverse pressure at every point of the intermediate line: which additional supposition is entirely useless, and quite unconnected with the essential conditions of the case. When the Newtonians were accused of introducing into philosophy an unknown cause which they termed attraction, they justly replied that they knew as much respecting attraction as their opponents did about impulse. In each case we have a knowledge of the conception in question so

* *Ancient Metaphysics*, vol. ii. p. 175.

far as we clearly apprehend it under the conditions of
those axioms of mechanical causation which form the
basis of our science on such subjects.

Having thus examined the degree of certainty and
generality to which our knowledge of the law of universal
gravitation has been carried, by the progress of mechanical
discovery and speculation up to the present time, we
might proceed to the other branches of science, and
examine in like manner their grounds and conditions.
But before we do this, it will be worth our while to
attend for a moment to the effect which the progress of
mechanical ideas among mathematicians and mechanical
philosophers has produced upon the minds of other persons, who share only in an indirect and derivative manner
in the influence of science.

CHAPTER X.

OF THE GENERAL DIFFUSION OF CLEAR MECHANICAL IDEAS.

1. WE have seen how the progress of knowledge
upon the subject of motion and force has produced, in
the course of the world's history, a great change in the
minds of acute and speculative men; so that such persons can now reason with perfect steadiness and precision
upon subjects on which, at first, their thoughts were vague
and confused; and can apprehend, as truths of complete
certainty and evidence, laws which it required great labour
and time to discover. This *complete* developement and
clear manifestation of mechanical ideas has taken place
only among mathematicians and philosophers. But yet a
progress of thought upon such subjects; an advance from
the obscure to the clear, and from error to truth; may be

traced in the world at large, and among those who have not directly cultivated the exact sciences. This diffused and collateral influence of science manifests itself, although in a wavering and fluctuating manner, by various indications, at various periods of literary history. The opinions and reasonings which are put forth upon mechanical subjects, and above all, the adoption into common language, of terms and phrases belonging to the prevalent mechanical systems, exhibit to us the most profound discoveries and speculations of philosophers in their effect upon more common and familiar trains of thought. This effect is by no means unimportant, and we shall point out some examples of such indications as we have mentioned.

2. The discoveries of the ancients in speculative mechanics were, as we have seen, very scanty; and hardly extended their influence to the unmathematical world. Yet the familiar use of the term "centre of gravity" preserved and suggested the most important part of what the Greeks had to teach. The other phrases which they employed, as momentum, energy, virtue, force, and the like, never had any exact meaning, even among mathematicians; and therefore never, in the ancient world, became the means of suggesting just habits of thought. I have pointed out, in the History of Science, several circumstances which appear to denote the general confusion of ideas which prevailed upon mechanical subjects during the times of the Roman empire. I have there taken as one of the examples of this confusion, the fable narrated by Pliny and others concerning the echineïs, a small fish, which was said to stop a ship merely by sticking to it*. This story was adduced as betraying the absence of any steady apprehension of the equality of action and reaction; since the fish, except it had some immoveable

* *Hist. Ind. Sci.*, i. 245.

obstacle to hold by, must be pulled forward by the ship, as much as it pulled the ship backward.   If the writers who speak of this wonder had shown any perception of the necessity of a reaction, either produced by the rapid motion of the fish's fins in the water, or in any other way, they would not be chargeable with this confusion of thought; but from their expressions it is, I think, evident that they saw no such necessity*.   Their idea of mechanical action was not sufficiently distinct to enable them to see the absurdity of supposing an intense pressure with no obstacle for it to exert itself against.

3. We may trace, in more modern times also, indications of a general ignorance of mechanical truths.   Thus the phrase of shooting at an object "point-blank," implies the belief that a cannon-ball describes a path of which the first portion is a straight line.   This error was corrected by the true mechanical principles which Galileo and his followers brought to light; but these principles made their way to popular notice, principally in consequence of their application to the motions of the solar system, and to the controversies which took place respecting those motions.   Thus by far the most powerful argument against the reception of the Copernican system of the universe, was that of those who asked, Why a stone dropt from a tower was not left behind by the motion of the earth?   The answer to this question, now universally

---

* See Prof. POWELL On the Nature and Evidence of the Laws of Motion. Reports of the Ashmolean Society. Oxford. 1837.   Professor Powell has made an objection to my use of this instance of confusion of thought; the remark in the text seems to me to justify what I said in the History.   As an evidence that the fish was not supposed to produce its effect by its muscular power acting on the water, we may take what Pliny says, Nat. Hist., xxxii. 1, "Domat mundi rabiem, nullo suo labore; non retinendo, aut alio modo quam adhærendo :" and also what he states in another place (ix. 41,) that when it is preserved in pickle, it may be used in recovering gold which has fallen into a deep well.   All this implies adhesion alone, with no conception of reaction.

familiar, involves a reference to the true doctrine of the composition of motions.   Again; Kepler's persevering and strenuous attempts* to frame a physical theory of the universe were frustrated by his ignorance of the first law of motion, which informs us that a body will retain its velocity without any maintaining force. He proceeded upon the supposition that the sun's force was requisite to *keep up* the motion of the planets, as well as to deflect and modify it; and he was thus led to a system which represented the sun as carrying round the planets in their orbits by means of a *vortex*, produced by his revolution. The same neglect of the laws of motion presided in the formation of Descartes' system of vortices.   Although Descartes had enunciated in words the laws of motion, he and his followers showed that they had not the practical habit of referring to these mechanical principles; and dared not trust the planets to move in free space without some surrounding machinery to support them†.

4. When at last mathematicians, following Newton, had ventured to consider the motion of each planet as a mechanical problem not different in its nature from the motion of a stone cast from the hand; and when the solution of this problem and its immense consequences had become matters of general notoriety and interest; the new views introduced, as is usual, new terms, which soon became extensively current. We meet with such phrases as "flying off in the tangent," and "deflexion from the tangent;" with antitheses between "centripetal" and "centrifugal force," or between "projectile" and "central force." "Centres of force," "disturbing forces,"

---

* *Hist. Ind. Sci.*, i. 408; ii. 129.

† I have, in the History, applied to Descartes the character which Bacon gives to Aristotle, "Audax simul et pavidus:" though he was bold enough to enunciate the laws of motion without knowing them aright, he had not the courage to leave the planets to describe their orbits by the agency of those laws, without the machinery of contact.

"perturbations," and "perturbations of higher orders," are not unfrequently spoken of: and the expression "to gravitate," and the term "universal gravitation," acquired a permanent place in the language.

Yet for a long time, and even up to the present day, we find many indications that false and confused apprehensions on such subjects are by no means extirpated. Arguments are urged against the mechanical system of the universe, implying in the opponents an absence of all clear mechanical notions. Many of this class of writers retrograde to Kepler's point of view. This is, for example, the case with Lord Monboddo, who, arguing on the assumption that force is requisite to maintain, as well as to deflect motion, produced a series of attacks upon the Newtonian philosophy; which he inserted in his *Ancient Metaphysics*, published in 1779 and the succeeding years. This writer (like Kepler), measures force by the velocity which the body *has* *, not by that which it *gains*. Such a use of language would prevent our obtaining any laws of motion at all. Accordingly, the author, in the very next page to that which I have just quoted, abandons this measure of force, and, in curvilinear motion, measures force by "the fall from the extremity of the arc." Again; in his objections to the received theory, he denies that curvilinear motion is compounded, although his own mode of considering such motion assumes this composition in the only way in which it was ever intended by mathematicians. Many more instances might be adduced to show that a want of cultivation of the mechanical ideas rendered this philosopher incapable of judging of a mechanical system.

The following extract from the *Ancient Metaphysics*, may be sufficient to show the value of the author's criticism on the subjects of which we are now speaking.

* *Anc. Met.*, vol. ii., b. v., c. 6., p. 413.

His object is to prove that there do not exist a centripetal and a centrifugal force in the case of elliptical motion. "Let any man move in a circular or elliptical line described to him; and he will find no tendency in himself either to the centre or from it, much less both. If indeed he attempt to make the motion with great velocity, or if he do it carelessly and inattentively, he may go out of the line, either towards the centre or from it: but this is to be ascribed, not to the nature of the motion, but to our infirmity; or perhaps to the animal form, which is more fitted for progressive motion in a right line than for any kind of curvilinear motion. But this is not the case with a sphere or spheroid, which is equally adapted to motion in all directions*." We need hardly remind the reader that the manner in which a man running round a small circle, finds it necessary to lean inwards, in order that there may be a centripetal inclination to counteract the centrifugal force, is a standard example of our mechanical doctrines; and this fact (quite familiar in practice as well as theory,) is in direct contradiction of Lord Monboddo's assertion.

5. A similar absence of distinct mechanical thought appears in some of the most celebrated metaphysicians of Germany. I have elsewhere noted† the opinion expressed by Hegel, that the glory which belongs to Kepler has been unjustly transferred to Newton; and I have suggested, as the explanation of this mode of thinking, that Hegel himself, in the knowledge of mechanical truth, had not advanced beyond Kepler's point of view. Persons who possess conceptions of space and number, but who have not learnt to deal with ideas of force and causation, may see more value in the discoveries of Kepler than in those of Newton. Another exemplification of this

* *Anc. Met.*, vol. i., b. ii., c. 19, p. 264.
† *Hist. Ind. Sc.*, ii., 181.

state of mind may be found in Mr. Schelling's speculations; for instance, in his *Lectures on the Method of Aca-
demical Study*. In the twelfth Lecture, on the Study of
Physics and Chemistry, he says, (p. 266,) " What the
mathematical natural philosophy has done for the knowledge of the laws of the universe since the time that they
were discovered by his (Kepler's) godlike genius, is, as
is well known, this: it has attempted a construction of
those laws which, according to its foundations, is altogether empirical.   We may assume it as a general rule, that
in any proposed construction, that which is not a pure
general form cannot have any scientific import or truth.
The foundation from which the centrifugal motion of the
bodies of the world is derived, is no necessary form, it is
an empirical fact.   The Newtonian attractive force, even
if it be a necessary assumption for a merely reflective
view of the subject, is still of no significance for the
Reason, which recognises only absolute relations.   The
grounds of the Keplerian laws can be derived, without
any empirical appendage, purely from the doctrine of
Ideas, and of the two Unities, which are in themselves
one Unity, and in virtue of which each being, while it is
absolute in itself, is at the same time in the absolute, and
reciprocally."

It will be observed, that in this passage our mechanical laws are objected to because they are not necessary
results of our ideas; which, however, as we have seen,
according to the opinion of some eminent mechanical
philosophers, they are.   But to assume this evident
necessity as a condition of every advance in science, is
to mistake the last, perhaps unattainable step, for the
first, which lies before our feet.   And, without inquiring
further about "the Doctrine of the two Unities," or the
manner in which from that doctrine we may deduce the
Keplerian laws, we may be well convinced that such a

doctrine cannot supply any sufficient reason to induce us to quit the inductive path by which all scientific truth up to the present time has been acquired.

6. But without going to schools of philosophy opposed to the Inductive School, we may find many loose and vague habits of thinking on mechanical subjects among the common classes of readers and reasoners. And there are some familiar modes of employing the phraseology of mechanical science, which are, in a certain degree, chargeable with inaccuracy, and may produce or perpetuate confusion. Among such cases we may mention the way in which the centripetal and centrifugal forces, and also the projectile and central forces of the planets, are often compared or opposed. Such antitheses sometimes proceed upon the false notion that the two members of these pairs of forces are of the same kind : whereas on the contrary the *projectile* force is a hypothetical impulsive force which may, at some former period, have caused the motion to begin ; while the *central* force is an actual force, which must act continuously and during the whole time of the motion, in order that the motion may go on in the curve. In the same manner the *centrifugal* force is not a distinct force in a strict sense, but only a certain result of the first law of motion, measured by the portion of *centripetal* force which counteracts it. Comparisons of quantities so heterogenous imply confusion of thought, and often suggest baseless speculations and imagined reforms of the received opinions.

7. I might point out other terms and maxims, in addition to those already mentioned, which, though formerly employed in a loose and vague manner, are now accurately understood and employed by all just thinkers; and thus secure and diffuse a right understanding of mechanical truths. Such are *momentum, inertia, quantity of matter, quantity of motion ;* that *force is proportional*

to its effects ; that *action and reaction are equal ;* that *what is gained in force by machinery is lost in time ;* that *the quantity of motion in the world cannot be either increased or diminished.* When the expression of the truth thus becomes easy and simple, clear and convincing, the meanings given to words and phrases by discoverers glide into the habitual texture of men's reasonings, and the effect of the establishment of true mechanical principles is felt far from the school of the mechanician. If these terms and maxims are understood with tolerable clearness, they carry the influence of truth to those who have no direct access to its sources. Many an extravagant project in practical machinery, and many a wild hypothesis in speculative physics, has been repressed by the general currency of such maxims as we have just quoted.

8. Indeed so familiar and evident are the elementary truths of mechanics when expressed in this simple form, that they are received as truisms ; and men are disposed to look back with surprise and scorn at the speculations which were carried on in neglect of them. The most superficial reasoner of modern times thinks himself entitled to speak with contempt and ridicule of Kepler's hypothesis concerning the physical causes of the celestial motions : and gives himself credit for intellectual superiority, because he sees, as self-evident, what such a man could not discover at all. It is well for such a person to recollect, that the real cause of his superior insight is not the pre-eminence of his faculties, but the successful labours of those who have preceded him. The language which he has learnt to use unconsciously, has been adapted to, and moulded on, ascertained truths. When he talks familiarly of accelerating forces, and deflexions from the tangent, he is assuming that which Kepler did not know, and which it cost Galileo and his disciples so much labour and thought to establish. Language is often called an

instrument of thought; but it is also the nutriment of thought; or rather, it is the atmosphere in which thought lives : a medium essential to the activity of our specu-lative power, although invisible and imperceptible in its operation; and an element modifying, by its qualities and changes, the growth and complexion of the faculties which it feeds.   In this way the influence of preceding discoveries upon subsequent ones, of the past upon the present, is most penetrating and universal, though most subtle and difficult to trace.   The most familiar words and phrases are connected by imperceptible ties with the reasonings and discoveries of former men and distant times.   Their knowledge is an inseparable part of ours; the present generation inherits and uses the scientific wealth of all the past.   And this is the fortune, not only of the great and rich in the intellectual world : of those who have the key to the ancient storehouses, and who have accumulated treasures of their own;—but the humblest inquirer, while he puts his reasonings into words, benefits by the labours of the greatest discoverers. When he counts his little wealth, he finds that he has in his hands coins which bear the image and superscription of ancient and modern intellectual dynasties; and that in virtue of this possession, acquisitions are in his power, solid knowledge within his reach, which none could ever have attained to, if it were not that the gold of truth, once dug out of the mine, circulates more and more widely among mankind.

9. Having so fully examined, in the preceding in-stances, the nature of the progress of thought which science implies, both among the peculiar cultivators of science, and in that wider world of general culture which receives only an indirect influence from scientific disco-veries, we shall not find it necessary to go into the same extent of detail with regard to the other provinces of

human knowledge. In the case of the Mechanical Sciences, we have endeavoured to show, not only that Ideas are requisite in order to form into a science the Facts which nature offers to us, but that we can advance, almost or quite, to a complete identification of the Facts with the Ideas. In the sciences to which we now proceed, we shall not seek to fill up the chasm by which Facts and Ideas are separated; but we shall endeavour to detect the Ideas which our knowledge involves, to show how essential these are; and in some respects to trace the mode in which they have been gradually developed among men.

10. The motions of the heavenly bodies, their laws, their causes, are among the subjects of the first division of the Mechanical Sciences; and of these sciences we formerly sketched the history, and have now endeavoured to exhibit the philosophy. If we were to take any other class of motions, *their* laws and causes might give rise to sciences which would be mechanical sciences in exactly the same sense in which Physical Astronomy is so. The phenomena of magnets, of electrical bodies, of galvanical apparatus, seem to form obvious materials for such sciences; and if they were so treated, the philosophy of such branches of knowledge would naturally come under our consideration at this point of our progress.

But on looking more attentively at the sciences of Electricity, Magnetism, and Galvanism, we discover cogent reasons for transferring them to another part of our arrangement; we find it advisable to associate them with Chemistry, and to discuss their principles when we can connect them with the principles of chemical science. For though the first steps and narrower generalizations of these sciences depend upon mechanical ideas, the highest laws and widest generalizations which we can reach respecting them, involve chemical relations. The progress of these portions of knowledge is in some respects opposite to the progress of Physical Astronomy. In this, we begin with phenomena which appear to indicate peculiar and various qualities in the bodies which we consider, (namely, the heavenly bodies,) and we find in the end that all these qualities resolve themselves into one common mechanical property, which exists alike in all bodies and parts of bodies. On the contrary, in studying magnetical and electrical laws, we appear at first to have a single extensive phenomenon, attraction and repulsion: but in our attempts to generalize this phenomenon, we find that it is governed by conditions depending upon something quite separate from the bodies themselves, upon the presence and distribution of peculiar and transitory agencies; and, so far as we can discover, the general laws óf these agencies are of a *chemical* nature, and are brought into action by peculiar properties of special substances. In cosmical phenomena, everything, in proportion as it is referred to mechanical principles, tends to simplicity,—to permanent uniform forces,—to one common, positive, property. In magnetical and electrical appearances, on the contrary, the application of mechanical principles leads only to a new complexity, which requires a new explanation; and this explanation involves changeable and various forces,—gradations and oppositions of qualities. The doctrine of the universal gravitation of matter is a simple and ultimate truth, in which the mind can acquiesce and repose. We rank gravity among the mechanical attributes of matter, and we see no necessity to derive it from any ulterior properties. Gravity belongs to matter, independent of any conditions. But the *conditions* of magnetic or electrical activity require investigation as much as the *laws* of their action. Of these conditions no mere mechanical explanation can be given; we are compelled to take along with us chemical properties and relations also : and thus magnetism, electricity, galvanism, are *mechanico-chemical sciences.*

12. Before considering these, therefore, I shall treat of what I shall call *Secondary Mechanical Sciences ;* by which expression I mean the sciences depending upon certain qualities which our senses discover to us in bodies ; *Optics,* which has visible phenomena for its subject ; *Acoustics,* the science of hearing ; the doctrine of *Heat,* a quality which our touch recognises ; to this last science I shall take the liberty of sometimes giving the name *Thermotics,* analogous to the names of the other two. If our knowledge of the phenomena of Smell and Taste had been successfully cultivated and systematized, the present part of our work would be the place for the philosophical discussion of those sensations as the subjects of science.

The branches of knowledge thus grouped in one class involve common Fundamental Ideas, from which their principles are derived in a mode analogous, at least in a certain degree, to the mode in which the principles of the mechanical sciences are derived from the fundamental ideas of causation and reaction. We proceed now to consider these Fundamental Ideas, their nature, development, and consequences.

BOOK IV.

THE PHILOSOPHY OF THE SECONDARY MECHANICAL SCIENCES.

CHAPTER I.

OF THE IDEA OF A MEDIUM AS COMMONLY EMPLOYED.

1. *Of Primary and Secondary Qualities.*—In the same way in which the mechanical sciences depend upon the Idea of Cause, and have their principles regulated by the development of that Idea, it will be found that the sciences which have for their subject Sound, Light, and Heat, depend for *their* principles upon the Fundamental Idea of Media by means of which we perceive those qualities. Like the idea of cause, this idea of a medium is unavoidably employed, more or less distinctly, in the common, unscientific operations of the understanding; and is recognised as an express principle in the earliest speculative essays of man. But here also, as in the case of the mechanical sciences, the developement of the idea, and the establishment of the scientific truths which depend upon it, was the business of a succeeding period, and was only executed by means of long and laborious researches, conducted with a constant reference to experiment and observation.

Among the most prominent manifestations of the influence of the idea of a medium of which we have now to speak, is the distinction of the qualities into *primary,* and *secondary* qualities. This distinction has been constantly spoken of in modern times : yet it has often been a subject of discussion among metaphysicians whether there be really such a distinction, and what the true difference is. Locke states it thus *: original or primary qualities of body are "such as are utterly inseparable from the body in what estate soever it may be,—such as sense constantly finds in every particle of matter which has bulk enough to be perceived, and the mind finds inseparable from every particle of matter, though less than to make itself singly perceived by our senses:" and he enumerates them as solidity, extension, figure, motion or rest, and number. Secondary qualities, on the other hand, are such "which in truth are nothing in the objects themselves, but powers to produce various sensations in us by their primary qualities, *i. e.*, by the bulk, figure, texture, and motion of their insensible parts, as colours, sounds, tastes, &c."

Dr. Reid†, reconsidering this subject, puts the difference in another way. There is, he says, a real foundation for the distinction of primary and secondary qualities, and it is this : "That our senses give us a direct and distinct notion of the primary qualities, and inform us what they are in themselves ; but of the secondary qualities, our senses give us only a relative and obscure notion. They inform us only that they are qualities that affect us in a certain manner, that is, produce in us a certain sensation ; but as to what they are in themselves, our senses leave us in the dark."

Dr. Brown‡ states the distinction somewhat otherwise. We give the name of matter, he observes, to that which has extension and resistance : these, therefore, are primary qualities of matter, because they compose our

---

* *Essay*, b. ii., ch. 8., s. 9, 10.   † *Essays*, b. ii., c. 17.
‡ *Lectures*, ii., 12.

definition of it.   All other qualities are secondary, since
they are ascribed to bodies only because we find them
associated with the primary qualities which form our
notion of those bodies.

It is not necessary to criticise very strictly these various distinctions.   If it were, it would be easy to cavil at
them.   Thus Locke, it may be observed, does not point
out any *reason* for believing that his secondary qualities
are produced by the primary.   How are we to learn that
the colour of a rose arises from the bulk, figure, texture,
and motion of its particles?   Certainly our senses do not
teach us this; and in what other way, on Locke's principles, can we learn it?   Reid's statement is not more
free from the same objection.   How does it appear that
our notion of warmth is relative to our own sensations
more than our notion of solidity?   And if we take
Brown's account, we may still ask whether our selection
of certain qualities to form our idea and definition of
matter be arbitrary and without reason?   If it be, how
can it make a real distinction; if it be not, what is the
reason?

I do not press these objections, because I believe that
any of the above accounts of the distinction of primary
and secondary qualities is right in the main, however imperfect it may be.   The difference between such qualities
as extension and solidity on the one hand, and colour or
fragrance on the other, is assented to by all, with a conviction so firm and indestructible, that there must be
some fundamental principle at the bottom of the belief,
however difficult it may be to clothe the principle in
words.   That successive efforts to express the real nature
of the difference were made by men so clear-sighted and
acute as those whom I have quoted, even if none of them
are satisfactory, shows how strong and how deeply-seated
is the perception of truth which impels us to such
attempts.

The most obvious mode of stating the difference of primary and secondary qualities, as it naturally offers itself to speculative minds, appears to be that employed by Locke, slightly modified. Certain of the qualities of bodies, as their bulk, figure, and motion, are perceived immediately in the bodies themselves. Certain other qualities as sound, colour, heat, are perceived by means of some medium. Our conviction that this is the case is spontaneous and irresistible; and this difference of qualities immediately and mediately perceived is the distinction of primary and secondary qualities. We proceed further to examine this conviction.

2. *The Idea of Externality.*—In reasoning concerning the secondary qualities of bodies, we are led to assume the bodies to be external to us, and to be perceived by means of some medium intermediate between us and them. These assumptions are fundamental conditions of perception, inseparable from it even in thought.

That objects are *external* to us, that they are *without* us, that they have *outness*, is as clear as it is that these words have any meaning at all. This conviction is, indeed, involved in the exercise of that faculty by which we perceive all things as existing in space; for by this faculty we place ourselves and other objects in one common space, and thus they are exterior to us. It may be remarked that this apprehension of objects as external to us, although it assumes the idea of space, is far from being implied in the idea of space. The objects which we contemplate are considered as existing in space, and by that means become invested with certain mutual relations of position; but when we consider them as existing without us, we make the additional step of supposing *ourselves* and the objects to exist in one common space. The question respecting the Ideal Theory of Berkeley has been mixed up with the recognition of this condition of the externality of objects. That philosopher maintained, as is well known, that the perceptible qualities of bodies have no existence except in a perceiving. mind. This system has often been understood as if he had imagined the world to be a kind of optical illusion, like the images which we see when we shut our eyes, appearing to be without us, though they are only in our organs ; and thus this Ideal System has been opposed to a belief in an external world. In truth, however, no such opposition exists. The Ideal System is an attempt to explain the mental process of perception, and to get over the difficulty of mind being affected by matter. But the author of that system did not deny that objects were perceived under the conditions of space and mechanical causation ; that they were external and material so far as those words describe perceptible qualities. Berkeley's system, however visionary or erroneous, did not prevent his entertaining views as just, concerning optics or acoustics, as if he had held any other doctrine of the nature of perception.

But when Berkeley's theory was understood as a denial of the existence of objects without us, how was it answered ? If we examine the answers which are given by Reid and other philosophers to this hypothesis, it will be found that they amount to this : that objects *are* without us, since we *perceive* that they are so ; that we perceive them to be external, by the same act by which we perceive them to be objects. And thus, in this stage of philosophical inquiry, the externality of objects is recognised as one of the inevitable conditions of our perception of them ; and hence the idea of externality is adopted as one of the necessary foundations of all reasoning concerning all objects whatever.

3. *Sensation by a Medium.*—Objects, as we have just seen, are necessarily apprehended as without us ; and in general, as removed from us by a great or small distance.

Yet they affect our bodily senses; and this leads us irresistibly to the conviction that they are perceived by means of something intermediate. Vision, or hearing, or smell, or the warmth of a fire, must be communicated to us by some medium of sensation. This unavoidable belief appears in all attempts, the earliest and the latest alike, to speculate upon such subjects. Thus, for instance, Aristotle says*, "Seeing takes place in virtue of some action which the sentient organ suffers: now it cannot suffer action from the colour of the object directly: the only remaining possible case then is, that it is acted upon by an intervening Medium; there must then be an intervening Medium." "And the same may be said," he adds, "concerning sounding and odorous bodies; for these do not produce sensation by touching the sentient organ, but the intervening Medium is acted on by the sound or the smell, and the proper organ, by the Medium....In sound the Medium is air; in smell we have no name for it." In the sense of taste, the necessity of a Medium is not at first so obviously seen, because the object tasted is brought into contact with the organ; but a little attention convinces us that the taste of a solid body can only be perceived when it is conveyed in some liquid vehicle. Till the fruit is crushed, and till its juices are pressed out, we do not distinguish its flavour. In the case of heat, it is still more clear that we are compelled to suppose some invisible fluid, or other means of communication, between the distant body which warms us and ourselves.

It may appear to some persons that the assumption of an intermedium between the object perceived and the sentient organ results from the principles which form the basis of our mechanical reasonings,—that every change must have a cause, and that bodies can act upon each other only by contact. It cannot be denied that this

* Περὶ Ψυχῆς. II. 7.

principle does offer itself very naturally as the ground of our belief in media of sensation; and it appears to be referred to for this purpose by Aristotle in the passage quoted above. But yet we cannot but ask, Does the principle, that matter produces its effect by contact only, manifestly apply here? When we so apply it, we include *sensation* among the *effects* which material contact produces;—a case so different from any merely mechanical effect, that the principle, so employed, appears to acquire a new signification. May we not, then, rather say that we have here a new axiom, That sensation implies a material cause immediately acting on the organ; than a new application of our former proposition, That all mechanical change implies contact?

The solution of this doubt is not of any material consequence to our reasonings; for whatever be the ground of the assumption, it is certain that we do assume the existence of media by which the sensations of sight, hearing, and the like, are produced; and it will be seen shortly that principles inseparably connected with this assumption are the basis of the sciences now before us.

This assumption makes its appearance in the physical doctrines of all the schools of philosophy. It is exhibited perhaps most prominently in the tenets of the Epicureans, who were materialists, and extended to all kinds of causation the axiom of the existence of a corporeal mechanism by which alone the effect is produced. Thus, according to them, vision is produced by certain images or material films which flow from the object, strike upon the eyes, and so become sensible. This opinion is urged with great detail and earnestness by Lucretius, the poetical expositor of the Epicurean creed among the Romans. His fundamental conviction of the necessity of a material medium is obviously the basis of his reasoning, though he attempts to show the existence of such a medium by facts.

Thus he argues*, that by shouting loud we make the throat sore; which shows, he says, that the voice must be material, so that it can hurt the passage in coming out.

Haud igitur dubium est quin voces verbaque constent
Corporeis e principiis ut lædere possint.

4. *The Process of Perception of Secondary Qualities.*
—The likenesses or representatives of objects by which they affect our senses were called by some writers *species,* or *sensible species,* a term which continued in use till the revival of science. It may be observed that the conception of these species as films cast off from the object, and retaining its shape, was different, as we have seen, from the view which Aristotle took, though it has sometimes been called the Peripatetic doctrine †. We may add that the expression was latterly applied to express the supposition of an emanation of any kind, and implied little more than that supposition of a medium of which we are now speaking. Thus Bacon, after reviewing the phenomena of sound, says‡, "Videntur motus soni fieri per *species spirituales :* ita enim loquendum donec certius quippiam inveniatur."

Though the fundamental principles of several sciences depend upon the assumption of a medium of perception, these principles do not at all depend upon any special view of the process of our perceptions. The mechanism of that process is a curious subject of consideration; but it belongs to physiology, more properly than either to metaphysics, or to those branches of physics of which we are now speaking. The general nature of the process is the same for all the senses. The object affects the appropriate intermedium; the medium, through the proper organ, the eye, the ear, the nose, affects the nerves of the par-

* Lib. iv. 529. † BROWN, vol. ii., p. 98.
‡ *Hist. Son. et Aud.*, vol. ix., p. 87.

ticular sense; and, by these, in some way, the sensation is conveyed to the mind. But to treat the *impression* upon the nerves as the *act* of sensation which we have to consider, would be to mistake our object, which is not the constitution of the human body, but of the human mind. It would be to mistake one link for the power which holds the end of the chain. No anatomical analysis of the corporeal conditions of vision, or hearing, or feeling warm, is necessary to the sciences of Optics, or Acoustics, or Thermotics.

Not only is this physiological research an extraneous part of our subject, but a partial pursuit of such a research may mislead the inquirer. We perceive objects *by means of* certain media, and *by means of* certain impressions on the nerves: but we cannot with propriety say that we perceive either the media or the impressions on the nerves. What person in the act of seeing is conscious of the little coloured spaces on the retina? or of the motions of the bones of the auditory apparatus whilst he is hearing? Surely, no one. This may appear obvious enough, and yet a writer of no common acuteness, Dr. Brown, has put forth several very strange opinions, all resting upon the doctrine that the coloured spaces on the retina are the *objects* which we perceive; and there are some supposed difficulties and paradoxes on the same subject which have become quite celebrated (as upright vision with inverted images), arising from the same confusion of thought.

As the consideration of the difficulties which have arisen respecting the philosophy of perception may serve still further to illustrate the principles on which we necessarily reason respecting the secondary qualities of bodies, I shall here devote a few pages to that subject.

Chapter II.

ON PECULIARITIES IN THE PERCEPTIONS OF THE DIFFERENT SENSES.

1. We cannot doubt that we perceive all secondary qualities by means of immediate impressions made, through the proper medium of sensation, upon our organs. Hence all the senses are sometimes vaguely spoken of as modifications of the sense of feeling. It will, however, be seen, on reflection, that this mode of speaking identifies in words things which in our conceptions have nothing in common. No impression on the organs of touch can be conceived as having any resemblance to colour or smell. No effort, no ingenuity, can enable us to describe the impressions of one sense in terms borrowed from another.

The senses have, however, each its peculiar powers, and these powers may be in some respects compared, so as to show their leading resemblances and differences, and the characteristic privileges and laws of each. This is what we shall do as briefly as possible.

(I.) *Prerogatives of Sight.*—The sight distinguishes colours, as the hearing distinguishes tones; the sight estimates degrees of brightness, the ear, degrees of loudness; but with several resemblances, there are most remarkable differences between these two senses.

2. *Position.*—The sight has this peculiar prerogative, that it apprehends the *place* of its objects directly and primarily. We see *where* an object is at the same instant that we see what it is. If we see two objects, we see their relative position. We cannot help perceiving that one is above or below, to the right or to the left of the other, if we perceive them at all.

There is nothing corresponding to this in sound. When we hear a noise, we do not necessarily assign a place to it. It may easily happen that we cannot tell from which side a thunder-clap comes. And though we often can judge in what direction a voice is heard, this is a matter of secondary impression, and of inference from con-comitant circumstances, not a primary fact of sensation. The judgments which we form concerning the position of sounding bodies are obtained by the conscious or uncon-scious comparison of the impressions made on the two ears, and on the bones of the head in general; they are not inseparable conditions of hearing. We may hear sounds, and be uncertain whether they are " above, around, or underneath;" but the moment any thing visible appears, however unexpected, we can say " see *where* it comes!"

Since we can see the relative position of things, we can see *figure*, which is but the relative position of the different parts of the boundary of the object. And thus the whole visible world exhibits to us a scene of various shapes, coloured and shaded according to their form and position, but each having relations of position to all the rest; and altogether, entirely filling up the whole range which the eye can command.

3. *Distance.*—The distance of objects from us is no matter of immediate perception, but is a judgment and inference formed from our sensations, in the same way as our judgment of position by the ear. That this is so, was most distinctly shown by Berkeley, in his *New Theory of Vision.* The elements on which we form our judgment are, the effort by which we fix both eyes on the same object, the effort by which we adjust each eye to distinct vision, and the known forms, colours, and parts of objects, as compared with their appearance. The right interpre-tation of the information which these circumstances give us respecting the true distances and forms of things, is

gradually learned by experience, the lesson being begun
in our earliest infancy, and inculcated upon us every hour
during which we use our eyes. The completeness with
which the lesson is learned is truly admirable; for we forget that our conclusion is obtained indirectly, and mistake
a judgment on evidence for an intuitive perception. This,
however, is not more surprising than the rapidity and
unconsciousness of effort with which we understand the
meaning of the speech that we hear, or the book that we
read. In both cases, the habit of interpretation is become
as familiar as the act of perception. And this is the case
with regard to vision. We see the breadth of the street
as clearly and readily as we see the house on the other
side of it. We see the house to be square, however
obliquely it be presented to us. Indeed the difficulty is,
to recover the consciousness of our real and original
sensations;—to discover what is the apparent relation of
the lines which appear before us. As we have already
said, in the common process of vision we suppose ourselves to see that which cannot be seen; and when we
would make a picture of an object, the difficulty is to
represent what is visible and no more.

But perfect as is our habit of interpreting what we
perceive, we could not interpret if we did not perceive.
If the eye did not apprehend visible position, it could not
infer actual position, which is collected as a consequence:
if we did not see apparent figure, we could not form any
opinion concerning real form. The perception of place,
which is the prerogative of the eye, is the basis of all its
other superiority.

The precision with which the eye can judge of apparent
position is remarkable. If we had before us two stars distant from each other by one-twentieth of the moon's diameter, we could easily decide the apparent direction of the
one from the other, as above or below, to the right or left.

Yet eight millions of stars might be placed in the visible hemisphere of the sky at such distances from each other; and thus the eye would recognise the relative position in a portion of its range not greater than one eight-millionth of the whole. Such is the accuracy of the sense of vision in this respect; and, indeed, we might with truth have stated it much higher. Our judgment of the position of distant objects in a landscape depends upon features far more minute than the magnitude we have here stated.

As our object is to point out principally the differences of the senses, we do not dwell upon the delicacy with which we distinguish tints and shades, but proceed to another sense.

(II.) *Prerogatives of Hearing.*—The sense of hearing has two remarkable prerogatives; it can perceive a definite and peculiar relation between certain tones, and it can clearly perceive two tones together; in both these circumstances it is distinguished from vision, and from the other senses.

4. *Musical Intervals.*—We perceive that two tones have, or have not, certain definite relations to each other, which we call *Concords:* one sound is a *Fifth*, an *Octave*, &c., above the other. And when this is the case, our perception of the relation is extremely precise. It is easy to perceive when a fifth is out of tune by one-twentieth of a tone; that is, by one-seventieth of itself. To this there is nothing analogous in vision. Colours have certain vague relations to one another; they look well together, by contrast or by resemblance; but this is an indefinite, and in most cases a casual and variable feeling The relation of *complementary* colours to one another, as of red to green, is somewhat more definite; but still has nothing of the exactness and peculiarity which belongs to a musical concord. In the case of the two sounds,

there is an exact point at which the relation obtains; when by altering one note we pass this point, the concord does not gradually fade away, but instantly becomes a discord; and if we go further still, we obtain another concord of quite a different character.

We learn from the theory of sound that concords occur when the times of vibration of the notes have exact simple ratios; an octave has these times as 1 to 2; a fifth, as 2 to 3. According to the undulatory theory of light, such ratios occur in colours, yet the eye is not affected by them in any peculiar way. The times of the undulations of certain red and violet rays are as 2 to 3, but we do not perceive any peculiar harmony or connexion between those colours.

5. *Chords.*—Again, the ear has this prerogative, that it can apprehend two notes together, yet distinct. If two notes, distant by a fifth from each other, are sounded on two wind instruments, both they and their musical relation are clearly perceived. There is not a mixture, but a concord, an interval. In colours, the case is otherwise. If blue and yellow fall on the same spot, they form green; the colour is simple to the eye; it can no more be decomposed by the vision than if it were the simple green of the prismatic spectrum: it is impossible for us, by sight, to tell whether it is so or not.

These are very remarkable differences of the two senses: two colours can be compounded into an apparently simple one; two sounds cannot: colours pass into each other by gradations and intermediate tints; sounds pass from one concord to another by no gradations: the most intolerable discord is that which is near a concord. We shall hereafter see how these differences affect the scales of sound and of colour.

6. *Rhythm.*—We might remark, that as we see objects in space, we hear sounds in time; and that we thus intro-

duce an arrangement among sounds which has several analogies with the arrangement of objects in space. But the conception of time does not seem to be peculiarly connected with the sense of hearing; a faculty of apprehending tone and time, or in musical phraseology *tune* and *rhythm*, are certainly very distinct. I shall not, therefore, here dwell upon such analogies.

The other Senses have not any peculiar prerogatives, at least none which bear on the formation of science. I may, however, notice, in the feeling of heat, this circumstance; that it presents us with two opposites, heat and cold, which graduate into each other. This is not quite peculiar, for vision also exhibits to us white and black, which are clearly opposites, and which pass into each other by the shades of gray.

7. *First Paradox of Vision. Upright Vision.*—All our senses appear to have this in common;—That they act by means of organs, in which a bundle of nerves receives the impression of the appropriate medium of the sense. In the construction of these organs there are great differences and peculiarities, corresponding, in part at least, to the differences in the information given. Moreover, in some cases, as we have noted in the case of audible position and visible distance, that which seems to be a perception is really a judgment founded on perceptions of which we are not directly aware. It will be seen, therefore, that with respect to the peculiar powers of each sense, it may be asked;—whether they can be explained by the construction of the peculiar organ;—whether they are acquired judgments and not direct perceptions;—or whether they are inexplicable in either of these ways, and cannot, at present at least, be resolved into anything but conditions of the intellectual act of perception.

Two of these questions with regard to vision, have

been much discussed by psychological writers: the cause of our seeing objects upright by inverted images on the retina; and of our seeing single with two such images.

Physiologists have very completely explained the exquisitely beautiful mechanism of the eye, considered as analogous to an optical instrument; and it is indisputable that by means of certain transparent lenses and humours, an inverted image of the objects which are looked at is formed upon the *retina*, or fine net-work of nerve, with which the back of the eye is lined. We cannot doubt that the impression thus produced on these nerves is essential to the act of vision; and so far as we consider the nerves themselves to feel or perceive by contact, we may say that they perceive this image, or the affections of light which it indicates. But we cannot with any propriety say that *we* perceive, or that our mind perceives, this image; for we are not conscious of it, and none but anatomists are aware of its existence: we perceive *by means of* it.

A difficulty has been raised, and dwelt upon in a most unaccountable manner, arising from the neglect of this obvious distinction. It has been asked, how is it that we see an object, a man for instance, upright, when the immediate object of our sensation, the image of the man on our retina, is inverted? To this we must answer, that we see him upright *because* the image is inverted; that the inverted image is the necessary means of seeing an upright object. This is granted, and where then is the difficulty? Perhaps it may be put thus: How is it that we do not judge the man to be inverted, since the sensible image is so? To this we may reply, that we have no notion of *upright* or inverted, except that which is founded on experience, and that all our experience, without exception, must have taught us that such a sensible image belongs to a man who is in an upright position. Indeed, the contrary judgment is not conceivable; a man is upright whose head is upwards and his feet downwards. But what are the sensible images of *upwards* and *downwards?* Whatever be our standard of up and down, the sensible representation of *up* will be an image moving on the retina towards the lower side, and the sensible representation of *down* will be a motion towards the upper side. The head of the man's image is towards the image of the sky, its feet are towards the image of the ground; how then should it appear otherwise than upright? But, perhaps, we expect that the whole world should appear inverted; but if the whole be inverted, how is the relation of the parts altered? or we expect that we should think our own persons inverted: yet this cannot be, for we look at them as we do at other objects: Or, perhaps we expect that things should appear to fall upwards; yet what do we know of upwards, except that it is the direction in which bodies do *not* fall? In short, the whole of this difficulty, though it has in no small degree embarrassed metaphysicians, appears to result from a very palpable confusion of ideas; from an attempt at comparison of what *we* see, with that which the retina feels, as if they were separately presentable. It is a sufficient explanation to say, that we do not see the image on the retina, but see by *means* of it. The perplexity does not require much more skill to disentangle, than it does to see that a word written in *black* ink, may signify *white*.

8. *Second Paradox of Vision. Single Vision.*—(1.) *Small or Distant Objects.*—The other difficulty, why with two images on the retina we see only one object, is of a much more real and important kind. This effect is manifestly limited by certain circumstances of a very precise nature; for if we direct our eyes at an object which is very near the eye, we see all other objects

double. The fact is not, therefore, that we are incapable of receiving two impressions from the two images, but that, under certain conditions, the two impressions form one. A little attention shows us that these conditions are, that with both eyes we should look at the same object; and again, we find that to look at an object with either eye, is to direct the eye so that the image falls on or near a particular point about the middle of the retina. Thus these middle points in the two retinas correspond, and we see an image single when the two images fall on the corresponding points.

Again, as each eye judges of position, and as the two eyes judge similarly, an object will be seen in the same place by one eye and by the other, when the two images which it produces are *similarly situated* with regard to the *corresponding points* of the retina.

This is the Law of Single Vision, at least so far as regards small objects; namely, objects so small that in contemplating them we consider their position only, and not their solid dimensions. The law is a distinct and original principle of our constitution; and it is a mistake to call in, as some have done, the influence of habit and of acquired judgments, in order to determine the result in such cases.

To ascribe the apparent singleness of objects to the impressions of vision corrected by the experience of touch*, would be to assert that a person who had not been in the habit of handling what he saw, would see all objects double; and also, to assert that a person beginning with the double world which vision thus offers to him, would, by the continued habit of handling objects, gradually and at last learn to see them single. But all the facts of the case show such suppositions to be utterly fantastical. No one can, in this case, go back from the habitual judgment of the singleness of objects,

* See BROWN, vol. ii. p. 81.

to the original and direct perception of their doubleness, as the draughtsman goes back from judgments to perception, in representing solid distances and forms by means of perspective pictures. No one can point out any case in which the habit is imperfectly formed; even children of the most tender age look at an object with both eyes, and see it as one.

In cases when the eyes are distorted (in squinting), one eye only is used, or if both are employed, there is double vision; and thus any derangement of the correspondence of motion in the two eyes will produce double-sightedness.

Brown is one of those* who assert that two images suggest a single object because we have *always found* two images to belong to a single object. He urges as an illustration, that the *two* words " he conquered," by custom excite exactly the same notion as the *one* Latin word " vicit;" and thus that two visual images, by the effect of habit, produce the same belief of a single object as one tactual impression. But in order to make this pretended illustration of any value, it ought to be true that when a person has thoroughly learnt the Latin language, he can no longer distinguish any separate meaning in " he" and in "conquered." We can by no effort perceive the double sensation, when we look *at* the object with the two eyes. Those who squint, learn by habit to see objects single : but the habit which they acquire is that of attending to the impressions of one eye only at once, not of combining the two impressions. It is obvious, that if each eye spreads before us the same visible scene, with the same objects and the same relations of place, then, if one object in each scene coincide, the whole of the two visible impressions will be coincident. And here the remarkable circumstance is,

* *Lectures*, vol. ii. p. 81.

that not only each eye judges for itself of the relations of position which come within its field of view; but that there is a superior and more comprehensive faculty which combines and compares the two fields of view ; which asserts or denies their coincidence; which contemplates, as in a relative position to one another, these two visible worlds, in which all other relative position is given. This power of confronting two sets of visible images and figured spaces before a purely intellectual tribunal, is one of the most remarkable circumstances in the sense of vision.

9. (2.) *Near Objects.*—We have hitherto spoken of the singleness of objects whose images occupy corresponding positions on the retina of the two eyes. But here occurs a difficulty. If an object of moderate size, a small thick book for example, be held at a little distance from the eyes, it produces an image on the retina of each eye; and these two images are perspective representations of the book from different points of view, (the positions of the two eyes,) and are therefore of different forms. Hence the two images cannot occupy corresponding points of the retina throughout their whole extent. If the central parts of the two images occupy corresponding points, the boundaries of the two will not correspond. How is it then consistent with the law above stated, that in this case the object appears single?

It may be observed, that the two images in such a case will differ most widely when the object is not a mere surface, but a solid. If a book, for example, be held with one of its edges towards the face, the right eye will see one side more directly than the left eye, and the left eye will see another side more directly, and the outline of the two images upon the two retinas will exhibit this difference. And it may be further observed, that this difference in the images received by the two eyes, is a plain and demonstrative evidence of the solidity of the object seen; since nothing but a solid object could (without some special contrivance) produce these different forms of the images in the two eyes.

Hence the absence of exact coincidence in the two images on the retina is the necessary condition of the solidity of the object seen, and must be one of the indications by means of which our vision apprehends an object as solid. And that this is so, Mr. Wheatstone has proved experimentally, by means of some most ingenious and striking contrivances. He has devised* an instrument by which two images (drawn in outline) differing exactly as much as the two images of a solid body seen near the face would differ, are conveyed, one to one eye, and the other to the other. And it is found that when this is effected, the object which the images represent is not only seen single, but is apprehended as solid with a clearness and reality of conviction quite distinct from any impression which a mere perspective representation can give.

At the same time it is found that the object is then only apprehended as single when the two images are such as are capable of being excited by one single object placed in solid space, and seen by the two eyes. If the images differ more or otherwise than this condition allows, the result is, that both are seen, their lines crossing and interfering with one another.

It may be observed, too, that if an object be of such large size as not to be taken in by a single glance of the eyes, it is no longer apprehended as single by a direct act of perception; but its parts are looked at separately and successively, and the impressions thus obtained are put together by a succeeding act of the mind. Hence the objects which are directly seen as solid, will be of mode-

* *Phil. Trans.*, 1839.

rate size; in which case it is not difficult to show that the outlines of the two images will differ from each other only slightly.

Hence we are led to the following, as the Law of Single Vision for *near* objects :—When the two images in the two eyes are situated (part for part) nearly, but not exactly, upon corresponding points, the object is apprehended as single, if the two images are such as are or would be given by a single solid object seen by the two eyes separately : and in this case the object is necessarily apprehended as solid.

This law of vision does not contradict that stated above for distant objects: for when an object is removed to a considerable distance, the images in the two eyes coincide exactly, and the object is seen as single, though without any direct apprehension of its solidity. The first law is a special case of the second. Under the condition of *exactly* corresponding points, we have the perception of singleness, but no evidence of solidity. Under the condition of *nearly* corresponding points, we may have the perception of singleness, and with it, of solidity.

We have before noted it as an important feature in our visual perception, that while we have two distinct impressions upon the sense, which we can contemplate separately and alternately, (the impressions on the two eyes,) we have a higher perceptive faculty which can recognise these two impressions, exactly similar to each other, as only two images of one and the same assemblage of objects. But we now see that the faculty by which we perceive visible objects can do much more than this: —it can not only unite two impressions, and recognise them as belonging to one object in virtue of their coincidence, but it can also unite and identify them, even when they do not exactly coincide. It can correct and adjust their small difference, so that they are both apprehended as representations of the same figure. It can infer from them a real form, not agreeing with either of them; and a solid space, which they are quite incapable of exemplifying. The visual faculty decides whether or not the two ocular images can be pictures of the same solid object, and if they can, it undoubtingly and necessarily accepts them as being so. This faculty operates as if it had the power of calling before it all possible solid figures, and of ascertaining by trial whether any of those will, at the same time, fit both the outlines which are given by the sense. It assumes the reality of solid space, and, if it be possible, reconciles the appearances with that reality. And thus an activity of the mind of a very remarkable and peculiar kind is exercised in the most common act of seeing.

10. It may be said that this doctrine, of such a visual faculty as has been described, is very vague and obscure, since we are not told what are its limits. It adjusts and corrects figures which *nearly* coincide, so as to identify them. But *how* nearly, it may be asked, must the figures approach each other, in order that this adjustment may be possible? What discrepance renders impossible the reconcilement of which we speak? Is it not impossible to give a definite answer to these questions, and therefore impossible to lay down definitely such laws of vision as we have stated? To this I reply, that the indefiniteness thus objected to us, is no new difficulty, but one with which philosophers are familiar, and to which they are already reconciled. It is, in fact, no other than the indefiniteness of the limits of distinct vision. How near to the face must an object be brought, so that we shall cease to see it distinctly? The distance, it will be answered, is indefinite: it is different for different persons; and for the same person, it varies with the degree of effort, attention, and habit. But this indefiniteness is

only the indefiniteness, in another form, of the deviation
of the two ocular images from one another: and in reply
to the question concerning them we must still say, as
before, that in doubtful cases, the power of apprehending
an object as single, when this *can* be done, will vary with
effort, attention, and habit.   The assumption that the
apparent object exists as a real figure, in real space, is to
be verified, if possible ; but, in extreme cases, from the
unfitness of the point of view, or from any other cause of
visual confusion or deception, the existence of a real
object corresponding to the appearance may be doubtful ;
as in any other kind of perception it may be doubtful
whether our senses, under disadvantageous circumstances,
give us true information.   The vagueness of the limits,
then, within which this visual faculty can be successfully
exercised, is no valid argument against the existence of
the faculty, or the truth of the law which we have stated
concerning its action.

11. *Visible Figure.* — There is one tenet on the
subject of vision which appears to me so extravagant
and unphilosophical, that I should not have thought it
necessary to notice it, if it had not been recently promulgated by a writer of great acuteness in a book which
has obtained, for a metaphysical work, considerable circulation.   I speak of Brown's opinion* that we have no
immediate perception of visible figure.   I confess myself
unable to comprehend fully the doctrine which he would
substitute in the place of the one commonly received.   He
states it thus†: " When the simple affection of sight is
blended with the ideas of suggestion [those arising from
touch, &c.] in what are termed the acquired perceptions
of vision, as, for example, in the perception of a sphere,
it is colour only which is blended with the large convexity, and not a small coloured plane." The doctrine

* *Lectures*, vol. ii., p. 82.   † *Ib.*, vol. ii., p. 90.

which Brown asserts in this and similar passages, appears
to be, that we do not by vision perceive *both* colour and
figure; but that the colour which we see is blended with
the figure which we learn the existence of by other
means, as by touch. But if this were possible when we
can call in other perceptions, how is it possible when we
cannot or do not touch the object? Why does the
moon appear round, gibbous, or horned? What sense
besides vision suggests to us the idea of *her* figure? And
even in objects which we can reach, what is that circumstance in the sense of vision which suggests to us that
the colour belongs to the sphere, except that we see the
colour *where* we see the sphere? If we do not see figure,
we do not see position; for figure is the relative position
of the parts of a boundary. If we do not see position,
why do we ascribe the yellow colour to the sphere on our
left, rather than to the cube on our right? We *associate*
the colour with the object, says Dr. Brown; but if his
opinion were true, we could not associate two colours
with two objects, for we could not apprehend the colours
as occupying two different places.

The whole of Brown's reasoning on this subject is so
irreconcileable with the first facts of vision, that it is
difficult to conceive how it could proceed from a person
who has reasoned with great acuteness concerning touch.
In order to prove his assertion, he undertakes to examine
the only reasons which, he says*, he can imagine for
believing the immediate perception of visible figure: (1)
That it is absolutely impossible, in our present sensations
of sight, to separate colour from extension; and (2) That
there are, in fact, figures on the retina corresponding to
the apparent figures of objects.

On the subject of the first reason, he says, that the
figure which we perceive as associated with colour, is the
real, and not the apparent figure. "Is there," he asks,

* *Lectures,* vol. ii., p. 83.

"the slightest consciousness of a perception of visible figure, corresponding to the affected portion of the retina?" To which, though he seems to think an affirmative answer impossible, we cannot hesitate to reply, that there is undoubtedly such a consciousness; that though obscured by being made the ground of habitual inference as to the real figure, this consciousness is constantly referred to by the draughtsmen, and easily recalled by any one. We may separate colour, he says again*, from the figures on the retina, as we may separate it from length, breadth, and thickness, which we do not see. But this is altogether false: we cannot separate colour from length, breadth, and thickness in *any other way*, than by transferring it to the visible figure which we do see. He cannot, he allows, separate the colour from the visible form of the trunk of a large oak; but just as little, he thinks, can he separate it from the convex mass of the trunk, which (it is allowed on all hands) he does not immediately see. But in this he is mistaken: for if he were to make a *picture* of the oak, he would separate the colour from the convex shape, which he does not imitate, but he could not separate it from the visible figure, which he does imitate; and he would then perceive that the fact that he *has not* an immediate perception of the convex form, is necessarily connected with the fact that he *has* an immediate perception of the apparent figure; so far is the rejection of immediate perception in the former case from being a reason for rejecting it in the latter.

Again, with regard to the second argument. It does not, he says, follow, that because a certain figured portion of the retina is affected by light, we should see such a figure; for if a certain figured portion of the olfactory organ were affected by odours, we should not acquire by smell any perception of such figure†. This is merely to

* *Lectures*, vol. ii., p. 84.   † *Ib.*, p. 87.

say, that because we do not perceive position and figure
by one sense, we cannot do so by another. But this
again is altogether erroneous. It is an office of our
sight to inform us of position, and consequently of figure;
for this purpose, the organ is so constructed that the
position of the object determines the position of the
point of the retina affected. There is nothing of this
kind in the organ of smell; objects in different positions
and of different forms do not affect different parts of the
olfactory nerve, or portions of different shape. Different
objects, remote from each other, if perceived by smell,
affect the same part of the olfactory organs. This is all
quite intelligible; for it is not the office of smell to
inform us of position. Of what use or meaning would
be the curious and complex structure of the eye, if it
gave us only such vague and wandering notions of the
colours and forms of the flowers in a garden, as we
receive from their odours when we walk among them
blindfold? It is, as we have said, the *prerogative* of
vision to apprehend position: the places of objects on
the retina give this information. We do not suppose
that the affection of a certain shape of nervous expanse
will necessarily and in all cases give us the impression of
figure; but we know that in vision it does; and it is
clear that if we did not acquire our acquaintance with
visible figure in this way, we could not acquire it in
any way*.

The whole of this strange mistake of Brown's appears
to arise from the fault already noticed;—that of considering the image on the retina as the *object* instead of

---

* When Brown says further (p. 87,) that we can indeed show the
image in the dissected eye; but that "it is not in the dissected eye
that vision takes place;" it is difficult to see what his drift is. Does
he doubt that there is an image formed in the living as completely as
in the dissected eye?

the *means* of vision. This indeed is what he says : "the true object of vision is not the distant body itself, but the light that has reached the expansive termination of the optic nerve*." Even if this were so, we do not see why we should not perceive the position of the impression on this expanded nerve. But as we have already said, the impression on the nerve is the means of vision, and enables us to assign a place, or at least a direction, to the object from which the light proceeds, and thus makes vision possible. Brown, indeed, pursues his own peculiar view till he involves the subject in utter confusion. Thus he says†, "According to the common theory [that figure can be perceived by the eye,] a visible sphere is at once to my perception convex and plane ; and if the sphere be a large one, it is perceived at once to be a sphere of many feet in diameter, and a plane circular surface of the diameter of a quarter of an inch." It is easy to deduce these and greater absurdities, if we proceed on his strange and baseless supposition that the object and the image on the retina are *both* perceived. But who is conscious of the image on the retina in any other way than as he sees the object by means of it?

Brown seems to have imagined that he was analysing the perception of figure in the same manner in which Berkeley had analysed the perception of distance. He ought to have recollected that such an undertaking, to be successful, required him to show what elements he analysed it *into*. Berkeley analysed the perception of real figure into the interpretation of visible figure according to certain rules which he distinctly stated. Brown analyses the perception of visible figure into no elements. Berkeley says, that we do not directly perceive distance, but that we perceive something else, from which we infer distance, namely, visible figure and colour, and our own

* *Lectures*, vol. ii, p. 57.   † *Ib.*, vol. ii., p. 89,

efforts in seeing; Brown says, that we do not see figure, but infer it; what then do we see which we infer it from? To this he offers no answer. He asserts the seeming perception of visible figure to be a result of "association;"—of "suggestion." But what meaning can we attach to this? Suggestion requires something which suggests; and not a hint is given what it is which suggests position. Association implies two things associated; what is the sensation which we associate with form? What is that visual perception which is not figure, and which we mistake for figure? What perception is it that suggests a square to the eye? What impressions are those which have been associated with a visible triangle, so that the revival of the impressions revives the notion of the triangle? Brown has nowhere pointed out such perceptions and impressions; nor indeed was it possible for him to do so; for the only visual perceptions which he allows to remain, those of colour, most assuredly do not suggest visible figures by their differences; red is not associated with square rather than with round, or with round rather than square. On the contrary, the eye, constructed in a very complex and wonderful manner in order that it may give to us directly the perception of position as well as of colour, has it for one of its prerogatives to give us this information; and the perception of the relative position of each part of the visible boundary of an object constitutes the perception of its apparent figure; which faculty we cannot deny to the eye without rejecting the plain and constant evidence of our senses, making the mechanism of the eye unmeaning, confounding the object with the means of vision, and rendering the mental process of vision utterly unintelligible.

Having sufficiently discussed the processes of perception, I now return to the consideration of the Ideas which these processes assume.

CHAPTER III.

SUCCESSIVE ATTEMPTS AT THE SCIENTIFIC
APPLICATION OF THE IDEA OF A
MEDIUM.

1. IN what precedes, we have shown by various considerations that we necessarily and universally assume the perception of secondary qualities to take place by means of a medium interjacent between the object and the person perceiving. Perception is affected by various peculiarities, according to the nature of the quality perceived: but in all cases a medium is equally essential to the process.

This principle, which, as we have seen, is accepted as evident by the common understanding of mankind, is confirmed by all additional reflection and discipline of the mind, and is the foundation of all the theories which have been proposed concerning the processes by which the perception takes place, and concerning the modifications of the qualities thus perceived. The medium, and the mode in which the impression is conveyed through the medium, seem to be different for different qualities; but the existence of the medium leads to certain necessary conditions or alternatives, which have successively made their appearance in science, in the course of the attempts of men to theorize concerning the principal secondary qualities, sound, light, and heat. We must now point out some of the ways, at first imperfect and erroneous, in which the consequences of the fundamental assumption were traced.

2. *Sound.*—In all cases the medium of sensation, whatever it is, is supposed to produce the effect of conveying secondary qualities to our perception by means of its primary qualities. It was conceived to operate by the

size, form, and motion of its parts. This is a fundamental principle of the class of sciences of which we have at present to speak.

It was assumed from the first, as we have seen in the passage lately quoted from Aristotle*, that in the conveyance of *sound*, the medium of communication was the air. But although the first theorists were right so far, that circumstance did not prevent their going entirely wrong when they had further to determine the nature of the process. It was conceived by Aristotle that the air acted after the manner of a rigid body;—like a staff, which, receiving an impulse at one end, transmits it to the other. Now this is altogether an erroneous view of the manner in which the air conveys the impulse by which sound is perceived. An approach was made to the true view of this process, by assimilating it to the diffusion of the little circular waves which are produced on the surface of still water when a stone is dropt into it. These little waves begin from the point thus disturbed, and run outwards, expanding on every side, in concentric circles, till they are lost. The propagation of sound through the air from the point where it is produced, was compared by Vitruvius to this diffusion of circular waves in water; and thus the notion of a propagation of impulse by the *waves* of a fluid was introduced, in the place of the former notion of the impulse of an unyielding body.

But though, taking an enlarged view of the nature of the progress of a wave, this is a just representation of the motion of air in conveying sound, we cannot suppose that the process was, at the period of which we speak, rightly understood. For the waves of water were contemplated only as affecting the surface of the water; and as the air has no surface, the communication must take place by means of an internal motion, which can bear only a remote and obscure resemblance to the waves which we

* *Supr.*, p. 271.

see. And even with regard to the waves of water, the mechanism by which they are produced and transferred was not at all understood; so that the comparison employed by Vitruvius must be considered rather as a loose analogy than as an exact scientific explanation. No correct account of such motions was given, till the formation of the science of mechanics in modern times had enabled philosophers to understand more distinctly the mode in which motion is propagated through a fluid, and to discern the forces which the process calls into play, so as to continue the motion once begun. Newton introduced into this subject the exact and rigorous conception of an *undulation*, which is the true key to the explanation of impulses conveyed through a fluid.

Even at the present day, the right apprehension of the nature of an undulation transmitted through a fluid is found to be very difficult for all persons except those whose minds have been duly disciplined by mathematical studies. When we see a wave run along the surface of water, we are apt to imagine at first that a portion of the fluid is transferred bodily from one place to another. But with a little consideration we may easily satisfy ourselves that this is not so: for if we look at a field of standing corn, when a breeze blows over it, we see waves like those of water run along its surface. Yet it is clear that in this case the separate stalks of corn only bend backwards and forwards, and no portion of the grain is really conveyed from one part of the field to the other. This is obvious even to popular apprehension. The poet speaks of

> . . . . The rye,
> That stoops its head when whirlwinds rave
> And springs again in eddying wave
> As each wild gust sweeps by.

Each particle of the mass in succession has a small

motion backwards and forwards; and by this means a large ridge made by many such particles runs along the mass to any distance. This is the general notion of an undulation.

Thus, when an undulation is propagated in a fluid, it is not matter, but form, which is transmitted from one place to another. The particles along the line of each wave assume a certain arrangement, and this arrangement passes from one part to another, the particles changing their places only within narrow limits, so as to lend themselves successively to the arrangements by which the successive waves, and the intervals between the waves, are formed.

When such an undulation is propagated through air, the wave is composed, not, as in water, of particles which are higher than the rest, but of particles which are closer to each other than the rest. The wave is not a ridge of elevation, but a line of condensation; and as in water we have alternately elevated and depressed lines, we have in air lines alternately condensed and rarefied. And the motion of the particles is not, as in water, up and down, in a direction transverse to that of the wave which runs forwards; in the motion of an undulation through air the motion of each particle is alternately forwards and backwards, while the motion of the undulation is constantly forwards.

This precise and detailed account of the undulatory motion of air by which sound is transmitted was first given by Newton. He further attempted to determine the motions of the separate particles, and to point out the force by which each particle affects the next, so as to continue the progress of the undulation once begun. The motions of each particle must be oscillatory; he assumed the oscillations to be governed by the simplest law of oscillation which had come under the notice of mathematicians, (that of small vibrations of a pendulum;) and he proved that in this manner the forces which are called into play by the contraction and expansion of the parts of the elastic fluid are such as the continuance of the motion requires.

Newton's proof of the exact law of oscillatory motion of the aërial particles was not considered satisfactory by succeeding mathematicians; for it was found that the same result, the development of forces adequate to continue the motion, would follow if any other law of the motion were assumed. Cramer proved this by a sort of *parody* of Newton's proof, in which, by the alteration of a few phrases in this formula of demonstration, it was made to establish an entirely different conclusion.

But the general conception of an undulation as presented by Newton was, as from its manifest mechanical truth it could not fail to be, accepted by all mathematicians; and in proportion as the methods of calculating the motions of fluids were further improved, the necessary consequences of this conception, in the communication of sound through air, were traced by unexceptionable reasoning. This was especially done by Euler and Lagrange, whose memoirs on such motions of fluids are some of the most admirable examples which exist, of refined mathematical methods applied to the solution of difficult mechanical problems.

But the great step in the formation of the theory of sound was undoubtedly that which we have noticed, the introduction of the conception of an undulation such as we have attempted to describe it:—a state, condition, or arrangement of the particles of a fluid, which is transferred from one part of space to another by means of small motions of the particles altogether distinct from the movement of the undulation itself. This is a conception which is not obvious to common apprehension.

It appears paradoxical at first sight to speak of a large *wave* (as the tide wave) running up a river at the rate of twenty miles an hour, while the *stream* of the river is all the while flowing downwards. Yet this is a very common fact. And the conception of such a motion must be fully mastered by all who would reason rightly concerning the transmission of impressions through a medium.

We have described the motion of sound as produced by small motions of the particles forwards and backwards, while the waves, or condensed and rarefied lines, move constantly forwards. It may be asked what right we have to suppose the motion to be of this kind, since when sound is heard no such motions of the particles of air can be observed, even by refined methods of observation. Thus Bacon declares himself against the hypothesis of such a vibration, since, as he remarks, it cannot be perceived in any visible impression upon the flame of a candle. And to this we reply, that the supposition of this vibration is made in virtue of a principle which is involved in the original assumption of a medium ; namely, That *a medium, in conveying secondary qualities, operates by means of its primary qualities*, the bulk, figure, motion, and other mechanical properties of its parts. This is an axiom belonging to the Idea of a Medium. In virtue of this axiom it is demonstrable that the motion of the air, when any how disturbed, must be such as is supposed in our acoustical reasonings. For the elasticity of the parts of the air, called into play by its expansion and contraction, lead, by a mechanical necessity, to such a motion as we have described. We may add that, by proper contrivances, this motion may be made perceptible in its visible effects. Thus the theory of sound, as an impression conveyed through air, is established upon evident general principles, although the mathematical calculations which are requisite to investigate its consequences are, some of them, of a very recondite kind.

3. *Light.*—The early attempts to explain vision represented it as performed by means of material rays proceeding *from* the eye, by the help of which the eye felt out the form and other visible qualities of an object, as a blind man might do with his staff. But this opinion could not keep its ground long: for it did not even explain the fact that light is necessary to vision. Light as a peculiar medium was then assumed as the machinery of vision; but the mode in which the impression was conveyed through the medium was left undetermined, and no advance was made towards sound theory, on that subject, by the ancients.

In modern times, when the prevalent philosophy began to assume a mechanical turn (as in the theories of Descartes), light was conceived to be a material substance which is emitted from luminous bodies, and which is also conveyed from all bodies to the eye, so as to render them visible. The various changes of direction by which the rays of light are affected, (reflection, refraction, &c.) Descartes explained, by considering the particles of light as small globules, which change their direction when they impinge upon other bodies, according to the laws of mechanics. Newton, with a much more profound knowledge of mechanics than Descartes possessed, adopted, in the most mature of his speculations, nearly the same view of the nature of light; and endeavoured to show that reflection, refraction, and other properties of light, might be explained as the effects which certain forces, emanating from the particles of bodies, produce upon the luminiferous globules.

But though some of the properties of light could thus be accounted for by the assumption of particles emitted from luminous bodies, and reflected or refracted by forces,

other properties came into view which would not admit
of the same explanation.   The phenomena of diffraction
(the fringes which accompany shadows) could never be
truly represented by such an hypothesis, in spite of many
attempts which were made.   And the colours of thin
plates, which show the rays of light to be affected by an
alternation of two different conditions at small intervals
along their length, led Newton himself to incline, often
and strongly, to some hypothesis of undulation.   The
double refraction of Iceland spar, a phenomenon in itself
very complex, could, it was found by Huyghens, be
expressed with great simplicity by a certain hypothesis of
undulations.

Two hypotheses of the nature of the luminiferous
medium were thus brought under consideration; the one
representing light as matter emitted from the luminous
object, the other, as undulations propagated through a fluid.
These two hypotheses remained in presence of each other
during the whole of the last century, neither of them
gaining any material advantage over the other, though the
greater part of mathematicians, following Newton, embraced the emission theory.   But at the beginning of the
present century, an additional class of phenomena, those
of the interference of two rays of light, were brought
under consideration by Dr. Young; and these phenomena
were strongly in favour of the undulatory theory, while
they were irreconcilable with the hypothesis of emission.
If it had not been for the original bias of Newton and his
school to the other side, there can be little doubt that
from this period light as well as sound would have been
supposed to be propagated by undulations; although in
this case it was necessary to assume as the vehicle of
such undulations a special medium or *ether*.   Several
points of the phenomena of vision no doubt remained
unexplained by the undulatory theory, as absorption, and
the natural colours of bodies; but such facts, though

they did not confirm, did not evidently contradict the theory of a luminiferous ether; and the facts which such a theory did explain, it explained with singular happiness and accuracy.

But before this undulatory theory could be generally accepted, it was presented in an entirely new point of view by being combined with the facts of polarization. The general idea of polarization must be illustrated hereafter; but we may here remark that Young and Fresnel, who had adopted the undulatory theory, after being embarrassed for some time by the new facts which were thus presented to their notice, at last saw that these facts might be explained by conceiving the vibrations to be transverse to the ray, the motions of the particles being not backwards and forwards in the line in which the impulse travels, but to the right and left of that line. This conception of *transverse vibrations*, though quite unforeseen, had nothing in it which was at all difficult to reconcile with the general notion of an undulation. We have described an undulation, or wave, as a certain condition or arrangement of the particles of the fluid successively transferred from one part of space to another: and it is easily conceivable that this arrangement or wave may be produced by a lateral transfer of the particles from their quiescent positions. This conception of transverse vibrations being accepted, it was found that the explanation of the phenomena of polarization and of those of interference led to the same theory with a correspondence truly wonderful; and this coincidence in the views collected from two quite distinct classes of phenomena was justly considered as an almost demonstrative evidence of the truth of this undulatory theory.

It remained to be considered whether the doctrine of transverse vibrations in a fluid could be reconciled with the principles of mechanics. And it was found that by making certain suppositions, in which no inherent improbability existed, the hypothesis of transverse vibrations would explain the laws, both of interference and of polarization of light, in air and in crystals of all kinds, with a surprising fertility and fidelity.

Thus the undulatory theory of light, like the undulatory theory of sound, is recommended by its conformity to the fundamental principle of the Secondary Mechanical Sciences, that the medium must be supposed to transmit its peculiar impulses according to the laws of mechanics. Although no one had previously dreamt of qualities being conveyed through a medium by such a process, yet when it is once suggested as the only mode of explaining some of the phenomena, there is nothing to prevent our accepting it entirely, as a satisfactory theory for all the known laws of light.

4. *Heat.*—With regard to heat as with regard to light, a fluid medium was necessarily assumed as the vehicle of the property. During the last century, this medium was supposed to be an emitted fluid. And many of the ascertained Laws of Heat, those which prevail with regard to its radiation more especially, were well explained by this hypothesis*. Other effects of heat, however, as for instance *latent heat*†, and the change of *consistence* of bodies‡, were not satisfactorily brought into connexion with the hypothesis; while *conduction*§, which at first did not appear to result from the fundamental assumption, was to a certain extent explained as internal radiation.

But it was by no means clear that an undulatory theory of heat might not be made to explain these phenomena equally well. Several philosophers inclined

* See the Account of the Theory of Exchanges, *Hist. Ind. Sci.*, ii. 474.
  † *Ib.*, ii. 499.   ‡ *Ib.*, 498.   § *Ib.*, 469.

to such a theory; and finally, Ampère showed that the doctrine that the heat of a body consists in the undulations of its particles propagated by means of the undulations of a medium, might be so adjusted as to explain all which the theory of emission could explain, and moreover to account for facts and laws which were out of the reach of that theory. About the same time it was discovered by Prof. Forbes and M. Nobili that radiant heat is, under certain circumstances, polarized. Now polarization had been most satisfactorily explained by means of transverse undulations in the case of light; while all attempts to modify the emission theory so as to include polarization in it, had been found ineffectual. Hence this discovery was justly considered as lending great countenance to the opinion that heat consists in the vibrations of its proper medium.

But what is this medium? Is it the same by which the impressions of light are conveyed? This is a difficult question; or rather it is one which we cannot at present hope to answer with certainty. No doubt the connexion between light and heat is so intimate and constant, that we can hardly refrain from considering them as affections of the same medium. But instead of attempting to erect our systems on such loose and general views of connexion, it is rather the business of the philosophers of the present day to determine the laws of the operation of heat, and its real relation to light, in order that we may afterwards be able to connect the theories of the two qualities. Perhaps in a more advanced state of our knowledge we may be able to state it as an axiom, that two secondary qualities, which are intimately connected in their causes and effects, must be affections of the same medium. But at present it does not appear safe to proceed upon such a principle, although many writers, in their speculations both concerning light and heat, and concerning other properties, have not hesitated to do so.

Some other consequences follow from the Idea of a Medium which must be the subject of another chapter.

Chapter IV.

OF THE MEASURE OF SECONDARY QUALITIES.

1. *Scales of Qualities in general.*—The ultimate object of our investigation in each of the Secondary Mechanical Sciences, is the nature of the processes by which the special impressions of sound, light, and heat, are conveyed, and the modifications of which these processes are susceptible. And of this investigation, as we have seen, the necessary basis is the principle, that these impressions are transmitted by means of a medium. But before we arrive at this ultimate object, we may find it necessary to occupy ourselves with several intermediate objects: before we discover the *cause*, it may be necessary to determine the *laws* of the phenomena. Even if we cannot immediately ascertain the mechanism of light or heat, it may still be interesting and important to arrange and measure the effects which we observe.

The idea of a medium affects our proceeding in this research also. We cannot measure secondary qualities in the same manner in which we measure primary qualities, by a mere addition of parts. There is this leading and remarkable difference, that while both classes of qualities are susceptible of changes of magnitude, primary qualities increase by addition of *extension*, secondary, by augmentation of *intensity*. A space is doubled when another equal space is placed by its side; one weight joined to another makes up the sum of the two. But when one degree of warmth is combined with another, or one shade of red colour with another, we cannot in like manner talk of the sum. The component parts do not evidently retain their separate existence; we cannot separate a strong green colour into two weaker ones, as we can separate a large force into two smaller. The increase is absorbed into the previous amount, and is no longer in evidence as a part of the whole. And this is the difference which has given birth to the two words *extended*, and *intense*. That is extended which has "partes extra partes," parts outside of parts: that is intense which becomes stronger by some indirect and unapparent increase of agency, like the stretching of the internal springs of a machine, as the term *intense* implies. Extended magnitudes can at will be resolved into the parts of which they were originally composed, or any other which the nature of their extension admits; their proportion is apparent; they are directly and at once subject to the relations of number. Intensive magnitudes cannot be resolved into smaller magnitudes; we can see that they differ, but we cannot tell in what proportion; we have no direct measure of their quantity. How many times hotter than blood is boiling water? The answer cannot be given by the aid of our feelings of heat alone.

This difference, as we have said, is connected with the fundamental principle that we do not perceive secondary qualities directly, but through a medium. We have no natural apprehension of light, or sound, or heat, as they exist in the bodies from which they proceed, but only as they affect our organs. We can only measure them, therefore, by some scale supplied by their effects. And thus while extended magnitudes, as space, time, are measurable directly and of themselves; intensive magnitudes, as brightness, loudness, heat, are measurable only by artificial means and conventional scales. Space, time, measure themselves: the repetition of a smaller space, or time, while it composes a larger one, measures it.   But for light and heat we must have photometers and thermometers, which measure something which is assumed to be an indication of the quality in question.   In one case, the mode of applying the measure, and the meaning of the number resulting, are seen by intuition; in the other, they are consequences of assumption and reasoning.   In the one case, they are *units*, of which the extension is made up; in the other, they are *degrees* by which the intensity ascends.

2. When we discover any property in a sensible quality, which at once refers us to number or space, we readily take this property as a measure; and thus we make a transition from quality to quantity.   Thus Ptolemy in the third chapter of the First Book of his *Harmonics* begins thus: " As to the differences which exist in sounds both in *quality* and in *quantity*, if we consider that difference which refers to the acuteness and graveness, we cannot at once tell to which of the above two classes it belongs, till we have considered the causes of such symptoms." But at the end of the chapter, having satisfied himself that grave sounds result from the magnitude of the string or pipe, other things being equal, he infers, " Thus the difference of acute and grave appears to be a difference of *quantity*."

In the same manner, in order to form Secondary Mechanical Sciences respecting any of the other properties of bodies, we must reduce these properties to a dependence upon quantity, and thus make them subject to measurement.   We cannot obtain any sciential truths respecting the comparison of sensible qualities, till we have discovered measures and scales of the qualities which we have to consider; and accordingly, some of the most important steps in such sciences have been the establishment of such measures and scales, and the invention of the requisite instruments.

The formation of the mathematical sciences which rest upon the measures of the intensity of sensible qualities took place mainly in the course of the last century. Perhaps we may consider Lambert, a mathematician who resided in Switzerland, and published about 1750, as the person who first clearly felt the importance of establishing such sciences. His Photometry, Pyrometry, Hygrometry, are examples of the systematic reduction of sensible qualities (light, heat, moisture) to modes of numerical measurement.

We now proceed to speak of such modes of measurement with regard to the most obvious properties of bodies.

3. (I.) *The Musical Scale.*—The establishment of the *Harmonic Canon*, that is, of a Scale and Measure of the musical place of notes, in the relation of *high* and *low*, was the first step in the science of Harmonics. The perception of the differences and relations of musical sounds is the office of the sense of hearing; but these relations are fixed, and rendered accurately recognisable by artificial means. "Indeed, in all the senses," as Ptolemy truly says in the opening of his Harmonics, "the sense discovers what is approximately true, and receives accuracy from another quarter: the reason receives the approximately-true from another quarter, and discovers the accurate truth." We can have no measures of sensible qualities which do not ultimately refer to the sense;—whether they do this immediately, as when we refer colours to an assumed standard; or mediately, as when we measure heat by expansion, having previously found by an appeal to sense that the expansion increases with the heat. Such relations of sensible qualities cannot be described in words, and can only be

apprehended by their appropriate faculty. The faculty by which the relations of sounds are apprehended is a *musical ear* in the largest acceptation of the term. In this signification the faculty is nearly universal among men; for all persons have musical ears sufficiently delicate to understand and to imitate the modulations corresponding to various emotions in speaking; which modulations depend upon the succession of acuter and graver tones. These are the relations now spoken of, and these are plainly perceived by persons who have very imperfect musical ears, according to the common use of the phrase. But the relations of tones which occur in speaking are somewhat indefinite; and in forming that musical scale which is the basis of our science upon the subject, we take the most definite and marked of such relations of notes; such as occur, not in speaking but in singing. Those musical relations of two sounds which we call the *octave*, the *fifth*, the *fourth*, the *third*, are recognised after a short familiarity with them. These *chords* or *intervals* are perceived to have each a peculiar character, which separates them from the relations of two sounds taken at random, and makes it easy to know them when sung or played on an instrument; and for most persons, not difficult to sing the sounds in succession exactly, or nearly correct. These musical relations, or *concords*, then, are the groundwork of our musical standard. But how are we to name these indescribable sensible characters? how to refer, with unerring accuracy, to a type which exists only in our own perceptions? We must have for this purpose a *Scale* and a *Standard*.

The Musical Scale is a series of eight notes, ascending by certain steps from the first or key-note to the octave above it, each of the notes being fixed by such distinguishable musical relations as we have spoken of above. We may call these notes c, d, e, f, g, a, b, c;

and we may then say that G is determined by its being a
fifth above C; D by its being a fourth below G; E by its
being a third above C; and similarly of the rest. It will
be recollected that the terms a *fifth*, a *fourth*, a *third*, have
hitherto been introduced as expressing certain simple and
indescribable musical relations among sounds, which
might have been indicated by any other names. Thus
we might call the fifth the *dominant*, and the fourth the
subdominant, as is done in one part of musical science.
But the names we have used, which are the common
ones, are in fact derived from the number of notes which
these intervals include in the scale obtained in the above
manner. The notes C, D, E, F, G, being five, the interval
from C to G is a fifth, and so of the rest. The fixation of
this scale gave the means of describing exactly any note
which occurs in the scale, and the method is easily applicable to notes above and below this range; for in a
series of sounds higher or lower by an octave than this
standard series, the ear discovers a recurrence of the
same relations so exact, that a person may sometimes
imagine he is producing the same notes as another when
he is singing the same air an octave higher.   Hence the
next eight notes may be conveniently denoted by a repetition of the same letters, as the first; thus, C, D, E, F, G, A, B,
c, d, e, f, g, a, b; and it is easy to devise a continuation
of such cycles.   And other admissible notes are designated by a further modification of the standard ones, as
by making each note *flat* or *sharp;* which modification it
is not necessary here to consider, since our object is only
to show how a standard is attainable, and how it serves
the ends of science.

We may observe, however, that the above is not an
exact account of the first, or early Greek scale; for this
scale was founded on a primary division of the interval of
two octaves (the extreme range which it admitted) into

five *tetrachords*, each tetrachord including the interval of a fourth. All the notes of this series had different names borrowed from this division*: thus *mese* was the middle or key-note; the note below it was *lichanos mesón*, the next below was *parypate mesón*, the next lower *hypate mesón*. The fifth above *mese* was *nete diazeugmenón*, the octave was *nete hyperbolæón*.

4. But supposing a complete system of such denominations established, how could it be with certainty and rigour applied? The human ear is fallible, the organs of voice imperfectly obedient; if this were not so, there would be no such thing as a *good* ear or a *good* voice. What means can be devised of finding at will a *perfect* concord, a fifth or a fourth? Or supposing such concords fixed by an acknowledged authority, how can they be referred to, and the authority adduced? How can we enact a Standard of sounds?

A Standard was discovered in the *Monochord*. A musical string properly stretched, may be made to produce different notes, in proportion as we intercept a longer or shorter portion, and make this portion vibrate. The relation of the length of the strings which thus sound the two notes G and C is fixed and constant, and the same is true of all other notes. Hence the musical interval of any notes of which we know the places in the musical scale, may be reproduced by measuring the lengths of string which are known to give them. If C be of the length 180, D is 169, E is 144, F is 135, G is 120; and thus the musical relations are reduced to numerical relations, and the monochord is a complete and perfect *tonometer*.

We have here taken the length of the string as the measure of the tone: but we may observe that there is in us a necessary tendency to assume that the ground of this mea-

* BURNEY's *History of Music*, vol. i. p. 28.

sure is to be sought in some ulterior cause; and when we consider the matter further, we find this cause in the frequency of these vibrations of. the string. The truth that the same note must result from the same frequency of vibration is readily assented to on a slight suggestion of experience. Thus Mersenne*, when he undertakes to determine the frequency of vibrations of a given sound, says "Supponendum est quoscunque nervos et quaslibet chordas unisonum facientes eundem efficere numerum recursuum eodem vel equali tempore, quod perpetuâ constat experientiâ." And he proceeds to apply it to cases where experience could not verify this assertion, or at least had not verified it, as to that of pipes.

The pursuit of these numerical relations of tones forms the science of Harmonics; of which here we do not pretend to give an account, but only to show, how the invention of a Scale and Nomenclature, a Standard and Measure of the tone of sounds, is its necessary basis. We will therefore now proceed to speak of another subject; *colour*.

5. (II.) *Scales of Colour.—The Prismatic Scale of Colour.*—A Scale of Colour must depend originally upon differences discernible by the eye, as a scale of notes depends on differences perceived by the ear. In one respect the difficulty is greater in the case of the visible qualities, for there are no relations of colour which the eye peculiarly singles out and distinguishes, as the ear selects and distinguishes an octave or a fifth. Hence we are compelled to take an arbitrary scale; and we have to find one which is fixed, and which includes a proper collection of colours. The prismatic spectrum, or coloured image produced when a small beam of light passes obliquely through any transparent surface (as the surface of a prism of glass,) offers an obvious Standard as far

* *Harmonia*, lib. ii. Prop. 19.

as it is applicable. Accordingly colours have, for various purposes, been designated by their place in the spectrum ever since the time of Newton; and we have thus a means of referring to such colours as are included in the series *red, orange, yellow, green, blue, violet, indigo*, and the intermediate tints.

But this scale is not capable of numerical precision. If the spectrum could be exactly defined as to its extremities, and if these colours occupied always the same proportional part of it, we might describe any colour in the above series by the measure of its position. But the fact is otherwise. The spectrum is too indefinite in its boundaries to afford any distinct point from which we may commence our measures; and moreover the spectra produced by different transparent bodies differ from each other. Newton had supposed that the spectrum and its parts were the same, so long as the refraction was the same; but his successors discovered that, with the same amount of refraction in different kinds of glass, there are different magnitudes of the spectrum; and what is still worse with reference to our present purpose, that the spectra from different glasses have the colours distributed in different proportions. In order, therefore, to make the spectrum the scale of colour, we must assume some fixed substance; for instance, we may take water, and thus the colours of the rainbow will be our standard. But we should still have an extreme difficulty in applying such a rule. The distinctions of colour which the terms of common language express, are not used with perfect unanimity or with rigorous precision. What one person calls *bluish green* another calls *greenish blue*. Nobody can say what is the precise boundary between red and orange. Thus the prismatic scale of colour was incapable of mathematical exactness, and this inconvenience was felt up to our own times.

But this difficulty was removed by a curious discovery of Fraunhofer; who found that there are, in the solar spectrum, certain fine black Lines which occupy a definite place in the series of colours, and can be observed with perfect precision. We have now no uncertainty as to what coloured light we are speaking of, when we describe it as that part of the spectrum in which Fraunhofer's Line c or d occurs. And thus, by this discovery, the prismatic spectrum of sunlight became, for certain purposes, an exact *Chromatometer*.

6. *Newton's Scale of Colours.*—Still, such a standard is arbitrary and seemingly anomalous. The lines A, B, C, D, &c., of Fraunhofer's spectrum are distributed without any apparent order or law; and we do not, in this way, obtain numerical measures, which is what, in all cases, we desire to have. Another discovery of Newton, however, gives us a spectrum containing the same colours as the prismatic spectrum, but produced in another way, so that the colours have a numerical relation. I speak of the *colours of thin plates.* The little rainbows which we sometimes see in the cracks of broken glass are governed by fixed and simple laws. The kind of colour produced at any point depends on the thickness of the thin plate of air included in the fissure. If the thickness be twelve-millionths of an inch, the colour is orange, if ten-millionths of an inch, we have green, and so on; and thus these numbers which succeed each other in a regular order from red to indigo, give a numerical measure of each colour; which measure, when we pursue the subject, we find is one of the bases of all optical theory. The series of colours obtained from plates of air of gradually increasing thickness is called *Newton's Scale of Colours;* but we may observe that this is not precisely what we are here speaking of, a scale of simple colours; it is a series produced by certain combinations, resulting from the

repetition of the first spectrum, and is mainly useful as a standard for similar phenomena, and not for colour in general. The real scale of colour is to be found, as we have said, in the numbers which express the thickness of the producing film;—in the length of a *fit* in Newton's phraseology, or the length of an undulation in the modern theory.

7. *Scales of Impure Colours.*—The standards just spoken of include (mainly at least) only pure and simple colours; and however complete they may be for certain objects of the science of optics, they are insufficient for other purposes. They do not enable us to put in their place mixed and impure colours. And there is, in the case of colour, a difficulty already noticed, which does not occur in the case of sound; two notes, when sounded together, are not necessarily heard as one; they are recognised as still two, and as forming a concord or a discord. But two colours form a single colour; and the eye cannot, in any way, distinguish between a green compounded of blue and yellow, and the simple, undecomposable green of the spectrum. By composition of three or more colours, innumerable new colours may be generated which form no part of the prismatic series; and by such compositions is woven the infinitely varied web of colour which forms the clothing of nature. How are we to classify and arrange all the possible colours of objects, so that each shall have a place and name? How shall we find a chromatometer for impure as well as for pure colour?

Though no optical investigations have depended on a scale of impure colours, such a scale has been wanted and invented for other purposes; for instance, in order to identify and describe objects of natural history. Not to speak of earlier essays, we may notice Werner's Nomenclature of Colours, devised for the purpose of describing minerals. This scale of colour was far superior to any which had previously been promulgated. It was, indeed, arbitrary in the selection of its degrees, and in a great measure in their arrangement; and the colours were described by the usual terms, though generally with some added distinction; as *blackish green, bluish green, apple green, emerald green*. But the great merit of the scale was its giving a *fixed* conventional meaning to these terms, so that they lost much of their usual vagueness. Thus *apple-green* did not mean the colour of any green apple casually taken; but a certain definite colour which the student was to bear in mind, whether or not he had ever seen an apple of that exact hue. The words were not a description, but a *record* of the colour: the memory was to retain a *sensation*, not a name.

The imperfection of the system (arising from its arbitrary form) was its incompleteness: however well it served for the reference of the colours which it did contain, it was applicable to no others; and thus, though Werner's enumeration extended to more than a hundred colours, there occur in nature a still greater number which cannot be exactly described by means of it.

In such cases the unclassed colour is, by the Wernerians, defined by stating it as intermediate between two others: thus we have an object described as *between emerald green and grass green*. The eye is capable of perceiving a gradation from one colour to another; such as may be produced by a gradual mixture in various ways. And if we image to ourselves such a mixture, we can compare with it a given colour. But in employing this method we have nothing to tell us in what part of the scale we must seek for an approximation to our unclassed colour. We have no rule for discovering where we are to look for the boundaries of the definition of a colour which the Wernerian series does not supply. For it is not always between contiguous members of the series that the undescribed colour is found. If we place emerald green between apple green and grass green, we may yet have a colour intermediate between emerald green and leek green; and, in fact, the Wernerian series of colours is destitute of a principle of self-arrangement and gradation; and is thus necessarily and incurably imperfect.

9. We should have a complete Scale of Colours, if we could form a series including all colours, and arranged so that each colour was intermediate in its tint between the adjacent terms of the series; for then, whether we took many or few of the steps of the series for our standard terms, the rest could be supplied by the law of continuity; and any given colour would either correspond to one of the steps of our scale or fall between two intermediate ones. The invention of a Chromatometer for Impure Colours, therefore, requires that we should be able to form all possible colours by such intermediation in a systematic manner; that is, by the mixture or combination of certain elementary colours according to a simple rule: and we are led to ask whether such a process has been shown to be possible.

The colours of the prismatic spectrum obviously do form a continuous series; green is intermediate between its neighbours yellow and blue, orange between red and yellow; and if we suppose the two ends of the spectrum bent round to meet each other, so that the arrangement of the colours may be circular, the violet and indigo will find their appropriate place between the blue and red. And all the interjacent tints of the spectrum, as well as the ones thus named, will result from such an arrangement. Thus all the pure colours are produced by combinations two and two of three primary colours, red, yellow, and blue; and the question suggests itself whether these three are not really the only primary colours, and whether all the impure colours do not arise from mixtures of the three in various proportions. There are various modes in which this suggestion may be applied to the construction of a scale of colours; but the simplest and the one which appears really to verify the conjecture that all possible colours may be so exhibited, is the following. A certain combination of red, yellow, and blue, will produce black, or pure grey, and when diluted, will give all the shades of grey which intervene between black and white. By adding various shades of grey, then, to pure colours, we may obtain all the possible ternary combinations of red, yellow, and blue; and in this way it is found that we exhaust the range of colours. Thus the circle of pure colours of which we have spoken may be accompanied by several other circles, in which these colours are tinged with a less or greater shade of grey; and in this manner it is found that we have a perfect chromatometer; every possible colour being exhibited either exactly or by means of approximate and contiguous limits. The arrangement of colours has been brought into this final and complete form by M. Merimée, whose chromatic scale is published by M. Mirbel in his *Elements of Botany*. We may observe that such a standard affords us a numerical exponent for every colour by means of the proportions of the three primary colours which compose it; or, expressing the same result otherwise, by means of the pure colour which is involved, and the proportion of grey by which it is rendered impure. In such a scale the fundamental elements would be the precise tints of red, yellow, and blue which are found or assumed to be primary; the numerical exponents of each colour would depend upon the arbitrary number of degrees which we interpose between each two primary colours; and between each pure colour and absolute blackness. No such numerical scale has, however, as yet, obtained general acceptation.

10. (III.) *Scales of Light. Photometer.*—Another instrument much needed in optical researches is a *Photometer*, a measure of the intensity of light. In this case, also, the organ of sense, the eye, is the ultimate judge; nor

has any effect of light, as light, yet been discovered which we can substitute for such a judgment. All instruments, such as that of Leslie, which employ the heating effect of light, or at least all that have hitherto been proposed, are inadmissible as photometers. But though the eye can judge of two surfaces illuminated by light of the same colour, and can determine when they are equally bright, or which is the brighter, the eye can by no means decide at sight the proportion of illumination. How much in such judgments we are affected by contrast, is easily seen when we consider how different is the apparent brightness of the moon at mid-day and at midnight, though the light which we receive from her is, in fact, the same at both periods. In order to apply a scale in this case, we must take advantage of the known numerical relations of light. We are certain that if all other illumination be excluded, two equal luminaries, under the same circumstances, will produce an illumination twice as great as one does; and we can easily prove, from mathematical considerations, that if light be not enfeebled by the medium through which it passes, the illumination on a given surface will diminish as the square of the distance of the luminary increases. If, therefore, we can by taking a fraction thus known of the illuminating effect of one luminary, make it equal to the total effect of another, of which equality the eye is a competent judge, we compare the effects of the two luminaries. In order to make this comparison we may, with Rumford, look at the shadows of the same object made by the two lights, or with Ritchie, we may view the brightness produced on two contiguous surfaces, framing an apparatus so that the equality may be brought about by proper adjustment; and thus a measure will become practicable. Or we may employ other methods as was done by Wollaston*, who reduced the light of the sun by observing it as reflected

* *Phil. Trans.*, 1829, p. 19.

from a bright globule, and thus found the light of the sun to be 10,000,000,000 times that of Sirius, the brightest fixed star. All these methods are inaccurate, even as methods of comparison; and do not offer any fixed or convenient numerical standard; but none better have yet been devised.

10. *Cyanometer.*—As we thus measure the brightness of a colourless light, we may measure the intensity of any particular colour in the same way; that is, by applying a standard exhibiting the gradations of the colour in question till we find a shade which is seen to agree with the proposed object. Such an instrument we have in the *Cyanometer*, which was invented by Saussure for the purpose of measuring the intensity of the blue colour of the sky. We may introduce into such an instrument a numerical scale, but the numbers in such a scale will be altogether arbitrary.

11. (1V.) *Scales of Heat.*—When we proceed to the sensation of heat, and seek a measure of that quality, we find, at first sight, new difficulties. Our sensations of this kind are more fluctuating than those of vision; for we know that the same object may feel warm to one hand and cold to another at the same instant, if the hands have been previously cooled and warmed respectively. Nor can we obtain here, as in the case of light, self-evident numerical relations of the heat communicated in given circumstances; for we know that the effect so produced will depend on the warmth of the body to be heated, as well as on that of the source of heat; the summer sun, which warms our bodies, will not augment the heat of a red-hot iron. The cause of the difference of these cases is, that bodies do not receive the whole of their heat, as they receive the whole of their light, from the immediate influence of obvious external agents. There is no readily-discovered absolute cold,

corresponding to the absolute darkness which we can easily produce or imagine. Hence we should be greatly at a loss to devise a *Thermometer*, if we did not find an indirect effect of heat sufficiently constant and measurable to answer this purpose. We discover. however, such an effect in the *expansion* of bodies by. the effect of heat.

12. Many obvious phenomena show that air, under given circumstances, expands by the effect of heat; the same is seen to be true of liquids, as of water, and spirit of wine; and the property is found to belong also to the metallic fluid, quicksilver. A more careful examination showed that the increase of bulk in some of these bodies by increase of heat was a fact of a nature sufficiently constant and regular to afford a means of measuring that previously intangible quality; and the Thermometer was invented. There were, however, many difficulties to overcome, and many points to settle before this instrument was fit for the purposes of science.

An explanation of the way in which this was done necessarily includes an important chapter of the history of Thermotics. We must now, therefore, briefly notice historically the progress of the Thermometer. The leading steps of this progress, after the first invention of the instrument, were—The establishment of *fixed points* in the thermometric scale—The *comparison of the scales* of different substances—And the reconcilement of these differences by some method of interpreting them as indications of the absolute *quantity of heat.*

13. It would occupy too much space to give in detail the history of the successive attempts by which these steps were effected. A thermometer is described by Bacon under the title *Vitrum Calendare;* this was an air thermometer. Newton used a thermometer of linseed oil, and he perceived that the first step requisite to give value to such an instrument was to fix its scale; accord-

ingly he proposed his *Scala Graduum Caloris**. But when thermometers of different liquids were compared, it appeared, from their discrepancies, that this fixation of the scale of heat was more difficult than had been supposed. It was, however, effected. Newton had taken freezing water, or rather thawing snow, as the zero of his scale, which is really a fixed point; Halley and Amontons discovered (in 1693 and 1702) that the heat of boiling water is another fixed point; and Daniel Gabriel Fahrenheit, of Dantzig, by carefully applying these two standard points, produced, about 1714, thermometers, which were constantly consistent with each other. This result was much admired at the time, and was, in fact, the solution of the problem just stated, the *fixation of the scale of heat*.

14. But the scale thus obtained is a conventional not a natural scale. It depends upon the fluid employed for the thermometer. The progress of expansion from the heat of freezing to that of boiling water is different for mercury, oil, water, spirit of wine, air. A degree of heat which is half-way between these two standard points according to a mercurial thermometer, will be below the half-way point in a spirit thermometer, and above it in an air thermometer. Each liquid has its own *march* in the course of its expansion. Deluc and others compared the marches of various liquids, and thus made what we may call a *concordance* of thermometers of various kinds.

15. Here the question further occurs : Is there not some *natural measure* of the degrees of heat? It appears certain that there must be such a measure, and that by means of it all the scales of different liquids must be reconciled. Yet this does not seem to have occurred at once to men's minds. Deluc, in speaking of the researches which we have just mentioned, says†, "When I undertook these experiments, it never once came into my

* *Phil. Trans.*, 1701.   † *Modif. de l'Atmosph.*, 1782, p. 303.

thoughts that they could conduct me with any probability
to a table of real degrees of heat   But hope grows with
success, and desire with hope." Accordingly he pursued
this inquiry for a long course of years.

What are the principles by which we are to be
guided to the true measure of heat? Here, as in all the
sciences of this class, we have the general principle, that
the secondary quality, heat, must be supposed to be perceived in some way by a material medium or fluid. If
we take that which is, perhaps, the simplest form of this
hypothesis, that the heat depends upon the *quantity* of this
fluid, or *caloric*, which is present, we shall find that we are
led to propositions which may serve as a foundation for a
natural measure of heat. The *Method of Mixtures* is
one example of such a result. If we mix together two
pints of water, one hot and one cold, is it not manifest
that the temperature of the mixture must be midway
between the two? Each of the two portions brings with it
its own heat. The whole heat, or caloric, of the mixture
is the sum of the two ; and the heat of each half must be
the half of this sum, and therefore its temperature must
be intermediate between the temperatures of the equal
portions which were mixed. Deluc made experiments
founded upon this principle, and was led by them to conclude that "the dilatations of mercury follow an accelerated march for successive equal augmentations of heat."

But there are various circumstances which prevent
this method of mixtures from being so satisfactory as at
first sight it seems to promise to be. The different *capa-
cities for heat* of different substances, and even of the
same substance at different temperatures, introduce much
difficulty into the experiments, and this path of inquiry
has not yet led to a satisfactory result.

16. Another mode of inquiring into the natural measure
of heat is to seek it by researches on the *law of cooling* of

hot bodies. If we assume that the process of cooling of hot bodies consists in a certain material heat flying off, we may, by means of certain probable hypotheses, determine mathematically the law according to which the temperature decreases as time goes on; and we may assume that to be the true measure of temperature which gives to the experimental law of cooling the most simple and probable form.

It appears evident from the most obvious conceptions which we can form of the manner in which a body parts with its superabundant heat, that the hotter a body is, the faster it cools; though it is not clear without experiment, by what law the rate of cooling will depend upon the heat of the body. Newton took for granted the most simple and seemingly natural law of this dependence: he supposed the rate of cooling to be *proportional* to the temperature, and from this supposition he could deduce the temperature of a hot iron, calculating from the original temperature and the time during which it had been cooling. By calculation founded on such a basis, he graduated his thermometer.

17. But a little further consideration showed that the rate of cooling of hot bodies depended upon the temperature of the surrounding bodies, as well as upon its own temperature. Prevost's *Theory of Exchanges** was propounded with a view of explaining this dependence, and was generally accepted. According to this theory, all bodies radiate heat to one another, and are thus constantly giving and receiving heat; and a body which is hotter than surrounding bodies, cools itself, and warms the surrounding bodies, by an exchange of heat for heat, in which they are the gainers. Hence if θ be the temperature of the bodies, or of the space, by which the hot body is surrounded, and $\theta + t$ the temperature of the hot

* *Recherches sur la Chaleur*, 1791. *Hist. Ind. Sci.*, ii. 474.

body, the rate of cooling will depend upon the excess of the radiation for a temperature $\theta + t$, above the radiation for a temperature θ.

Accordingly, in the admirable researches of MM. Dulong and Petit upon the cooling of bodies, it was assumed that the rate of cooling of the hot body was represented by the excess of $F(\theta+t)$ above $F(\theta)$; where F represented some mathematical *function*, that is, some expression obtained by arithmetical operations from the temperatures $\theta+t$ and θ; although what these operations are to be, was left undecided, and was in fact determined by the experiments. And the result of their investigations was, that the function is of this kind: when the temperature increases by equal intervals, the function increases in a continued geometric proportion*. This was, in fact, the same law which had been assumed by Newton and others, with this difference, that they had neglected the term which depends upon the temperature of the surrounding space.

18. This law falls in so well with the best conceptions we can form of the mechanism of cooling upon the supposition of a radiant fluid caloric, that it gives great probability to the scale of temperature on which the simplicity of the result depends. Now the temperatures in the formulæ just referred to were expressed by means of the *air thermometer*. Hence MM. Dulong and Petit justly state that while all different substances employed as thermometers give different laws of thermotical phenomena, their own success in obtaining simple and general laws by means of the air thermometer, is a strong recommendation of that as the *natural scale of heat*. They add†,

---

* The formula for the rate of cooling is $m a^{\theta+t} - m a^{\theta}$, where the quantity m depends upon the nature of the body, the state of it surface, and other circumstances.—*Ann. Chim.* vii. 150.

† *Annales de Chimie*, vii. 153.

" The well-known uniformity of the principal physical properties of all gases, and especially the perfect identity of their laws of dilatation by heat, [a very important discovery of Dalton and Gay Lussac*,] make it very probable that in this class of bodies the disturbing causes have not the same influence as in solids and liquids; and consequently that the changes of bulk produced by the action of heat are here in a more immediate dependence on the force which produces them."

19. Still we cannot consider this point as settled till we obtain a more complete theoretical insight into the nature of heat itself. If it be true that heat consists in the vibrations of a fluid, then, although, as Ampère has shown†, the laws of radiation will, on mathematical grounds, be the same as they are on the hypothesis of emission, we cannot consider the natural scale of heat as determined, till we have discovered some means of measuring the caloriferous vibrations as we measure luminiferous vibrations. We shall only know what the quantity of heat is when we know what heat itself is ;—when we have obtained a theory which satisfactorily explains the manner in which the substance or medium of heat produces its effects. When we see how radiation and conduction, dilatation and liquefaction are all produced by mechanical changes of the same fluid, we shall then see what the nature of that change is which dilatation really measures, and what relation it bears to any more proper standard of heat.

We may add, that while our thermotical theory is still so imperfect as it is, all attempts to divine the true nature of the relation between light and heat are premature, and must be in the highest degree insecure and visionary. Speculations in which, from the general assumption of a caloriferous and luminiferous medium, and from a few facts arbitrarily selected and loosely analysed, a general theory of light and heat is asserted, are entirely foreign to the course of inductive science, and cannot lead to any stable and substantial truth.

20. *Other Instruments for measuring Heat.*—It does not belong to our present purpose to speak of instruments of which the object is to measure, not sensible qualities, but some effect or modification of the cause by which such qualities are produced: such, for instance, are the *Calorimeter*, employed by Lavoisier and Laplace, in order to compare the *specific heat* of different substances; and the *Actinometer*, invented by Sir John Herschel, in order to determine the *effect of the sun's rays* by means of the heat which they communicate in a given time; which effect is, as may readily be supposed, very different under different circumstances of atmosphere and position. The laws of such effects may be valuable contributions to our knowledge of heat, but the interpretation of them must depend on a previous knowledge of the relations which temperature bears to heat, according to the views just explained.

21. (V.) *Scales of other Qualities.*—Before quitting the subject of the measures of sensible qualities, we may observe that there are several other such qualities for which it would be necessary to have scales and means of measuring, in order to make any approach to science on such subjects. This is true, for instance, of tastes and smells. Indeed some attempts have been made towards a classification of the tastes of sapid substances, but these have not yet assumed any satisfactory or systematic character; and I am not aware that any instruments have been suggested for *measuring* either the flavour or the odour of bodies which possess such qualities.

22. *Quality of Sounds.*—The same is true of that kind of difference in sounds which is peculiarly termed their

quality; that character by which, for instance, the sound of a flute differs from that of a hautbois, when the note is the same; or a woman's voice from a boy's.

23. *Articulate Sounds.*—There is also in sounds another difference, of which the nature is still obscure, but in reducing which to rule, and consequently to measure, some progress has nevertheless been made. I speak of the differences of sound considered as *articulate.* Classifications of the sounds of the usual alphabets have been frequently proposed; for instance, that which arranges the *consonants* into the following groups :—

Sharp.	Flat.	Sharp Aspirate.	Flat Aspirate.	Nasal.
p	b	ph (f)	bh (v)	m
k	g (hard)	kh	gh	ng
t	d	th (sharp)	th (flat)	n
s	z	sh	zh	

It is easily perceived that the relations of the sounds in each of these horizontal lines are analogous; and accordingly the rules of derivation and modification of words in several languages proceed upon such analogies. In the same manner the *vowels* may be arranged in an order depending on their sound. But to make such arrangements fixed and indisputable, we ought to know the mechanism by which such modifications are caused. Instruments have been invented by which some of these sounds can be imitated; and if such instruments could be made to produce the above series of articulate sounds, by connected and regular processes, we should find, in the process, a *measure* of the sound produced. This has been in a great degree effected for the Vowels by Professor Willis's artificial mode of imitating them. For he finds that if a musical reed be made to sound through a cylindrical pipe, we obtain by gradually lengthening the cylindrical pipe, the series of vowels I, E, A, O, U, with intermediate sounds*. In this instrument, then, the

* *Camb. Trans.,* vol. iii., p. 239.

length of the pipe would determine the vowel, and might be used numerically to express it. Such an instrument so employed would be a measure of vowel quality.

Our business at present, however, is not with instruments which might be devised for measuring sensible qualities, but with those which have been so used, and have thus been the basis of the sciences in which such qualities are treated of; and this we have now done sufficiently for our present purpose.

24. There is another Idea which, though hitherto very vaguely entertained, has had considerable influence in the formation, both of the sciences spoken of in the present Book, and on others which will hereafter come under our notice: namely, the Idea of Polarity. This Idea will be the subject of the ensuing Book. And although this Idea forms a part of the basis of various other extensive portions of science, as Optics and Chemistry, it occupies so peculiarly conspicuous a place in speculations belonging to what I have termed the Mechanico-Chemical Sciences, (Magnetism and Electricity,) that I shall designate the discussion of the Idea of Polarity as the Philosophy of those Sciences.

BOOK V.

OF THE PHILOSOPHY OF THE MECHANICO-CHEMICAL SCIENCES.

CHAPTER I.

ATTEMPTS AT THE SCIENTIFIC APPLICATION OF THE IDEA OF POLARITY.

1. IN some of the mechanical sciences, as Magnetism and Optics, the phenomena are found to depend upon position (the position of the magnet, or of the ray of light,) in a peculiar alternate manner. This dependence, as it was first apprehended, was represented by means of certain conceptions of space and force, as for instance by considering the two *poles* of a magnet. But in all such modes of representing these alternations by the conceptions borrowed from other ideas, a closer examination detected something superfluous and something defective ; and in proportion as the view which philosophers took of this relation was gradually purified from these incongruous elements, and was rendered more general and abstract by the discovery of analogous properties in new cases, it was perceived that the relation could not be adequately apprehended without considering it as involving a peculiar and independent Idea, which we may designate by the term *Polarity*.

We shall trace some of the forms in which this Idea has manifested itself in the history of science. In doing so we shall not begin, as in other Books of this work we have done, by speaking of the notion as it is employed in common use: for the relation of polarity is of so abstract and technical a nature, that it is not employed, at least in any distinct and obvious manner, on any ordinary or practical occasions. The idea belongs peculiarly to the region of speculation: in persons of common habits of thought it is probably almost or quite undeveloped; and even most of those whose minds have been long occupied by science, find a difficulty in apprehending it in its full generality and abstraction, and stript of all irrelevant hypothesis.

2. *Magnetism.*—The name and the notion of *Poles* were first adopted in the case of a magnet. If we have two magnets, their extremities attract and repel each other alternatively. If the first end of the one attract the first end of the other, it repels the second end, and conversely. In order to express this rule conveniently, the two ends of each magnet are called the *north pole* and the *south pole* respectively, the denominations being borrowed from the poles of the earth and heavens. "These poles," as Gilbert says*, "regulate the motions of the celestial spheres and of the earth. In like manner the magnet has its poles, a northern and a southern one; certain and determined points constituted by nature in the stone, the primary terms of its motions and effects, the limits and governors of many actions and virtues."

The nature of the opposition of properties of which we speak may be stated thus.

The North pole of one magnet attracts the South pole of another magnet.

The North pole of one magnet *repels* the North pole of another magnet.

The South pole of one magnet repels the South pole of another magnet.

* *De Magn.*, lib. i. c. 3.

The South pole of one magnet *attracts* the North pole of another magnet.

It will be observed that the contrariety of position which is indicated by putting the South pole for the North pole in either magnet, is accompanied by the opposition of mechanical effect which is expressed by changing attraction into repulsion and repulsion into attraction: and thus we have the general feature of polarity:—A contrast of properties corresponding to a contrast of positions.

3. *Electricity.*—When the phenomena of electricity came to be studied, it appeared that they involved relations in some respects analogous to those of magnetism.

Two kinds of electricity were distinguished, the positive and the negative; and it appeared that two bodies electrized positively or two electrized negatively, repelled each other, like two north or two south magnetic poles; while a positively and a negatively electrized body attracted each other, like the north and south poles of two magnets. In conductors of an oblong form, the electricity could easily be made to distribute itself so that one end should be positively and one end negatively electrized; and then such conductors acted on each other exactly as magnets would do.

But in conductors, however electrized, there is no peculiar point which can permanently be considered as the *pole*. The distribution of electricity in the conductor depends upon external circumstances: and thus, although the phenomena offer the general character of *polarity*—alternative results corresponding to alternative positions,—they cannot be referred to poles. Some other mode of representing the forces must be adopted than that which makes them emanate from permanent points as in a magnet.

The phenomena of attraction and repulsion in electrized bodies were conveniently represented by means of the hypothesis of *two* electric *fluids*, a positive and a negative one, which were supposed to be distributed in the bodies. Of these fluids, it was supposed that each repelled its own parts and attracted those of the opposite fluid : and it was found that this hypothesis explained all the obvious laws of electric action. Here then we have the phenomena of polarization explained by a new kind of machinery :—two opposite fluids distributed in bodies, and supplying them, so to speak, with their polar forces. This hypothesis not only explains electrical attraction, but also the electrical spark : when two bodies, of which the neighbouring surfaces are charged with the two opposite fluids, approach near to each other, the mutual attraction of the fluids becomes more and more intense, till at last the excess of fluid on the one body breaks through the air and rushes to the other body, in a form accompanied by light and noise. When this transfer has taken place, the attraction ceases, the positive and the negative fluid having neutralized each other. Their effort was to unite ; and this union being effected, there is no longer any force in action. Bodies in their natural unexcited condition may be considered as occupied by a combination of the two fluids : and hence we see how the production of either kind of electricity is necessarily accompanied with the production of an equivalent amount of the opposite kind.

4. *Voltaic Electricity.*—Such is the case in Franklinic electricity,—that which is excited by the common electrical machine. In studying Voltaic electricity, we are led to the conviction that the fluid which is in a condition of momentary equilibrium in electrized conductors, exists in the state of *current* in the voltaic circuit. And here we find polar relations of a new kind existing among the forces. Two voltaic currents attract each other when

they are moving in the same, and repel each other when they are moving in opposite, directions.

But we find, in addition to these, other polar relations of a more abstruse kind, and which the supposition of two fluids does not so readily explain. For instance, if such fluids existed, distinct from each other, it might be expected that it would be possible to exhibit one of them separate from the other. Yet in all the phenomena of electromotive currents, we attempt in vain to obtain one kind of electricity separately. "I have not," says Mr. Faraday*, "been able to find a single fact which could be adduced to prove the theory of two electricities rather than one, in electric currents; or, admitting the hypothesis of two electricities, have I been able to perceive the slightest grounds that one electricity can be more powerful than the other,—or that it can be present without the other,—or that it can be varied or in the slightest degree affected without a corresponding variation in the other." "Thus," he adds, "the polar character of the powers is rigorous and complete." Thus, we too may remark, all the superfluous and precarious parts gradually drop off from the hypothesis which we devise in order to represent polar phenomena; and the abstract notion of polarity—of equal and opposite powers called into existence by a common condition—remains unincumbered with extraneous machinery.

5. *Light.*—Another very important example of the application of the idea of polarity is that supplied by the discovery of the polarization of light. A ray of light may, by various processes, be modified, so that it has different properties according to its different *sides*, although this difference is not perceptible by any common effects. If, for instance, a ray thus modified, pass perpendicularly

* *Researches*, 516.

through a circular glass, and fall upon the eye, we may turn the glass round and round its frame, and we shall make no difference in the brightness of the spot which we see. But if, instead of a glass, we look through a longitudinal slice of tourmaline, the spot is alternately dark and bright as we turn the crystal through successive quadrants. Here we have a contrast of properties (dark and bright) corresponding to a contrast of positions, (the position of a line east and west being contrasted with the position north and south,) which, as we have said, is the general character of polarity. It was with a view of expressing this character that the term *polarization* was originally introduced. Malus was forced by his disco-veries into the use of this expression. "We find," he says, in 1811, "that light acquires properties which are relative only to the sides of the ray,—which are the same for the north and south sides of the ray, (using the points of the compass for description's sake only,) and which are different when we go from the north and south to the east or to the west sides of the ray. I shall give the name of *poles* to these sides of the ray, and shall call *polarization* the modification which gives to light these properties relative to these poles. I have *put off* hitherto the admission of this term into the description of the physical phenomena with which we have to do: I did not *dare* to introduce it into the Memoirs in which I published my last observations: but the variety of forms in which this new phenomenon appears, and the difficulty of describing them, compel me to admit this new expres-sion; which signifies simply the modification which light has undergone in acquiring new properties which are not relative to the direction of the ray, but only to its sides considered at right angles to each other, and in a plane perpendicular to its direction."

The theory which represents light as an emission of

particles was in vogue at the time when Malus published his discoveries; and some of his followers in optical research conceived that the phenomena which he thus described rendered it necessary to ascribe poles and an axis to each particle of light. On this hypothesis, light would be polarized when the axes of all the particles were in the same direction: and, making such a supposition, it may easily be conceived capable of transmission through a crystal whose axis is parallel to that of the luminous particles, and intransmissible when the axis of the crystal is in a position transverse to that of the particles.

The hypothesis of particles possessing *poles* is a rude and arbitrary assumption, in this as in other cases; but it serves to convey the general notion of polarity, which is the essential feature of the phenomena. The term "polarization of light" has sometimes been complained of in modern times as hypothetical and obscure. But the real cause of obscurity was, that the Idea of Polarity was, till lately, very imperfectly developed in men's minds. As we have seen, the general notion of polarity,—opposite properties in opposite directions,—exactly describes the character of the optical phenomena to which the term is applied.

It is to be recollected that in optics we never speak of the *poles*, but of the *plane of polarization* of a ray. The word *sides*, which Newton and Malus have used, neither of them appears to have been satisfied with; Newton, in employing it, had recourse to the strange Gallicism of speaking of the *coast* of usual and of unusual refraction of a crystal.

The modern theory of optics represents the plane of polarization of light as depending, not on the position in which the axes of the luminiferous particles lie, but on the direction of those transverse vibrations in which light consists. This theory is, as we have stated in the History, recommended by an extraordinary series of successes in accounting for the phenomena. And this hypothesis of transverse vibrations shows us another mechanical mode, (besides the hypothesis of particles with axes,) by which we may represent the polarity of a ray. But we may remark that the general notion of polarity, as applied to light in such cases, would subsist, even if the undulatory theory were rejected. The idea is, as we have before said, independent of all hypothetical machinery.

I need not here refer to the various ways in which light may be polarized, as, for instance, by being reflected from the surface of water or of glass at certain angles, by being transmitted through crystals, and in other ways. In all cases the modification produced, the polarization, is identically the same property. Nor need I mention the various kinds of phenomena which appear as contrasts in the result; for these are not merely light and dark, or white and black, but red and green, and generally, a colour and its *complementary* colour, exhibited in many complex and varied configurations. These multiplied modes in which polarized light presents itself add nothing to the original conception of polarization: and I shall therefore pass on to another subject.

6. *Crystallization.*—Bodies which are perfectly crystallized exhibit the most complete regularity and symmetry of form; and this regularity not only appears in their outward shape, but pervades their whole texture, and manifests itself in their cleavage, their transparency, and in the uniform and determinate optical properties which exist in every part, even the smallest fragment of the mass. If we conceive crystals as composed of particles, we must suppose these particles to be arranged in the most regular manner; for example, if we suppose

each particle to have an axis, we must suppose all these axes to be parallel; for the direction of the axis of the particles is indicated by the physical and optical properties of the crystal, and therefore this direction must be the same for every portion of the crystal. This parallelism of the axes of the particles may be conceived to result from the circumstance of each particle having poles, the opposite poles attracting each other. In virtue of forces acting as this hypothesis assumes, a collection of small magnetic particles would arrange themselves in parallel positions; and such a collection of magnetic particles offers a sort of image of a crystal. Thus we are led to conceive the particles of crystals as polarized, and as determined in their crystalline positions by polar forces. This mode of apprehending the constitution of crystals has been adopted by some of our most eminent philosophers. Thus Berzelius says*, "It is demonstrated, that the regular forms of bodies presuppose an effort of their atoms to touch each other by preference in certain points; that is, they are founded upon a Polarity;"—he adds, "a polarity which can be no other than an electric or magnetic polarity." In this latter clause we have the identity of different kinds of polarity asserted; a principle which we shall speak of in the next chapter. But we may remark, that even without dwelling upon this connexion, any notion which we can form of the structure of crystals necessarily involves the idea of polarity. Whether this polarity necessarily requires us to believe crystals to be composed of atoms which exert an effort to touch each other in certain points by preference, is another question. And, in agreement with what has been said respecting other kinds of polarity, we shall probably find, on a more profound examination of the subject, that while the idea of polarity is essential,

* *Essay on the Theory of Chemical Properties*, 1820, p. 113.

the machinery by which it is thus expressed is precarious and superfluous.

7. *Chemical Affinity.*—We shall have, in the next Book, to speak of Chemical Affinity at some length ; but since the ultimate views to which philosophers have been led, induce them to consider the forces of affinity as polar forces, we must enumerate these among the examples of polarity. In chemical processes, opposites tend to unite, and to neutralize each other by their union. Thus an *acid* or an *alkali* combine with vehemence, and form a compound, a neutral salt, which is neither acid nor alkaline.

This conception of contrariety and mutual neutralization, involves the idea of polarity. In the conception, as entertained by the earlier chemists, the idea enters very obscurely : but in the attempts which have more recently been made to connect this relation (of acid and base,) with other relations, the chemical elements have been conceived as composed of particles which possess poles ; *like* poles repelling, and *unlike* attracting each other, as they do in magnetic and electric phenomena. This is, however, a rude and arbitrary way of expressing polarity, and, as may be easily shown, involves many difficulties which do not belong to the idea itself. Mr. Faraday, who has been led by his researches to a conviction of the polar nature of the forces of chemical affinity, has expressed their character in a more general manner, and without any of the machinery of particles indued with poles. According to his view, chemical synthesis and analysis must always be conceived as taking place in virtue of equal and opposite forces, by which the particles are united or separated. These forces, by the very circumstance of their being polar, may be transferred from point to point. For if we conceive a string of particles, and if the positive force of the first particle be liberated and brought into action, its negative force also must be set free: this negative force neutralizes the positive force of the next particle, and therefore the negative force of this particle (before employed in neutralizing its positive force,) is set free: this is in the same way transferred to the next particle, and so on. And thus we have a positive force active at one extremity of a line of particles, corresponding to a negative force at the other extremity, all the intermediate particles reciprocally neutralizing each other's action. This conception of the transfer of chemical action was indeed at an earlier period introduced by Grotthus*, and confirmed by Davy. But in Mr. Faraday's hands we see it divested of all that is superfluous, and spoken of, not as a line of particles, but as "an axis of power, having [at every point,] contrary forces, exactly equal, in opposite directions."

8. *General Remarks.*—Thus, as we see, the notion of polarity is applicable to many large classes of phenomena. Yet the idea in a distinct and general form is only of late growth among philosophers. It has gradually been abstracted and refined from many extraneous hypotheses which were at first supposed to be essential to it. We have noticed some of these hypotheses;—as the poles of a body; the poles of the particles of a fluid; two opposite fluids; a single fluid in excess and defect; transverse vibrations. To these others might be added. Thus Dr. Prout† assumes that the polarity of molecules results from their rotation on their axes, the opposite motions of contiguous molecules being the cause of opposite (positive and negative) polarities.

But none of these hypotheses can be proved by the fact of polarity alone; and they have been in succession rejected when they had been assumed on that ground.

---

* Dumas, *Leçons sur la Philosophie Chimique*, p. 401.
† *Bridgwater Treatise*, p. 559.

Thus Davy, in 1826, speaking of chemical forces says[*], " In assuming the idea of two ethereal, subtile, elastic fluids, attractive of the particles of each other, and repulsive as to their own particles, capable of combining in different proportions with bodies, and according to their proportions giving them their specific qualities and rendering them equivalent masses, it would be natural to refer the action of the poles to the repulsions of the substances combined with the excess of one fluid, and the attractions of those united to the excess of the other fluid ; and a history of the phenomena, not unsatisfactory to the reason, might in this way be made out. But as it is possible likewise to take an entirely different view of the subject, on the idea of the dependence of the results upon the primary attractive powers of the parts of the combination on a single subtile fluid, I shall not enter into any discussion on this obscure part of the theory." Which of these theories will best represent the case, will depend upon the consideration of other facts, in combination with the polar phenomena, as we see in the history of optical theory. In like manner Mr. Faraday proved by experiment[†] the error of all theories which ascribe electro-chemical decomposition to the attraction of the poles of the voltaic battery.

In order that they may distinctly image to themselves the idea of polarity, men clothe it in some of the forms of machinery above spoken of; yet every new attempt shows them the unnecessary difficulties in which they thus involve themselves. But on the other hand it is difficult to apprehend this idea divested of all machinery; and to entertain it in such a form that it shall apply at the same time to magnetism and electricity, galvanism and chemistry, crystalline structure and light. The Idea of *Polarity* becomes most pure and genuine, when we entirely reject the conception of *Poles,* as Faraday has taught us to do in considering electro-chemical decomposition; but it is only by degrees and by effort that we can reach this point of abstraction and generality.

9. There is one other remark which we may here make. It was a maxim commonly received in the ancient schools of philosophy, that "like attracts like:" but as we have seen, the universal maxim of polar phenomena is, that like *repels* like, and attracts unlike. The north pole attracts the south pole, the positive fluid attracts the negative fluid; opposite elements rush together; opposite motions reduce each other to rest. The permanent and stable course of things is that which results from the balance and neutralization of contrary tendencies. Nature is constantly labouring after repose by the effect of such tendencies; and so far as polar forces enter into her economy, she seeks harmony by means of discord, and unity by opposition.

Although the Idea of Polarity is still somewhat vague and obscure, even in the minds of the cultivators of physical science, it has still given birth to some general principles which have been accepted as evident, and have had great influence on the progress of science. These we shall now consider.

CHAPTER II.

OF THE CONNEXION OF POLARITIES.

1. IT has appeared in the preceding chapter that in cases in which the phenomena suggest to us the idea of polarity, we are also led to assume some material machinery as the mode in which the polar forces are exerted. We assume, for instance, globular particles which possess

poles, or the vibrations of a fluid, or two fluids attracting each other; in every case, in short, some hypothesis by which the existence and operation of the polarity is embodied in geometrical and mechanical properties of a medium; nor is it possible for us to avoid proceeding upon the conviction that some such hypothesis must be true; although the nature of the connexion between the mechanism and the phenomena must still be indefinite and arbitrary.

But since each class of polar phenomena is thus referred to an ulterior cause, of which we know no more than that it has a polar character, it follows that different polarities may result from the same cause manifesting its polar character under different aspects. Taking, for example, the hypothesis of globular particles, if electricity result from an action dependent upon the *poles* of each globule, magnetism may depend upon an action in the *equator* of each globule; or taking the supposition of transverse vibrations, if polarized light result directly from such vibrations, crystallization may have reference to the axes of the elasticity of the medium by which the vibrations are rendered transverse,—so far as the polar character only of the phenomena is to be accounted for. I say this *may* be so, *in so far* only as the polar character of the phenomena is concerned; for whether the relation of electricity to magnetism, or of crystalline forces to light, can really be explained by such hypotheses, remains to be determined by the facts themselves. But since the first necessary feature of the hypothesis is, that it shall give polarity, and since an hypothesis which does this may, by its mathematical relations, give polarities of different kinds and in different directions, any two co-existent kinds of polarity may' result from the same cause, manifesting itself in various manners.

The conclusion to which we are led by these general

considerations is, that two co-existing classes of polar phenomena *may* be effects of the same cause. But those who have studied such phenomena more deeply and attentively have, in most or in all cases, arrived at the conviction that the various kinds of polarity in such cases *must* be connected and fundamentally identical. As this conviction has exercised a great influence, both upon the discoveries of new facts and upon the theoretical speculations of modern philosophers, and has been put forward by some writers as a universal principle of science, I will consider some of the cases in which it has been thus applied.

2. *Connexion of Magnetic and Electric Polarity.*— The polar phenomena of electricity and magnetism are clearly analogous in their laws : and obvious facts showed at an early period that there was some connexion between the two agencies. Attempts were made to establish an evident and definite relation between the two kinds of force, which attempts proceeded upon the principle now under consideration ;—namely, that in such cases, the two kinds of polarity must be connected. Professor Œrsted, of Copenhagen, was one of those who made many trials founded upon this conviction : yet all these were long unsuccessful. At length, in 1820, he discovered that a galvanic current, passing at right angles near to a magnetic needle, exercises upon it a powerful deflecting force. The connexion once detected between magnetism and galvanism was soon recognised as constant and universal. It was represented in different hypothetical modes by different persons ; some considering the galvanic current as the primitive axis, and the magnet as constituted of galvanic currents passing round it at right angles to the magnetic axis ; while others conceived the magnetic axis as the primitive one, and the electric current as implying a magnetic current round the wire.

So far as many of the general relations of these two kinds
of force were concerned, either mode of representation
served to express them; and thus the assumption that
the two polarities, the magnetic and the electric, were
fundamentally identical, was verified, so far as the phenomena of magnetic attraction, and the like, were concerned.

I need not here mention how this was further confirmed by the experiments in which, by means of the
forces thus brought into view, a galvanic wire was made
to revolve round a magnet, and a magnet round a galvanic wire; in which artificial magnets were constructed
of coils of galvanic wire; and finally, in which the galvanic spark was obtained from the magnet. The identity
which sagacious speculators had divined even before it
was discovered, and which they had seen to be universal as
soon as it was brought to light, was completely manifested
in every imaginable form.

The relation of the electric and magnetic polarities
was found to be, that they were *transverse* to each other,
and this relation exhibited under various conditions of
form and position of the apparatus, gave rise to very
curious and unexpected perplexities. The degree of complication which this relation may occasion, may be judged
of from the number of constructions and modes of conception offered by Œrsted, Wollaston, Faraday, and others,
for the purpose of framing a technical memory of the
results. The magnetic polarity gives us the north and
south poles of the needle; the electric polarity makes the
current positive and negative; and these pairs of opposites
are connected by relations of situation, as above and below,
right and left; and give rise to the resulting motion of
the needle one way or the other.

3. Ampère, by framing his hypotheses of the action of
voltaic currents and the constitution of magnets, reduced

all these technical rules to rigorous deductions from one general principle. And thus the vague and obscure persuasion that there *must* be *some* connexion between electricity and magnetism, so long an idle and barren conjecture, was unfolded into a complete theory, according to which magnetic and electromotive actions are only two different manifestations of the same forces; and all the above-mentioned complex relations of polarities are reduced to one single polarity, that of the electro-dynamic current.

4. As the idea of polarity was thus firmly established and clearly developed, it became an instrument of reasoning. Thus it led Ampère to maintain that the original or elementary forces in electro-dynamic action could not be as M, Biot thought they were, a statical *couple*, but must be directly opposite to each other. The same idea enabled Mr. Faraday to carry on with confidence such reasonings as the following * : "No other known power has like direction with that exerted between an electric current and a magnetic pole; it is tangential, while all other forces acting at a distance are direct. Hence if a magnetic pole on one side of a revolving plate follow its course by reason of its obedience to the tangential force exerted upon it by the very current of electricity which it has itself caused; a similar pole on the other side of the plate should immediately set it free from this force; for the currents which have to be formed by the two poles are in contrary directions." And in Article 1114 of his Researches, the same eminent philosopher infers that if electricity and magnetism are considered as the results of a peculiar agent or condition, exerted in determinate directions perpendicular to each other, one must be by some means convertible into the other; and this he was afterwards able to prove to be the case in fact.

* *Researches*, 244.

Thus the principle that the co-existent polarities of magnetism and electricity are connected and fundamentally identical, is not only true, but is far from being either vague or barren. It has been a fertile source both of theories which have, at present, a very great probability, and of the discovery of new and striking facts. We proceed to consider other similar cases.

5. *Connexion of Electrical and Chemical Polarities.*—The doctrine that the chemical forces by which the elements of bodies are held together or separated, are identical with the polar forces of electricity, is a great discovery of modern times; so great and so recent, indeed, that probably men of science in general have hardly yet obtained a clear view and firm hold of this truth. This doctrine is now, however, entirely established in the minds of the most profound and philosophical chemists of our time. The complete developement and confirmation of this as of other great truths, was preceded by more vague and confused opinions gradually tending to this point; and the progress of thought and of research was impelled and guided, in this as in similar cases, by the persuasion that these co-existent polarities could not fail to be closely connected with each other. While the ultimate and exact theory to which previous incomplete and transitory theories tended is still so new and so unfamiliar, it must needs be a matter of difficulty and responsibility for a common reader to describe the steps by which truth has advanced from point to point. I shall, therefore, in doing this, guide myself mainly by the historical sketches of the progress of this great theory, which, fortunately for us, have been given us by the two philosophers who have played by far the most important parts in the discovery, Davy and Faraday.

It will be observed that we are concerned here with the progress of theory, and not of experiment, except so

far as it is confirmatory of theory. In Davy's Memoir*
of 1826, on the Relations of Electrical and Chemical
Changes, he gives the historical details to which I have
alluded. Already in 1802 he had conjectured that all
chemical decompositions might be polar. In 1806 he
attempted to confirm this conjecture, and succeeded, to
his own satisfaction, in establishing† that the combinations and decompositions by electricity were referable
to the law of electrical attractions and repulsions; and
advanced the hypothesis (as he calls it,) that chemical and
electrical attractions were produced by the same cause,
acting in one case on particles, in the other on masses.
This hypothesis was most strikingly confirmed by the
author's being able to use electrical agency as a more
powerful means of chemical decomposition than any
which had yet been applied. "Believing," he adds, "that
our philosophical systems are exceedingly imperfect, I
never attached much importance to this hypothesis; but
having formed it after a copious induction of facts, and
having gained by the application of it a number of practical results, and considering myself as much the author
of it as I was of the decomposition of the alkalies, and
having developed it in an elementary work as far as the
present state of chemistry seemed to allow, I have never,"
he says, "criticized or examined the manner in which
different authors have adopted or explained it, contented,
if in the hands of others, it assisted the arrangements of
chemistry or mineralogy, or became an instrument of discovery." When the doctrine had found an extensive
acceptance among chemists, attempts were made to show
that it had been asserted by earlier writers: and though
Davy justly denies all value to these pretended anticipations, they serve to show, however dimly, the working of
that conviction of the connexion of co-existent proper-

* *Phil. Trans.*, 1826, p. 383.   † P. 389.

ties which all along presided in men's minds during this course of investigation. " Ritter and Winterl have been quoted," Davy says*, "among other persons, as having imagined or anticipated the relation between electrical powers and chemical affinities before the discovery of the pile of Volta. But whoever will read with attention Ritter's ' Evidence that Galvanic action exists in orga- nized nature,' and Winter's *Prolusiones ad Chemiam sæculi decimi noni,* will find nothing to justify this opi- nion." He then refers to the Queries of Newton at the end of his Optics. "These," he says, "contain more grand and speculative views that might be brought to bear upon this question than any found in the works of modern electricians; but it is very unjust to the experi- mentalists who by the laborious application of new in- struments, have discovered novel facts and analogies, to refer them to any such suppositions as that all attractions, chemical, electrical, magnetical, and gravitative, may de- pend upon the same cause." It is perfectly true, that such vague opinions, though arising from that tendency to generalize which is the essence of science, are of no value except so far as they are both rendered intelligible, and confirmed by experimental research.

The phenomena of chemical decomposition by means of the voltaic pile, however, led other persons to views very similar to those of Davy. Thus Grotthus in 1805† published an hypothesis of the same kind. "The pile of Volta," he says, " is an electrical magnet, of which each element, that is, each pair of plates, has a positive and a negative pole. The consideration of this polarity sug- gested to me the idea that a similar polarity may come into play between the elementary particles of water when acted upon by the same electrical agent; and I avow that this thought was for me a flash of light."

* *Phil. Trans.,* 1826, p. 384.   † *Ann. Chim.,* lxviii., 54.

6. The thought, however, though thus brought into being, was very far from being as yet freed from vagueness, superfluities, and errors. I have elsewhere noticed* Faraday's remark on Davy's celebrated Memoir of 1806; that " the mode of action by which the effects take place is stated very generally, so generally, indeed, that probably a dozen precise schemes of electro-chemical action might be drawn up, differing essentially from each other, yet all agreeing with the statement there given." When Davy and others proceeded to give a little more definiteness and precision to the statement of their views, they soon introduced into the theory features which it was afterwards found necessary to abandon. Thus† both Davy, Grotthus, Riffault, and Chompré, ascribed electrical decomposition to the action of the poles, and some of them even pretended to assign the proportion in which the force of the pole diminishes as the distance from it increases. Faraday, as I have already stated, showed that the polarity must be considered as residing not only in what had till then been called the poles, but at every point of the circuit. He ascribed‡ electro-chemical decomposition to internal forces, residing in the particles of the matter under decomposition, not to external forces, exerted by the poles. Hence he shortly afterwards§ proposed to reject the word poles altogether, and to employ instead, the term *electrode*, meaning the doors or passages (of whatever surface formed,) by which the decomposed elements pass out. What have been called the positive and negative poles he further termed the *anode* and *cathode;* and he introduced some other changes in nomenclature connected with these. He then, as I have

* *Hist. Ind. Sci.*, iii. 161.
† See Faraday's Historical Sketch, *Researches*, 481—492.
‡ Art. 524.
§ In 1834. Eleventh Series of *Researches*. Art. 662.

related in the History*, invented the Volta-electrometer, which enabled him to measure the quantity of voltaic action, and this he found to be identical with the quantity of chemical affinity; and he was thus led to the clearest view of the truth towards which he and his predecessors had so long been travelling, that electrical and chemical forces are identical †.

7. It will, perhaps, be said that this beautiful train of discovery was entirely due to experiment, and not to any *à priori* conviction that co-existent polarities must be connected. I trust I have sufficiently stated that such an *à priori* principle could not be proved, nor even understood, without a most laborious and enlightened use of experiment; but yet I think that the doctrine when once fully unfolded, exhibited clearly, and established as true, takes possession of the mind with a more entire conviction of its certainty and universality, in virtue of the principle we are now considering. When the theory has assumed so simple a form, it appears to derive immense probability (to say the least) from its simplicity. Like the laws of motion, when stated in its most general form, it appears to carry with it its own evidence. And thus this great theory borrows something of its character from the Ideas which it involves, as well as from the experiments by which it was established.

8. We may find in many of Mr. Faraday's subsequent reasonings, clear evidence that this idea of the connexion of polarities, as now developed, is not limited in its application to facts already known experimentally, but, like other ideas, determines the philosopher's researches into the unknown, and gives us the *form* of knowledge even before we possess the *matter*. Thus, he says, in his Thirteenth Series‡, "I have long sought, and still seek, for an effect or condition which shall be to statical electricity

---

* *Hist. Ind. Sci.*, iii., 168.   † Art. 915, 916, 917.   ‡ Art. 1658.

what magnetic force is to current electricity; for as the lines of discharge are associated with a certain transverse effect, so it appeared to me impossible but that the lines of tension or of inductive action, which of necessity precede the discharge, should also have their correspondent transverse condition or effect." Other similar passages might be found.

I will now consider another case to which we may apply the principle of connected polarities.

9. *Connexion of Chemical and Crystalline Polarities.* —The close connexion between the chemical affinity and the crystalline attraction of elements cannot be overlooked. Bodies never crystallize but when their elements combine chemically; and solid bodies which combine, when they do it most completely and exactly, also crystallize. The forces which hold together the elements of a crystal of alum are the same forces which make it a crystal. There is no distinguishing between the two sets of forces.

Both chemical and crystalline forces are polar, as we stated in the last chapter; but the polarity in the two cases is of a different kind. The polarity of chemical forces is then put in the most distinct form, when it is identified with electrical polarity; the polarity of the particles of crystals has reference to their geometrical form. And it is clear that these two kinds of polarity must be connected. Accordingly, Berzelius expressly asserts* the necessary identity of these two polarities. "The regular forms of bodies suppose a polarity which can be no other than an electric or magnetic polarity." This being so seemingly inevitable, we might expect to find the electric forces manifesting some relation to the definite directions of crystalline forms. Mr. Faraday tried, but in vain, to detect some such relation. He attempted

* *Essay on Chemical Prop.*, 113.

to ascertain* whether a cube of rock crystal transmitted the electrical force of tension with different intensity along and across the axis of the crystal. In the first specimen there seemed to be some difference; but in other experiments, made both with rock crystal and with calc spar, this difference disappeared. Although therefore we may venture to assert that there must be some very close connexion between electrical and crystalline forces, we are, as yet, quite ignorant what the nature of the connexion is, and in what kind of phenomena it will manifest itself.

10. *Connexion of Crystalline and Optical Polarities.*— Crystals present to us optical phenomena which have a manifestly polar character. The double refraction, both of uniaxal and of biaxal crystals, is always accompanied with opposite polarization of the two rays; and in this and in other ways light is polarized in directions dependent upon the axes of the crystalline form, that is, on the directions of the polarities of the crystalline particles. The identity of these two kinds of polarity (crystalline and optical) is too obviou to need insisting on; and it is not necessary for us here to decide by what hypothesis this identity may most properly be represented. We may hereafter perhaps find ourselves justified in considering the crystalline forces as determining the elasticity of the luminiferous ether to be different in different directions within the crystal, and thus as determining the refraction and polarization of the light which the crystal transmits. But at present we merely note this case as an additional example of the manifest connexion and fundamental identity of two co-existent polarities.

11. *Connexion of Polarities in general.*—Thus we find that the connexion of different kinds of polarities, magnetic, electric, chemical, crystalline, and optical, is certain

---

* *Researches.* Art. 1689.

as a truth of experimental science. We have attempted to show further that in the minds of several of the most eminent discoverers and philosophers, such a conviction is something more than a mere empirical result: it is a principle which has regulated their researches while it was still but obscurely seen and imperfectly unfolded, and has given to their theories a character of generality and self-evidence which experience alone cannot bestow.

It will, perhaps, be said that these doctrines,—that scientific researches may usefully be directed by principles in themselves vague and obscure;—that theories may have an evidence superior to and anterior to experience;—are doctrines in the highest degree dangerous, and utterly at variance with the soundest maxims of modern times respecting the cultivation of science.

To the justice and wisdom of this caution I entirely agree : and although I have shown that this principle of the *connexion of polarities*, rightly interpreted and established in each case by experiment, involves profound and comprehensive truths; I think it no less important to remark that, at least in the present stage of our knowledge, we can make no use of this principle without taking care, at every step, to determine by clear and decisive experiments, its proper meaning and application. All endeavours to proceed otherwise have led, and must lead, to ignorance and confusion. Attempts to deduce from our bare idea of polarity, and our fundamental convictions respecting the connexion of polarities, theories concerning the forces which really exist in nature, can hardly have any other result than to bewilder men's minds, and to misdirect their efforts.

So far, indeed, as this persuasion of a connexion among apparently different kinds of agencies impels men, engaged in the pursuit of knowledge, to collect observations, to multiply, repeat, and vary experiments, and to contemplate the result of these in all aspects and relations, it may be an occasion of the most important discoveries. Accordingly we find that the great laws of phenomena which govern the motions of the planets about the sun, were first discovered by Kepler, in consequence of his scrutinizing the recorded observations with an intense conviction of the existence of geometrical and arithmetical harmonies in the solar system. Perhaps we may consider the discovery of the connexion of magnetism and electricity by Professor Œrsted in 1820, as an example somewhat of the same kind; for he also was a believer in certain comprehensive but undefined relations among the properties of bodies; and in consequence of such views entertained great admiration for the *Prologue to the Chemistry of the Nineteenth Century*, of Winterl, already mentioned. M. Œrsted, in 1803, published a summary of this work; and in so doing, praised the views of Winterl as far more profound and comprehensive than those of Lavoisier. Soon afterwards a Review of this publication appeared in France*, in which it was spoken of as a work only fit for the dark ages, and as the indication of a sect which had for some time "ravaged Germany," and inundated that country with extravagant and unintelligible mysticism. It was, therefore, a kind of triumph to M. Œrsted to be, after some years' labour, the author of one of the most remarkable and fertile physical discoveries of his time.

12. It was not indeed without some reason that certain of the German philosophers were accused of dealing in doctrines vast and profound in their aspect, but, in reality, indefinite, ambiguous, and inapplicable. And the most prominent of such doctrines had reference to the principle now under our consideration; they represented the properties of bodies as consisting in certain polarities,

* *Ann. Chim.*, tom. 50 (1804), p. 191.

and professed to deduce, from the very nature of things, with little or no reference to experiment, the existence and connexion of these polarities. Thus Schelling, in his *Ideas towards a Philosophy of Nature*, published in 1803, says*, "Magnetism is the universal act of investing Multiplicity with Unity; but the universal form of the reduction of Multiplicity to Unity is the Line, pure Longitudinal Extension: hence Magnetism is determination of pure Longitudinal Extension; and as this manifests itself by absolute Cohesion, Magnetism is the determination of absolute Cohesion." And as Magnetism was, by such reasoning, conceived to be proved as a universal property of matter, Schelling asserted it to be a confirmation of his views when it was discovered that other bodies besides iron are magnetic. In like manner he used such expressions as the following†. "The threefold character of the Universal, the Particular, and the Indifference of the two,—as expressed in their Identity, is Magnetism, as expressed in their Difference, is Electricity, and as expressed in the Totality, is Chemical Process. Thus these forms are only one form; and the Chemical Process is a mere transfer of the three Points of Magnetism into the Triangle of Chemistry."

It was very natural that the chemists should refuse to acknowledge, in this fanciful and vague language, (delivered, however, it is to be recollected, in 1803,) an anticipation of Davy's doctrine of the identity of electrical and chemical forces, or of Œrsted's electro-magnetic agency. Yet it was perhaps no less natural that the author of such assertions should look upon every great step in the electro-chemical theory as an illustration of his own doctrines. Accordingly we find Schelling welcoming, with a due sense of their importance, the discoveries of Faraday. When he heard of the experiment

* P. 223.   † P. 486.

in which electricity was produced from common magnetism, he fastened with enthusiasm upon the discovery, even before he knew any of its details, and proclaimed it at a public meeting of a scientific body* as one of the most important advances of modern science. We have (he thus reasoned) three effects of polar forces;—electrochemical Decomposition, electrical Action, Magnetism. Volta and Davy had confirmed experimentally the identity of the two former agencies: Œrsted showed that a closed voltaic circuit acquired magnetic properties: but in order to exhibit the identity of electric and magnetic action it was requisite that electric forces should be extricated from magnetic. This great step Faraday, he remarked, had made, in producing the electric spark by means of magnets.

13. Although conjectures and assertions of the kind thus put forth by Schelling involve a persuasion of the pervading influence and connexion of polarities, which persuasion has already been confirmed in many instances, they involve this principle in a manner so vague and ambiguous that it can rarely, in such a form, be of any use or value. Such views of polarity can never teach us in what cases we are and in what we are not to expect to find polar relations; and indeed tend rather to diffuse error and confusion, than to promote knowledge. Accordingly we cannot be surprised to find such doctrines put forward by their authors as an evidence of the small value and necessity of experimental science. This is done by the celebrated metaphysician Hegel, in his *Encyclopædia* †. "Since," says he, "the plane of incidence and of reflection in simple reflection is the same plane, when a second reflector is introduced which further distributes the illumination reflected from the

* UEBER FARADAY's *Neueste Entdeckung. München.* 1832.
† Sec. 278.

first, the position of the first plane with respect to the second plane, containing the direction of the first reflection and of the second, has its influence upon the position, illumination or darkening of the object as it appears by the second reflection. This influence must be the strongest when the two planes are what we must call *negatively* related to each other:—that is, when they are at right angles." "But," he adds, "when men infer (as Malus has done) from the modification which is produced by this situation, in the illumination of the reflection, that the molecules of light in themselves, that is, on their different sides, possess different physical energies; and when on this foundation, along with the phenomena of entoptical colours therewith connected, a wide labyrinth of the most complex theory is erected; we have then one of the most remarkable examples of the *inferences* of physics from experiment." If Hegel's reasoning prove anything, it must prove that polarization always accompanies reflection under such circumstances as he describes: yet all physical philosophers know that in the case of metals, in which the reflection is most complete, light is not completely polarized at any angle; and that in other substances the polarization depends upon various circumstances which show how idle and inapplicable is the account he thus gives of the property. His self-complacent remark about the inferences of physics from experiment, is intended to recommend by comparison his own method of considering the nature of things in themselves; a mode of obtaining physical truth which had been more than exhausted by Aristotle, and out of which no new attempts have extracted anything of value since his time.

14. Thus the general conclusion to which we are led on this subject is, that the persuasion of the existence and connexion or identity of various polarities in nature,

although very naturally admitted, and in many cases interpreted and confirmed by observed facts, is of itself, so far as we at present possess it, a very insecure guide to scientific doctrines. When it is allowed to dictate our theories, instead of animating and extending our experimental researches, it leads only to error, confusion, obscurity, and mysticism.

This Fifth Book, on the subject of Polarities, is a short one compared with most of the others. This arises in a great measure from the circumstance that the Idea of Polarity has only recently been apprehended and applied, with any great degree of clearness, among physical philosophers; and is even yet probably entertained in an obscure and ambiguous manner by most experimental inquirers. I have been desirous of not attempting to bring forward any doctrines upon the subject, except such as have been fully illustrated and exemplified by the acknowledged progress of the physical sciences. If I had been willing to discuss the various speculations which have been published respecting the universal prevalence of polarities in the universe, and their results in every province of nature, I might easily have presented this subject in a more extended form; but this would not have been consistent with my plan of tracing the influence of scientific ideas only so far as they have really aided in disclosing and developing scientific truths. And as the influence of this idea is clearly distinguishable both from those which precede and those which follow in the character of the sciences to which it gives rise, and appears likely to be hereafter of great extent and consequence, it seemed better to treat of it in a separate Book, although of a brevity disproportioned to the rest.

BOOK VI.

THE PHILOSOPHY OF CHEMISTRY.

CHAPTER I.

ATTEMPTS TO CONCEIVE ELEMENTARY COMPOSITION.

1. WE have now to bring into view, if possible, the ideas and general principles which are involved in Chemistry,—the science of the composition of bodies. For in this as in other parts of human knowledge, we shall find that there are certain ideas, deeply seated in the mind, though shaped and unfolded by external observation, which are necessary conditions of the existence of such a science. These ideas it is which impel man to such a knowledge of the composition of bodies, which give meaning to facts exhibiting this composition, and universality to special truths discovered by experience. These are the Ideas of *Element* and of *Substance.*

Unlike the idea of polarization, of which we treated in the last Book, these ideas have been current in men's minds from very early times, and formed the subject of some of the first speculations of philosophers. It happened however, as might have been expected, that in the first attempts they were not clearly distinguished from other notions, and were apprehended and applied in an obscure and confused manner. We cannot better exhibit the peculiar character and meaning of these ideas than by tracing the form which they have assumed and the efficacy which they have exerted in these successive essays. This, therefore, I shall endeavour to do, beginning with the Idea of Element.

2. That bodies are composed or made up of certain parts, elements, or principles, is a conception which has existed in men's minds from the beginning of the first attempts at speculative knowledge. The doctrine of the four elements, earth, air, fire and water, of which all things in the universe were supposed to be constituted, is one of the earliest forms in which this conception was systematized; and this doctrine is stated by various authors to have existed as early as the times of the ancient Egyptians*. The words usually employed by Greek writers to express these elements are ἀρχή, a principle or beginning, and στοιχεῖον, which probably meant a letter (of a word) before it meant an element of a compound. For the resolution of a word into its letters is undoubtedly a remarkable instance of a successful analysis performed at an early stage of man's history; and might very naturally supply a metaphor to denote the analysis of substances into their intimate parts, when men began to contemplate such an analysis as a subject of speculation. The Latin word *elementum* itself, though by its form it appears to be a derivative abstract term, comes from some root now obsolete; probably† from a word signifying to grow or spring up.

The mode in which elements form the compound bodies and determine their properties was at first, as might be expected, vaguely and variously conceived. It will, I trust, hereafter be made clear to the reader that the relation of the elements to the compound involves a

---

* GILBERT's *Phys.*, l. i. c. 3.

† VOSSIUS *in voce.* "Conjecto esse ab antiquâ voce *eleo* pro *oleo*, id est *cresco:* à qua significatione *proles, suboles, adolescens:* ut ab *juratum, juramentum;* ab *adjutum, adjumentum:* sic ab *eletum, elementum:* quia inde omnia crescunt ac nascuntur."

peculiar and appropriate Fundamental Idea, not susceptible of being correctly represented by any comparison or
combination of other ideas, and guiding us to clear and
definite results only when it is illustrated and nourished
by an abundant supply of experimental facts.   But at first
the peculiar and special notion which is required in a just
conception of the constitution of bodies was neither discerned nor suspected; and up to a very late period in the
history of chemistry, men went on attempting to apprehend the constitution of bodies more clearly by substitu-
ting for this obscure and recondite idea of elementary
composition, some other idea more obvious, more luminous, and more familiar, such as the ideas of resemblance,
position, and mechanical force.   We shall briefly speak of
some of these attempts, and of the errors which were
thus introduced into speculations on the relations of
elements and compounds.

3. *Compounds assumed to resemble their Elements.*—
The first notion was that compounds derive their qualities
from their elements by *resemblance:*—they are hot in
virtue of a hot element, heavy in virtue of a heavy
element, and so on.   In this way the doctrine of the *four
elements* was framed; for every body is either hot or
cold, moist or dry; and by combining these qualities in
all possible ways, men devised four elementary substances, as has been stated in the History*.

This assumption of the derivation of the qualities of
bodies from similar qualities in the elements was, as we
shall see, altogether baseless and unphilosophical, yet it
prevailed long and universally.   It was the foundation of
medicine for a long period, both in Europe and Asia;
disorders being divided into hot, cold, and the like; and
remedies being arranged according to similar distinctions.
Many readers will recollect, perhaps, the story† of the

* *Hist. Ind. Sci.,* i. 47.   † See *Hadji Baba.*

indignation which the Persian physicians felt towards the European, when he undertook to cure the ill effects of cucumber upon the patient, by means of mercurial medicine: for cucumber, which is cold, could not be counteracted, they maintained, by mercury, which in their classification is cold also. Similar views of the operation of medicines might easily be traced in our own country. A moment's reflection may convince us that when drugs of any kind are subjected to the chemistry of the human stomach and thus made to operate on the human frame, it is utterly impossible to form the most remote conjecture what the result will be from any such vague notions of their qualities as the common use of our senses can give. And in like manner the common operations of chemistry give rise in almost every instance to products which bear no resemblance to the materials employed. The results of the furnace, the alembic, the mixture frequently bear no visible resemblance to the ingredients operated upon. Iron becomes steel by the addition of a little charcoal; but what visible trace of the charcoal is presented by the metal thus modified? The most beautiful colours are given to glass and earthenware by minute portions of the ores of black or dingy metals, as iron and manganese. The worker in metal, the painter, the dyer, the vintner, the brewer, all the artisans in short who deal with practical chemistry, are able to teach the speculative chemist that nothing can be so false as to expect that the qualities of the elements shall be still discoverable, in an unaltered form, in the compound. This first rude notion of an element, that it determines the properties of bodies by resemblance, must be utterly rejected and abandoned before we can make any advance towards a true apprehension of the constitution of bodies.

4. This step accordingly was made, when the hypo-

thesis of the four elements was given up, and the doctrine of the *three principles*, salt, sulphur and mercury, was substituted in its place.   For in making this change, as I have remarked in the History*, the real advance was the acknowledgment of the changes produced by the chemist's operations as results to be accounted for by the union and separation of substantial elements, however great the changes, and however unlike the product might be to the materials.   And this step once made, chemists went on constantly advancing towards a truer view of the nature of an element, and consequently, towards a more satisfactory theory of chemical operations.

5. Yet we may, I think, note one instance, even in the works of eminent modern chemists, in which this maxim, that we have no right to expect any resemblance between the elements and the compound, is lost sight of.   I speak of certain classifications of mineral substances.   Berzelius, in his System of Mineral Arrangement, places *sulphur* next to the *sulphurets*.   But surely this is an error, involving the ancient assumption of the resemblance of elements and compounds; as if we were to expect the sulphurets to bear a resemblance to sulphur.   All classifications are intended to bring together things resembling each other: the sulphurets of metals have certain general resemblances which make them a tolerably distinct, well determined, class of bodies.   But sulphur has no resemblances with these, no analogies with them, either in physical or even in chemical properties.   It is a simple body; and both its resemblances and its analogies direct us to place it along with other simple bodies, (selenium, and phosphorus,) which, united with metals, produce compounds not very different from the sulphurets.   Sulphur cannot be, nor approach to being, a sulphuret; we must not confound what it *is* with what it *makes*.   Sulphur has

* *Hist. Ind. Sci.*, iii. 100.

its proper influence in determining the properties of the compound into which it enters; but it does not do this according to resemblance of qualities, or according to any principle which properly leads to propinquity in classification.

6. *Compounds assumed to be determined by the Figure of Elements.*—I pass over the fanciful modes of representing chemical changes which were employed by the Alchemists; for these strange inventions did little in leading men towards a juster view of the relations of elements to compounds. I proceed for an instant to the attempt to substitute another obvious conception for the still obscure notion of elementary composition. It was imagined that all the properties of bodies and their mutual operations might be accounted for by supposing them constituted of *particles* of various *forms*, round or angular, pointed or hooked, straight or spiral. This is a very ancient hypothesis, and a favourite one with many casual speculators in all ages. Thus Lucretius undertakes to explain why wine passes rapidly through a sieve and oil slowly, by telling us that the latter substance has its particles either larger than those of the other, or more hooked and interwoven together. And he accounts for the difference of sweet and bitter by supposing the particles in the former case to be round and smooth, in the latter sharp and jagged[*]. Similar assumptions prevailed in modern times on the revival of the mechanical philosophy, and constitute a large part of the physical schemes of Descartes and Gassendi. They were also adopted to a considerable extent by the chemists. Acids were without hesitation assumed to consist of sharp pointed particles; which, "I hope," Lemery says[†], "no one will dispute, seeing every one's experience does demonstrate it: he needs but taste an acid to be satisfied of it, for it pricks the tongue like

* *De Rerum Natura*, ii. 390 sqq.   † *Chemistry*, p. 25.

anything keen and finely cut." Such an assumption is not only altogether gratuitous and useless, but appears to be founded in some degree upon a confusion in the metaphorical and literal use of such words as *keen* and *sharp*. The assumption once made, it was easy to accommodate it, in a manner equally arbitrary, to other facts. "A demonstrative and convincing proof that an acid does consist of pointed parts is, that not only all acid salts do crystallize into edges, but all dissolutions of different things, caused by acid liquors, do assume this figure in their crystallization. These crystals consist of points differing both in length and bigness one from another, and this diversity must be attributed to the keener or blunter edges of the different sorts of acids: and so likewise this difference of the points in subtilty is the cause that one acid can penetrate and dissolve with one sort of *mixt*, that another can't rarify at all: Thus *vinegar* dissolves *lead*, which *aqua fortis* can't : *aqua fortis* dissolves *quicksilver*, which *vinegar* will not touch ; *aqua regalis* dissolves *gold*, whenas *aqua fortis* cannot meddle with it ; on the contrary, *aqua fortis* dissolves *silver*, but can do nothing with gold, and so of the rest."

The leading fact of the vehement combination and complete union of acid and alkali readily suggested a fit form for the particles of the latter class of substances. "This effect," Lemery adds, "may make us reasonably conjecture that an alkali is a terrestrious and solid matter whose forms are figured after such a manner that the acid points entering in do strike and divide whatever opposes their motion." And in a like spirit are the speculations in Dr. MEAD's *Mechanical Account of Poisons* (1745). Thus he explains the poisonous effect of *corrosive sublimate* of mercury by saying* that the particles of the salt are a kind of lamellæ or blades to which the

* P. 199.

mercury gives an additional weight. If resublimed with three-fourths the quantity of mercury, it loses its corrosiveness, (becoming *calomel*,) which arises from this, that in sublimation " the crystalline blades are divided every time more and more by the force of the fire ;" and " the broken pieces of the crystals uniting into little masses of differing figures from their former make, those cutting points are now so much smaller that they cannot make wounds deep enough to be equally mischievous and deadly: and therefore do only vellicate and twitch the sensible membranes of the stomach."

7. Among all this very fanciful and gratuitous assumption we may notice one true principle clearly introduced, namely, that the suppositions which we make respecting the forms of the elementary particles of bodies and their mode of combination must be such as to explain the facts of crystallization,‐ as well as of mere chemical change. This principle we shall hereafter have occasion to insist upon further.

I now proceed to consider a more refined form of assumption respecting the constitution of bodies, yet still one in which a vain attempt is made to substitute for the peculiar idea of chemical composition a more familiar mechanical conception.

8. *Compounds assumed to be determined by the Mechanical Attraction of the Elements.*—When, in consequence of the investigations and discoveries of Newton and his predecessors, the conception of mechanical force had become clear and familiar, so far as the action of external forces upon a body was concerned, it was very natural that the mathematicians who had pursued this train of speculation should attempt to apply the same conception to that mutual action of the internal parts of a body by which they are held together. Newton himself had pointed the way to this attempt. In the Preface to the

Principia, after speaking of what he has done in calculating the effects of forces upon the planets, satellites, &c., he adds, " Would it were permitted us to deduce the other phenomena of nature from mechanical principles by the same kind of reasoning. For many things move me to suspect that all these phenomena depend upon certain forces, by which the particles of bodies, through causes not yet known, are either urged towards each other, and cohere according to regular figures, or are repelled and recede from each other; which forces being unknown, philosophers have hitherto made their attempts upon nature in vain." The same thought is at a later period followed out further in one of the Queries at the end of the Opticks*. " Have not the small particles of bodies certain Powers, Virtues, or Forces by which they act at a distance, not only upon the rays of light for reflecting, refracting and inflecting them, but also upon one another for producing a great part of the phenomena of nature?" And a little further on he proceeds to apply this expressly to chemical changes. " When Salt of Tartar runs *per deliquium* [or as we now express it, deliquesces] is not this done by an *attraction* between the particles of the Salt of Tartar and the particles of the water which float in the air in the form of vapours? And why does not common salt, or saltpetre, or vitriol, run *per deliquium,* but for want of such an attraction? or why does not Salt of Tartar draw more water out of the air than in a certain proportion to its quantity, but for want of an attractive force after it is saturated with water?" He goes on to put a great number of similar cases, all tending to the same point, that chemical combinations cannot be conceived in any other way than as an attraction of particles.

9. Succeeding speculators in his school attempted to

* Query 31.

follow out this view. Dr. Frend, of Christ Church, in
1710, published his *Prælectiones Chymicæ, in quibus omnes
fere Operationes Chymicæ ad vera Principia ex ipsius
Naturæ Legibus rediguntur. Oxonii habitæ.* This book is
dedicated to Newton, and in the dedication, the promise
of advantage to chemistry from the influence of the
Newtonian discoveries is spoken of somewhat largely,—
much more largely, indeed, than has yet been justified by
the sequel. After declaring in strong terms that the
only prospect of improving science consists in following
the footsteps of Newton, the author adds, "That force
of attraction, of which you first so successfully traced
the influence in the heavenly bodies, operates in the most
minute corpuscles, as you long ago hinted in your *Principia,* and have lately plainly shown in your *Opticks;*
and this force we are only just beginning to perceive and
to study. Under these circumstances I have been desirous of trying what is the result of this view in chemistry."
The work opens formally enough, with a statement of
general mechanical principles, of which the most peculiar
are these :—That there exists an attractive force by which
particles when at very small distances from each other,
are drawn together ;—that this force is different, according to the different figure and density of the particles ;
—that the force may be greater on one side of a particle than on the other ;—that the force by which particles cohere together arises from attraction, and is variously modified according to the quantity of contacts."
But these principles are not applied in any definite
manner to the explanation of specific phenomena. He
attempts, indeed, the question of special solvents*. Why
does *aqua fortis* dissolve silver and not gold, while *aqua
regia* dissolves gold and not silver? which, he says, is
the most difficult question in chemistry, and which is

* P. 54.

certainly a fundamental one in the formation of chemical theory. He solves it by certain assumptions respecting the forces of attraction of the particles, and also the diameter of the particles of the acids and the pores of the metals, all which suppositions are gratuitous.

10. We may observe further, that by speaking, as I have stated that he does, of the figure of particles, he mixes together the assumption of the last section with the one which we are considering in this. This combination is very unphilosophical, or, to say the least, very insufficient, since it makes a new hypothesis necessary. If a body be composed of cubical particles, held together by their mutual attraction, by what force are the parts of each cube held together? In order to understand their structure, we are obliged again to assume a cohesive force of the second order, binding together the particles of each particle. And therefore Newton himself says*, very justly, "The parts of all homogeneal hard bodies which fully touch each other, stick together very strongly: and for explaining how this is, some have invented hooked atoms, *which is begging the question.*" For (he means to imply,) how do the parts of the hook stick together?

The same remark is applicable to all hypotheses in which particles of a complex structure are assumed as the constituents of bodies : for while we suppose bodies and their known properties to result from the mutual actions of these particles, we are compelled to suppose the parts of each particle to be held together by forces still more difficult to conceive, since they are disclosed only by the properties of these particles, which as yet are unknown. Yet Newton himself has not abstained from such hypotheses: thus he says†, "A particle of a salt may be compared to a chaos, being dense, hard, dry, and earthy in the centre, and moist and watery in the circumference."

* *Opticks*, p. 364.   † *Ib.*, p. 362.

Since Newton's time the use of the term attraction, as expressing the cause of the union of the chemical elements of bodies, has been familiarly continued; and has, no doubt, been accompanied in the minds of many persons with an obscure notion that chemical attraction is, in some way, a kind of mechanical attraction of the particles of bodies. Yet this view has never, so far as I am aware, been worked out into a system of chemical theory; nor even applied with any distinctness as an explanation of any particular chemical phenomena. Any such attempt, indeed, could only tend to bring more clearly into view the entire inadequacy of such a mode of explanation. For the leading phenomena of chemistry are all of such a nature that no mechanical combination can serve to express them, without an immense accumulation of additional hypotheses. If we take as our problem the changes of colour, transparency, texture, taste, odour, produced by small changes in the ingredients, how can we expect to give a mechanical account of these, till we can give a mechanical account of colour, transparency, texture, taste, odour, themselves? And if our mechanical hypothesis of the elementary constitution of bodies does not explain *such* phenomena as those changes, what can it explain, or what can be the value of it? I do not here insist upon a remark which will afterwards come before us, that even crystalline form, a phenomenon of a far more obviously mechanical nature than those just alluded to, has never yet been in any degree explained by such assumptions as this, that bodies consist of elementary particles exerting forces of the same nature as the central forces which we contemplate in Mechanics.

When therefore Newton asks, "When some stones, as spar of lead, dissolved in proper menstruums, become salts, do not these things show that salts are dry earth and watery acid united by *attraction?*" we may answer, that this mode of expression appears to be intended to identify chemical combination with mechanical attraction;—that there would be no objection to any such identification if we could, in that way, explain, or even classify well, a collection of chemical facts; but that this has never yet been done by the help of such expressions. Till some advance of this kind can be pointed out, we must necessarily consider the power which produces chemical combination as a peculiar principle, a special relation of the elements, not rightly expressed in mechanical terms. And we now proceed to consider this relation under the name by which it is most familiarly known.

CHAPTER II.

ESTABLISHMENT AND DEVELOPMENT OF THE IDEA OF CHEMICAL AFFINITY.

1. THE earlier chemists did not commonly involve themselves in the confusion into which the mechanical philosophers ran, of comparing chemical to mechanical forces. Their attention was engaged, and their ideas were moulded, by their own pursuits. They saw that the connexion of elements and compounds with which they had to deal, was a peculiar relation which must be studied directly; and which must be understood, if understood at all, in itself, and not by comparison with a different class of relations. At different periods of the progress of chemistry, the conception of this relation, still vague and obscure, was expressed in various manners; and at last this conception was clothed in tolerably consistent phraseology, and the principles which it involved were, by the united force of thought and experiment, brought into view.

2. The power by which the elements of bodies combine chemically, being, as we have seen, a peculiar agency, different from mere mechanical connexion or attraction, it is desirable to have it designated by a distinct and peculiar name; and the term *affinity* has been employed for that purpose by most modern chemists. The word "affinity" in common language means, sometimes resemblance, and sometimes relationship and ties of family. It is from the latter sense that the metaphor is borrowed when we speak of chemical affinity. By the employment of this term we do not indicate resemblance, but disposition to unite. Using the word in a common unscientific manner, we might say that chlorine, bromine, and iodine have a great natural affinity with each other, for there are considerable resemblances and analogies among them; but these bodies have very little *chemical* affinity for each other. The use of the word in the former sense, of resemblance, can be traced in earlier chemists; but it does not appear to have acquired its peculiar chemical meaning till after Boerhaave's time. Boerhaave, however, is the writer in whom we first find a due apprehension of the peculiarity and importance of the Idea which it now expresses. When we make a chemical solution*, he says, not only are the particles of the dissolved body separated from each other, but they are closely united to the particles of the solvent. When *aqua regia* dissolves gold, do you not see, he says to his hearers, that there must be between each particle of the solvent and of the metal, a mutual virtue by which each loves, unites with, and holds the other (*amat, unit, retinet*)? The opinion previously prevalent had been that the solvent merely separates the parts of the body dissolved: and most philosophers had conceived this separation as performed by mechanical operations of the par-

* *Elementa Chemiæ. Lugd. Bat.* 1732, p. 677.

tioles, resembling, for instance, the operation of wedges breaking up a block of timber. But Boerhaave forcibly and earnestly points out the insufficiency of the conception. This, he says, does not account for what we see. We have not only a separation, but a new combination. There is a force by which the particles of the solvent associate to themselves the parts dissolved, not a force by which they repel and dissever them. We are here to imagine not mechanical action, not violent impulse, not antipathy, but love, at least if love be the desire of uniting. (Non igitur hic etiam actiones mechanicæ, non propulsiones violentæ, non inimicitiæ cogitandæ, sed amicitiæ, si amor dicendus copulæ cupido.) The novelty of this view is evidenced by the mode in which he apologizes for introducing it. " Fateor, paradoxa hæc assertio." To Boerhaave, therefore, (especially considering his great influence as a teacher of chemistry,) we may assign the merit of first diffusing a proper view of chemical affinity as a peculiar force, the origin of almost all chemical changes and operations.

3. To Boerhaave is usually assigned also the credit of introducing the *word* " affinity" among chemists ; but I do not find that the word is often used by him in this sense ; perhaps not at all*. But however this may be, the term is

---

* See Dumas, *Leçons de Philos. Chim.*, p. 364. Rees' *Cyclopædia*, Art. Chemistry. In the passage of Boerhaave to which I refer above, *affinitas* is rather opposed to, than identified with, chemical combination. When, he says, the parts of the body to be dissolved are dissevered by the solvent, why do they remain united to the particles of the solvent, and why do not rather both the particles of the solvent and of the dissolved body collect into homogeneous bodies by their *affinity ?* denuo se affinitate suæ naturæ colligant in corpora homogenea ? And the answer is, because they possess another force which counteracts this affinity of homogeneous particles, and makes compounds of different elements. Affinity, in chemistry, now means the tendency of *different* kinds of matter to unite : but it appears, as I have said, to have acquired this sense since Boerhaave's time.

on many accounts well worthy to be preserved, as I shall endeavour to show. Other terms were used in the same sense during the early part of the eighteenth century. Thus when Geoffroy, in 1718, laid before the Academy of Paris his Tables of Affinities, which perhaps did more than any other event to fix the idea of affinity, he termed them "Tables of the Relations of Bodies;" "*Tables des Rapports:*" speaking however, also, of their "disposition to unite," and using other phrases of the same import.

The term *attraction*, having been recommended by Newton as a fit word to designate the force which produces chemical combination, continued in great favour in England, where the Newtonian philosophy was looked upon as applicable to every branch of science. In France, on the contrary, where Descartes still reigned triumphant, "attraction," the watch-word of the enemy, was a sound never uttered but with dislike and suspicion. In 1718 (in the notice of Geoffroy's Tables,) the Secretary of the Academy, after pointing out some of the peculiar circumstances of chemical combinations says, "Sympathies and attractions would suit well here, if there were such things." "Les sympathies, les attractions conviendroient bien ici, si elles etaient quelque chose." And at a later period, in 1731, having to write the *éloge* of Geoffroy after his death, he says, "He gave, in 1718, a singular system, and a Table of *Affinities*, or Relations of the different substances in chemistry. These affinities gave uneasiness to some persons, who feared that they were *attractions in disguise*, and all the more dangerous in consequence of the seductive forms which clever people have contrived to give them. It was found in the sequel that this scruple might be got over."

This is the earliest published instance, so far as I am aware, in which the word "affinity" is distinctly used for the cause of chemical composition; and taking into account the circumstances, the word appears to have been adopted in France in order to avoid the word *attraction*, which had the taint of Newtonianism. Accordingly we find the word *affinité* employed in the works of French chemists from this time. Thus, in the *Transactions of the French Academy* for 1746, in a paper of Macquer's upon Arsenic, he says*, " On peut facilement rendre raison de ces phenomènes par le moyen des affinités que les differens substances qui entrent dans ces combinaisons, ont les uns avec les autres :" and he proceeds to explain the facts by reference to Geoffroy's Table. And in Macquer's *Elements of Chemistry*, which appeared a few years later, the " affinity of composition " is treated of as a leading part of the subject, much in the same way as has been practised in such books up to the present time. From this period the word appears to have become familiar to all European chemists in the sense of which we are now speaking. Thus, in the year 1758, the Academy of Sciences at Rouen offered a prize for the best dissertation on Affinity. The prize was shared between M. Limbourg of Theux, near Liege, and M. Le Sage of Geneva†. About the same time other persons (Manherr‡, Nicolai §, and others) wrote on the same subject, employing the same name.

Nevertheless, in 1775, the Swedish chemist Bergman, pursuing still further this subject of chemical affinities, and the expression of them by means of tables, returned again to the old Newtonian term; and designated the disposition of a body to combine with one rather than

---

* A. P. 1746, p. 201.

† Thomson's *Chemistry*, iii. 10.   Limbourg's Dissertation was published at Liege, in 1761 ; and Le Sage's at Geneva.

‡ *Dissertatio de Affinitate Corporum.*   Vindob. 1762.

§ *Progr.* I. II. *de Affinitate Corporum Chimica.*   Jen. 1775, 1776.

another of two others as *elective attraction*. And as his work on *Elective Attractions* had great circulation and great influence, this phrase has obtained a footing by the side of affinity, and both one and the other are now in common use among chemists.

4. I have said above that the term *Affinity* is worthy of being retained as a technical term. If we use the word attraction in this case, we identify or compare chemical with mechanical attraction; from which identification and comparison, as I have already remarked, no one has yet been able to extract the means of expressing any single scientific truth. If such an identification or comparison be not intended, the use of the same word in two different senses can only lead to confusion: and the proper course, recommended by all the best analogies of scientific history, is to adopt a peculiar term for that peculiar relation on which chemical composition depends. The word affinity, even if it were not rigorously proper according to its common meaning, still, being simple, familiar, and well established in this very usage, is much to be preferred before any other.

But further, there are some analogies drawn from the common meaning of this word, which appear to recommend it as suitable for the office which it has to discharge. For common mechanical attractions and repulsions, the forces by which one body considered as a *whole* acts upon another external to it, are, as we have said, to be distinguished from those more intimate ties by which the *parts* of each body are held together. Now this difference is implied, if we compare the former relations, the attractions and repulsions, to alliances and wars between states, and the latter, the internal union of particles, to those bonds of affinity which connect the citizens of the same state with one another, and especially to the ties of family. We have seen that Boerhaave compares the union of two elements of a compound to their marriage; "we must allow," says an eminent chemist of our own time*, "that there is some truth in this poetical comparison." It contains this truth, that the two become one to most intents and purposes, and that the unit thus formed (the family) is not a mere juxtaposition of the component parts. And thus the idea of Affinity as the peculiar principle of chemical composition, is established among chemists, and designated.by a familiar and appropriate name.

5. *Analysis is possible.*—We must, however, endeavour to obtain a further insight into this idea, thus fixed and named. We must endeavour to extricate, if not from the idea itself, from the processes by which it has obtained acceptation and currency among chemists, some principles which may define its application, some additional specialties in the relations which it implies. This we shall proceed to do.

The idea of affinity, as already explained, implies a disposition to combine. But this combination is to be understood as admitting also of a possibility of separation. Synthesis implies analysis as conceivable: or to recur to the image which we have already used, divorce is possible when the marriage has taken place.

That there is this possibility, is a conviction implied in all the researches of chemists, ever since the true notion of composition began to predominate in their investigations. One of the first persons who clearly expressed this conviction was Mayow, an English physician, who published his *Medico-Physical Tracts* in 1674. The first of them, *De Sale-Nitro et Spiritu Nitro-Aerio*, contains a clear enunciation of this principle. After showing how, in the combinations of opposite elements, as acid and alkali, their properties entirely disappear, and a new substance

* DUMAS, *Leçons de Phil. Chim.*, p. 363.

is formed not at all resembling either of the ingrédients, he adds*, "Although these salts thus mixed appear to be destroyed, it is still possible for them to be separated from each other, with their powers still entire." He proceeds to exemplify this, and illustrates it by the same image which I have already alluded to: "Salia acida a salibus volatilibus discedunt, ut cum sale fixo tartari, tanquam sponso magis idoneo, conjugium strictius ineunt." This idea of a synthesis which left a complete analysis still possible, was opposed to a notion previously current, that when two heterogeneous bodies united together and formed a third body, the two constituents were entirely destroyed, and the result formed out of their ruins †. And this conception of synthesis and analysis, as processes which are possible successively and alternately, and each of which supposes the possibility of the other, has been the fundamental and regulative principle of the operations and speculations of analytical chemistry from the time of Mayow to the present day.

6. *Affinity is elective.*—When the idea of chemical affinity, or disposition to unite, was brought into view by the experiments and reasonings of chemists, they found it necessary to consider this disposition as elective;— each element *chose* one rather than another of the elements which were presented to it, and quitted its union with one to unite with another which it preferred. This has already appeared in the passage just quoted from Mayow. He adds in the same strain, "I have no doubt that fixed salts choose one acid rather than another, in order that they may coalesce with it in a more intimate union."—"Nullus dubito salia fixa acidum unum præ aliis *eligere*, ut cum eodem arctiore unione coalescant." The same thought is expressed and exemplified by other chemists: they notice innumerable cases in which, when

* Cap. xiv., p. 233.   † Thomson's *Chemistry*, iii. 8.

an ingredient is combined with a liquid, if a new substance be immersed which has a greater affinity for the
liquid, the liquid combines with the new substance by
election, and the former ingredient is *precipitated*. Thus
Stahl says[*], "In spirit of nitre dissolve silver; put in
copper and the silver is thrown down; put in iron and
the copper goes down; put in zinc, the iron precipitates;
put in volatile alkali, the zinc is separated; put in fixed
alkali, the volatile quits its hold."—As may be seen in
this example, we have in such cases, not only a preference, but a long gradation of preferences. The spirit of
nitre will combine with silver, but it prefers copper;
prefers iron more; zinc still more; volatile alkali yet
more; fixed alkali the most.

The same thing was proved to obtain with regard to
each element; and when this was ascertained, it became
the object of chemists to express these degrees of preference, by lists in which substances were arranged according to their disposition to unite with another substance.
In this manner was formed Geoffroy's Table of Affinities
(1718), which we have already mentioned. This Table
was further improved by other writers, as Gellert (1751)
and Limbourg (1761). Finally Bergman improved these
Tables still further, taking into account not only the
order of affinities of each element for others, but the *sum*
of the tendencies to unite of each two elements, which
sum, he held, determined the resulting combination when
several elements were in contact with each other.

7. As we have stated in the History[†], when the doctrine of elective affinities had assumed this very definite
and systematic form, it was assailed by Berthollet, who
maintained, in his *Essai de Statique Chimique*, (1803,)
that chemical affinities are *not* elective:—that, when
various elements are brought together, their combinations

---

[*] *Zymotechnia*, 1697, p. 117.   [†] *Hist. Ind. Sci.*, iii 115.

do not depend upon the kind of elements alone, but upon the quantity of each which is present, that which is most abundant always entering most largely into the resulting compounds. It may seem strange that it should be possible, at so late a period of the science, to throw doubt upon a doctrine which had presided over and directed its progress so long. Proust answered Berthollet, and again maintained that chemical affinity is elective. I have, in the History, given the judgment of Berzelius upon this controversy. " Berthollet," he says, " defended himself with an acuteness which makes the reader hesitate in his judgment; but the great mass of facts finally decided the point in favour of Proust." I may here add the opinion pronounced upon this subject by Dr. Turner*. " Bergman erred in supposing the result of chemical action to be in every case owing to elective affinity [for this power is modified in its effects by various circumstances]: but Berthollet ran into the opposite extreme in declaring that the effects formerly ascribed to that power are never produced by it. That chemical attraction is exerted between different bodies with different degrees of energy, is, I apprehend, indisputable." And he then proceeds to give many instances of differences in affinity which cannot be accounted for by the operation of any modifying causes. Still more recently, M. Dumas has taken a review of this controversy; and, speaking with enthusiasm of the work of Berthollet, as one which had been of inestimable service to himself in his early study of chemistry, he appears at first disposed to award to him the victory in this dispute. But his final verdict leaves undamaged the general principle now under our consideration, that chemical affinity is elective. " For my own part," he says†, " I willingly admit the no-

* *Chemistry*, p. 199.   6th edition.
† *Leçons de Philosophie Chimique*, p. 386.

tions of Berthollet when we have to do with acids or with
bases, of which the energy is nearly equal: but when
bodies endued with very energetic affinities are in presence of other bodies of which the affinities are very
feeble, I propose to adopt the following rule: In a solution, everything remaining dissolved, the strong affinities
satisfy themselves, leaving the weak affinities to arrange
matters with one another. The strong acids take the
strong bases, and the weak acids can only unite with the
weak bases. The known facts are perfectly in accordance
with this practical rule." It is obvious that this recognition of a distinction between strong and weak affinities
which operates to such an extent as to determine entirely
the result, is a complete acknowledgement of the elective
nature of affinity as far as any person acquainted with
chemical operations could contend for it. For it must
be allowed by all, that solubility, and other collateral circumstances, influence the course of chemical combinations, since they determine whether or not there shall
take place that contact of elements without which affinity
cannot possibly operate.

8. *Affinity is Definite as to Quantity.*—In proportion
as chemists obtained a clearer view of the products of the
laboratory as results of the composition of elements,
they saw more and more clearly that these results were
definite; that one element not only preferred to combine
with another of a certain kind, but also would combine
with it to a certain extent and no further, thus giving to
the result not an accidental and variable, but a fixed and
constant character. Thus salts being considered as the
result of the combination of two opposite principles, acid
and alkali, and being termed *neutral* when these principles exactly balanced each other, Rouelle (who was
Royal Professor at Paris in 1742,) admits of neutral
salts with excess of acid, neutral salts with excess of

base, and perfect neutral salts. Beaume maintained*
against him that there were no salts except those perfectly neutral, the other classes being the results of mixture and imperfect combination. But this question was
not adequately treated till chemists made every experiment with the balance in their hands. When this was
done, they soon discovered that, in each neutral salt, the
proportional weights of the ingredients which composed it
were always the same. This was ascertained by Wenzel,
whose Doctrine of the Affinities of Bodies appeared in
1777. He not only ascertained that the proportions of
elements in neutral chemical compounds are definite, but
also that they are reciprocal; that is, that if A, a certain
weight of a certain acid, neutralize *m*, a certain weight of
a certain base, and B, a certain weight of a certain other
acid, neutralize *n*, a certain weight of a certain other base;
the compound of A and *n* will also be neutral; as also that
of B and *m*. The same views were again presented by
Richter in 1792, in his *Principles of the Measure of Chemical Elements*. And along with these facts, that of the
combination of elements in multiple proportions being
also taken into account, the foundations of the Atomic
Theory were laid; and that theory was propounded in
1803 by Mr. Dalton. That theory, however, rests upon
the idea of substance, as well as upon that idea of chemical affinity which we are here considering; and the discussion of its evidence and truth must be for the present
deferred.

9. The two principles just explained, that affinity is
definite as to the kind, and as to the quantity of the elements which it unites, have here been stated as results of
experimental investigation. That they could never have
been clearly understood, and therefore never firmly established, without laborious and exact experiments, is

---

* DUMAS, *Phil. Chim.*, p. 198.

certain; but yet we may venture to say that being once known, they possess an evidence beyond that of mere experiment. For how, in fact, can we conceive combinations, otherwise than as definite in kind and quantity? If we were to suppose each element ready to combine with any other indifferently, and indifferently in any quantity, we should have a world in which all would be confusion and indefiniteness. There would be no fixed kinds of bodies; salts, and stones, and ores, would approach to and graduate into each other by insensible degrees. Instead of this, we know that the world consists of bodies distinguishable from each other by definite differences, capable of being classified and named, and of having general propositions asserted concerning them. And as we cannot conceive a world in which this should not be the case, it would appear that we cannot conceive a state of things in which the laws of the combination of elements should not be of that definite and measured kind which we have above asserted.

This will, perhaps, appear more clearly by stating our fundamental convictions respecting chemical composition in another form, which I shall, therefore, proceed to do.

10. *Chemical Composition determines Physical Properties.*—However obscure and incomplete may be our conception of the internal powers by which the ultimate particles of bodies are held together, it involves, at least, this conviction:—that these powers are what determine bodies to *be* bodies, and therefore contain the reason of all the properties which, as bodies, they possess. The forces by which the particles of a body are held together, also cause it to be hard or soft, heavy or light, opake or transparent, black or red; for if these forces are not the cause of these peculiarities, what can be the cause? By the very supposition which we make respecting these forces, they include all the relations by which the parts

are combined into a whole, and therefore they, and they only, must determine all the attributes of the whole. The foundation of all our speculations respecting the intimate constitution of bodies must be this, that their composition determines their properties.

Accordingly we find our chemists reasoning from this principle with great confidence, even in doubtful cases. Thus Davy, in his researches concerning the diamond, says: "That some chemical difference must exist between the hardest and most beautiful of the gems and charcoal, between a non-conductor and a conductor of electricity, it is scarcely possible to doubt: and it seems reasonable to expect that a very refined or perfect chemistry will confirm *the analogies of nature;* and show that bodies cannot be the same in their composition or chemical nature, and yet totally different in their chemical properties." It is obvious that the principle here assumed is so far from being a mere result of experience, that it is here appealed to to prove that all previous results of experience on this subject must be incomplete and inaccurate; and that there must be some chemical difference between charcoal and diamond, though none had hitherto been detected.

11. In what manner, according to what rule, the chemical composition shall determine the kind of the substance, we cannot reasonably expect to determine by mere conjecture or assumption, without a studious examination of natural bodies and artificial compounds. Yet even in the most recent times, and among men of science, we find that an assumption of the most arbitrary character has in one case been mixed up with this indisputable principle, that the elementary composition determines the kind of the substance. In the classification of minerals, one school of mineralogists have rightly taken it as their fundamental principle that the chemical composition shall decide the position of the mineral in the system. But

they have appended to this principle, arbitrarily and unjustifiably, the maxim that the element which is *largest in quantity* shall fix the class of the substance. To make such an assumption is to renounce, at once, all hope of framing a system which shall be governed by the resemblances of the things classified; for how can we possibly know beforehand that fifty-five per cent. of iron shall give a substance its predominant properties, and that forty-five per cent. shall not? Accordingly, the systems of mineralogical arrangement which have been attempted in this way, (those of Haüy, Phillips, and others,) have been found inconsistent with themselves, ambiguous, and incapable of leading to any general truths.

12. Thus the physical properties of bodies depend upon their chemical composition, but in a manner which a general examination of bodies with reference to their properties and their composition can alone determine. We may, however, venture to assert further, that the more definite the properties are, the more distinct may we expect to find this dependence. Now the most definite of the properties of bodies are those constant properties which involve relations of space: that is, their figure. We speak not, however, of that external figure, derived from external circumstances, which, so far from being constant and definite, is altogether casual and arbitrary; but of that figure which arises from their internal texture, and which shows itself not only in the regular forms which they spontaneously assume, but in the disposition of the parts to separate in definite directions and no others. In short, the most definite of the properties of perfect chemical compounds is their crystalline structure; and therefore it is evident that the crystalline structure of each body, and the forms which it affects, must be in a most intimate dependence upon its chemical composition.

Here again we are led to the brink of another theory; —that of crystalline structure, which has excited great interest among philosophers ever since the time of Haüy. But this theory involves, besides that idea of chemical composition with which we are here concerned, other conceptions which enter into the relations of figure. These conceptions, governed principally by the idea of Symmetry, must be unfolded and examined before we can venture to discuss any theory of crystallization: and we shall proceed to do this as soon as we have first duly considered the Idea of Substance and its consequences.

<hr>

CHAPTER III.

OF THE IDEA OF SUBSTANCE.

1. *Axiom of the Indestructibility of Substance.*—We now come to an Idea of which the history is very different from those of which we have lately been speaking. Instead of being gradually and recently brought into a clear light, as has been the case with the Ideas of Polarity and Affinity, the Idea of Substance has been entertained in a distinct form from the first periods of European speculation. That this is so, is proved by our finding a principle depending upon this idea current as an axiom among the early philosophers of Greece:—namely, that *nothing can be produced out of nothing.* Such an axiom, more fully stated, amounts to this: that the substance of which a body consists is incapable of being diminished (and consequently incapable of being augmented) in quantity, whatever apparent changes it may undergo. Its form, its distribution, its qualities may vary, but the substance itself is identically the same under all these variations.

The axiom just spoken of was the great principle of the physical philosophy of the Epicurean school, as it must be of every merely material philosophy. The reader of Lucretius will recollect the emphasis with which it is repeatedly asserted in his poem:

E nilo nil gigni, in nilum nil posse reverti;
Nought comes of nought, nor ought returns to nought.

Those who engaged in these early attempts at physical speculation were naturally much pleased with the clearness which was given to their notions of change, composition, and decomposition, by keeping steadily hold of the Idea of Substance, as marked by this fundamental axiom. Nor has its authority ever ceased to be acknowledged. A philosopher was asked*, What is the weight of smoke? He answered, Subtract the weight of the ashes from the weight of the wood which is burnt, and you have the weight of the smoke. This reply would be assented to by all; and it assumes as incontestable that even under the action of fire, the material, the substance, does not perish, but only changes its form.

This principle of the indestructibility of substance might easily be traced in many reasonings and researches, ancient and modern. For instance, when the chemist works with the *retort*, he places the body on which he operates in one part of an inclosed cavity, which, by its bendings and communications, separates at the same time that it confines, the products which result from the action of fire: and he assumes that this process is an analysis of the body into its ingredients, not a creation of anything which did not exist before, or a destruction of anything which previously existed. And he assumes further, that the total quantity of the substance thus analysed is the sum of the quantities of its ingredients. This principle is the very basis of chemical speculation as we shall hereafter explain more fully.

* KANT, *Kritik, der R. V.*, p. 167

2. *The Idea of Substance.*—The axiom above spoken of depends upon the Idea of Substance, which is involved in all our views of external objects. We unavoidably assume that the qualities and properties which we observe are properties of *things*;—that the adjective implies a substantive;—that there is, besides the external characters of things, something *of which* they are the characters. An apple which is red, and round, and hard, is not merely redness, and roundness, and hardness: these circumstances may all alter while the apple remains the same apple. Behind the appearances which we see, we conceive something of which we think; or to use the metaphor which obtained currency among the ancient philosophers, the attributes and qualities which we observe are supported by and inherent in something: and this something is hence called a substratum or *substance*, that which stands beneath the apparent qualities and supports them.

That we have such an *Idea*, using the term in the sense in which I have employed it throughout these disquisitions, is evident from what has been already said. The Axiom of the indestructibility of substance proves the existence of the Idea of Substance, just as the Axioms of Geometry and Arithmetic prove the existence of the Ideas of Space and Number. In the case of substance, as of space or number, the ideas cannot be said to be borrowed from experience, for the axioms have an authority of a far more comprehensive and demonstrative character than any which experience can bestow. The axiom that nothing can be produced from nothing and nothing destroyed, is so far from being a result of experience, that it is apparently contradicted by the most obvious observation. It has, at first, the air of a paradox, and by those who refer to it, it is familiarly employed to show how fallacious common observation is. The assertion is usually made in this form; that nothing is created and nothing annihilated, *notwithstanding* that the common course of our experience appears to show the contrary. The principle is not an empirical, but a necessary and universal truth: is collected, not from the evidence of our senses, but from the operation of our ideas. And thus the universal and undisputed authority of the axiom proves the existence of the Idea of Substance.

3. *Locke's Denial of the Idea of Substance.*—I shall not attempt to review the various opinions which have been promulgated respecting this idea: but it may be worth our while to notice briefly the part it played in the great controversy concerning the origin of our ideas which LOCKE's *Essay* occasioned. Locke's object was to disprove the existence of all ideas not derived from Sensation or Reflection: and since the idea of substance as distinct from external qualities, is manifestly not derived directly from sensation, nor by any very obvious or distinct process from reflection, Locke was disposed to exclude the idea as much as possible. Accordingly, in his argumentation against Innate Ideas*, he says plainly, "the idea of substance, which we neither have nor can have by sensation or reflection." And the inference which he draws is, "that we have no such clear idea at all." What then, it may be asked, do we mean by the word substance? This also he answers, though somewhat strangely, "We signify nothing by the word substance, but only an uncertain supposition of we know not what, *i. e.*, of something whereof we have no particular distinct positive idea, which we take to be the substratum, or support, of those ideas we know." That while he indulged in this tautological assertion of our ignorance and uncertainty, he should still have been compelled to acknowledge that the word substance had

* *Essay*, b. i., ch, 4., s. 18,

some meaning, and should have been driven to explain it by the identical metaphors of substratum and support, is a curious proof how impossible it is entirely to reject this idea.

But as we have already seen, the supposition of the existence of substance is so far from being uncertain, that it carries with it irresistible conviction, and substance is necessarily conceived as something which cannot be produced or destroyed. It may be easily supposed, therefore, that when the controversy between Locke and his assailants came to this point, he would be in some difficulty. And, indeed, though with his accustomed skill in controversy, he managed to retain a triumphant tone, he was driven from his main points. Thus he repels the charge that he took the being of substance to be doubtful*. He says, " Having everywhere affirmed and built upon it that man is a substance, I cannot be supposed to question or doubt of the being of substance, till I can question or doubt of my own being." He attempts to make a stand by saying that *being* of things does not depend upon our *ideas;* but if he had been asked how, without having an *idea* of substance, he knew substance *to be*, it is difficult to conceive what answer he could have made. Again, he had said that our idea of substance arises from our accustoming ourselves to suppose a substratum of qualities. Upon this his adversary, Bishop Stillingfleet, very properly asks, Is this custom grounded upon true reason or no? To which Locke replies, that it is grounded upon this: That we cannot conceive how simple ideas of sensible qualities should subsist alone; and therefore we suppose them to exist in, and to be supported by some common subject, which support we denote by the name substance. Thus he allows, not only that we necessarily assume the reality of substance, but that we cannot conceive qualities

* *Essay*, b. ii., ch. 2, and *First Letter to the Bishop of Worcester.*

without substance; which are concessions so ample as
almost to include all that any advocate for the Idea of
Substance need desire.

Perhaps Locke, and the adherents of Locke, in deny-
ing that we have an idea of substance in general, were
latently influenced by finding that they could not, by any
effort of mind, call up any *image* which could be considered as an image of substance in general. That in
this sense we have no idea of substance, is plain enough;
but in the same sense we have no idea of space in
general, or of time, or number, or cause, or resemblance.
Yet we certainly have such a power of representing to
our minds space, time, number, cause, resemblance, as to
arrive at numerous truths by means of such representations. These general representations I have all along
called Ideas, nor can I discover any more appropriate
word; and in this sense, we have also, as has now been
shown, an Idea of Substance.

4. *Is all Material Substance heavy?*—The principle
that the quantity of the substance of any body remains
unchanged by our operations upon it, is, as we have said,
of universal validity. But then the question occurs, how
are we to ascertain the quantity of substance, and thus
to apply the principle in particular cases. In the case
above mentioned, where smoke was to be weighed, it
was manifestly assumed that the quantity of the substance
might be known by its weight; and that the total
quantity being unchanged, the total weight also would
remain the same. Now on what grounds do we make
this assumption? Is all material substance heavy? and
if we can assert this to be so, on what grounds does the
truth of the assertion rest? These are not idle questions
of barren curiosity; for in the history of that science
(Chemistry) to which the idea of substance is principally
applicable, nothing less than the fate of a comprehensive

and long established theory (the Phlogiston theory)
depended upon the decision of this question. When it
was urged that the reduction of a metal from a calcined
to a metallic form could not consist in the addition of
phlogiston, because the metal was lighter than the calx
had been; it was replied by some, that this was not con-
clusive, for that phlogiston was a principle of levity,
diminishing the weight of the body to which it was
added. This reply was, however, rejected by all the
sounder philosophers, and the force of the argument
finally acknowledged. But why was this suggestion of a
substance having no weight, or having absolute levity,
repudiated by the most reflective reasoners? It is assumed, it appears, that all matter must be heavy; what is
the ground of this assumption?

The ground of such an assumption appears to be the
following. Our idea of substance includes in it this:
that substance is a quantity capable of addition; and
thus capable of making up, by composition, a sum equal
to all its parts. But substance, and the quantity of substance, can be known to us only by its attributes and qualities. And the qualities which are capable constantly
and indefinitely of increase and diminution by increase
and diminution of the parts, must be conceived inseparable from the substance. For the qualities, if removable
from the substance at all, must be removable by some
operation performed upon the substance; and by the
idea of substance, all such operations are only equivalent
to separation, junction, and union of parts. Hence those
characters which thus universally increase and diminish
by addition and subtraction of the things themselves,
belong to the substance of the things. They are measures
of quantity, and not merely separable qualities.

The weight of bodies is such a character. However
we compound or divide bodies, we compound and divide

their weight in the same manner. We may dismember a body into the minutest parts; but the sum of the weights of the parts is always equal to the whole weight of the body. The weight of a body can be in no way increased or diminished except by adding something to it or taking something from it. If we bake a brick, we do not conceive that the change of colour or of hardness, implies that anything has been created or destroyed. It may easily be that the parts have only assumed a new arrangement; but if the brick have lost weight, we suppose that something (moisture for instance,) has been removed elsewhere.

Thus weight is apprehended as essential to matter. In considering the dismemberment or analysis of bodies, we assume that there must be some criterion of the quantity of substance; and this criterion can possess no other properties than their weight possesses. If we assume an element which has no weight, or the weight of which is negative, as some of the defenders of phlogiston attempted to do, we put an end to all speculation on such subjects. For if weight is not the criterion of the quantity of one element, phlogiston for instance, why is weight the criterion of the quantity of any other element? We may, by the same right, assume any other real or imaginary element to have levity instead of gravity; or to have a peculiar intensity of gravity which makes its weight no index of its quantity. In short, if we do this, we deprive of all possibility of application our notions of element, analysis, and composition; and violate the postulates on which the questions are propounded which we thus attempt to decide.

We must, then, take a constant and quantitative property of matter, such as weight is, to be an index of the quantity of matter or of substance to which it belongs. I do not here speak of the question which has sometimes been proposed, whether the *weight* or the *inertia* of bodies be the more proper measure of the quantity of matter. For the measure of inertia is regulated by the same assumption as that of substance:—that the quantity of the whole must be equal to the quantity of all the parts: and inertia is measured by weight, for the same reason that substance is so.

Having thus established the certainty, and ascertained the interpretation of the fundamental principle which the Idea of Substance involves, we are prepared to consider its application in the science upon which it has a peculiar bearing.

CHAPTER IV.

APPLICATION OF THE IDEA OF SUBSTANCE IN CHEMISTRY.

1. *A Body is Equal to the Sum of its Elements.*— From the earliest periods of chemistry the balance has been familiarly used to determine the proportions of the ingredients and of the compound; and soon after the middle of the last century, this practice was so studiously followed, that Wenzel and Richter were thereby led to the doctrine of definite proportions. But yet the full value and significance of the balance, as an indispensable instrument in chemical researches, was not understood till the gaseous, as well as solid and fluid ingredients were taken into the account. When this was done, it was found that the principle, that the whole is equal to the sum of its parts, of which, as we have seen, the necessary truth, in such cases, flows from the idea of substance, could be applied in the most rigorous manner. And conversely, it was found that by the use of the balance, the chemist could decide, in doubtful cases, which was a whole, and which were parts.

For it may be observed that chemistry considers all the changes which belong to her province as compositions and decompositions of elements: but still the question may occur, whether an observed change be the one or the other. How can we distinguish whether the process which we contemplate be composition or decomposition? Whether the new body be formed by addition of a new, or subtraction of an old element? Again; in the case of decomposition, we may inquire, what are the ultimate limits of our analysis? If we decompound bodies into others more and more simple, how far can we carry this succession of processes? How far can we proceed in the road of analysis? And in our actual course, what evidence have we that our progress, as far as it has gone, has carried us from the more complex to the more simple? To this we reply, that the criterion which enables us to distinguish, decidedly and finally, whether our process have been a mere analysis of the proposed body into its ingredients, or a synthesis of some of them with some new element, is the principle stated above, that the weight of the whole is equal to the weight of all the parts. And no process of chemical analysis or synthesis can be considered complete till it has been verified by this fact;—by finding that the weight of the compound is the weight of its supposed ingredients; or, that if there be an element which we think we have detached from the whole, its loss is betrayed by a corresponding diminution of weight.

I have already noticed what an important part this principle has played in the great chemical controversy which ended in the establishment of the oxygen theory The calcination of a metal was decided to be the union of oxygen with the metal, and not the separation of phlogiston from it, because it was found that in the process of calcination, the weight of the metal increased, and increased exactly as much as the weight of ambient air diminished. When oxygen and hydrogen were exploded together, and a small quantity of water was produced, it was held that this was really a synthesis of water, because, when very great care was taken with the process, the weight of the water which resulted was equal to the weight of the gases which disappeared.

2. *Lavoisier.*—It was when gases came to be considered as entering largely into the composition of liquid and solid bodies, that extreme accuracy in weighing was seen to be so necessary to the true understanding of chemical processes. It was in this manner discovered by Lavoisier and his contemporaries that oxygen constitutes a large ingredient of calcined metals, of acids, and of water. A countryman of Lavoisier[*] has not only given most just praise to that great philosopher for having constantly tested all his processes by a careful and skilful use of the balance, but has also claimed for him the merit of having introduced the maxim, that in chemical operations nothing is created and nothing lost. But I think it is impossible to deny that this maxim is assumed in all the attempts at analysis made by his contemporaries, as well as by him. This maxim is indeed included in any clear notion of analysis: it could not be the result of the researches of any one chemist, but was the governing principle of the reasonings of all. Lavoisier, however, employed this principle with peculiar assiduity and skill. In applying it, he does not confine himself to mere additions and subtractions of the quantities of ingredients; but often obtains his results by more complex processes. In one of his investigations he says, " I may consider the ingredients which are brought together, and the result which is obtained as an algebraical equation; and if I successively suppose each of the quantities of this equation to be unknown, I can obtain its value from the rest: and

---

[*] M. Dumas, *Leçons de la Philosophie Chimique.* 1837. p. 157.

thus I can rectify the experiment by the calculation, and the calculation by the experiment. I have often taken advantage of this method, in order to correct the first results of my experiments, and to direct me in repeating them with proper precautions."

The maxim that the whole is equal to the sum of all its parts, is thus capable of most important and varied employment in chemistry. But it may be applied in another form to the exclusion of a class of speculations which are often put forwards.

3. *Maxim respecting Imponderable Elements.*—Several of the phenomena which belong to bodies, as heat, light, electricity, magnetism, have been explained hypothetically by assuming the existence of certain fluids; but these fluids have never been shown to have weight. Hence such hypothetical fluids have been termed *imponderable elements.* It is however plain, that so long as these fluids appear to be without weight, they are not elements of bodies in the same sense as those elements of which we have hitherto been speaking. Indeed we may with good reason doubt whether those phenomena depend upon transferable fluids at all. We have seen strong reason to believe that light is not matter, but only motion; and the same thing appears to be probable with regard to heat. Nor is it at all inconceivable that a similar hypothesis respecting electricity and magnetism should hereafter be found tenable. Now if heat, light, and those other agents, be not matter, they are not elements in such a sense as to be included in the principle referred to above, that the body is equal to the sum of its elements. Consequently the maxim just stated, that in chemical operations nothing is created, nothing annihilated, does not apply to light and heat. They are not *things.* And whether heat can be produced where there was no heat before, and light struck out from darkness,

the ideas of which we are at present treating do not enable us to say. In reasoning respecting chemical synthesis and analysis therefore, we shall only make confusion by attempting to include in our conception the light and heat which are produced and destroyed. Such phenomena may be very proper subjects of study, as indeed they undoubtedly are ; but they cannot be studied to advantage by considering them as sharing the nature of composition and decomposition.

Again: in all attempts to explain the processes of nature, the proper course is, first to measure the facts with precision, and then to endeavour to understand their cause. Now the facts of chemical composition and decomposition, the weights of the ingredients and of the compounds, are facts measurable with the utmost precision and certainty. But it is far otherwise with the light and heat which accompany chemical processes. When combustion, deflagration, explosion, takes place, how can we measure the light or the heat? Even in cases of more tranquil action, though we can apply the thermometer, what does the thermometer tell us respecting the *quantity* of the heat? Since then we have no measure which is of any value as regards such circumstances in chemical changes, if we attempt to account for these phenomena *on chemical principles*, we introduce, into investigations in themselves perfectly precise and mathematically rigorous, another class of reasonings, vague and insecure, of which the only possible effect is to vitiate the whole reasoning, and to make our conclusions inevitably erroneous.

We are led then to this maxim : that *imponderable fluids are not to be admitted as chemical elements of bodies**.

---

* Since we are thus warned by a sound view of the nature of science, from considering chemical affinity as having any hold upon imponderable elements, we are manifestly still more decisively pro-

4. It appears, I think, that our best and most philosophical chemists have proceeded upon this principle in their investigations. In reasoning concerning the constitution of bodies and the interpretation of chemical changes, the attempts to include in these interpretations the heat or cold produced, by the addition or subtraction of a certain hypothetical caloric, have become more and more rare among men of science. Such statements, and the explanations often put forwards of the light and heat which appear under various circumstances in the form of fire, must be considered as unessential parts of any sound theory. Accordingly we find Mr. Faraday gradually relinquishing such views. In January, 1834, he speaks generally of an hypothesis of this kind*. "I cannot refrain from recalling here the beautiful idea put forth, I believe by Berzelius, in his developement of his views of the electro-chemical theory of affinity, that the heat and light evolved during cases of powerful combination are the consequence of the electric discharge which is at that moment taking place." But in April of the same year†, he observes, that in the combination of oxygen and hydrogen to produce water, electric powers to a most enormous amount are for the time active, but that the flame which is produced gives but feeble traces of such powers. "Such phenomena," therefore, he adds, "may not, cannot be, taken as evidences of the nature of the action; but are merely incidental results, incomparably small in relation to the forces concerned, and supplying no information of the way in which the particles are

hibited from supposing mechanical impulse or pressure to have any effect upon such elements. To make this supposition, is to connect the most subtle and incorporeal objects which we know in nature by the most material ties. This remark seems to be applicable to M. Poisson's hypothesis that the electric fluid is retained at the surface of bodies by the pressure of the atmosphere.

* *Researches*, 870.　　　　　　　† *Ib.* 960.

active on each other, or in which their forces are finally arranged."

In pursuance of this maxim, we must consider as unessential parts of the oxygen theory that portion of it, much insisted upon by its author at the time, in which when sulphur, for instance, combined with oxygen to produce sulphuric acid, the combustion was accounted for by means of the *caloric* which was supposed to be *liberated* from its combination with oxygen.

5. *Controversy of the Composition of Water.*—There is another controversy of our times to which we may with great propriety apply the maxim now before us. After the glory of having first given a true view of the composition of water had long rested tranquilly upon the names of Cavendish and Lavoisier, a claim was made in favour of James Watt as the real author of this discovery by his son, (Mr. J. Watt,) and his eulogist, (M. Arago*.) It is not to our purpose here to discuss the various questions which have arisen on this subject respecting priority of publication, and respecting the translation of opinions published at one time into the language of another period. But if we look at Watt's own statement of his views, given soon after those of Cavendish had been published, we shall perceive that it is marked by a violation of this maxim: we shall find that he does admit imponderable fluids as chemical elements; and thus shows a great vagueness and confusion in his idea of chemical composition. With such imperfection in his views, it is not surprising that Watt, not only did not anticipate, but did not fully appreciate the discovery of Cavendish and Lavoisier. Watt's statement of his views is as follows†:—"Are we not authorized to conclude that water is composed of

---

* Eloge de James Watt, *Annuaire du Bur. des Long.*, 1839.
† *Phil. Trans.*, 1784, p. 332.

dephlogisticated air and phlogiston deprived of part of their latent or elementary heat; that dephlogisticated or pure air is composed of water deprived of its phlogiston and united to elementary heat and light; and that the latter are contained in it in a latent state, so as not to be sensible to the thermometer or to the eye; and if light be only a modification of heat, or a circumstance attending it, or a component part of the inflammable air, then pure or dephlogisticated air is composed of water deprived of its phlogiston and united to elementary heat?"

When we compare this doubtful and hypothetical statement, involving so much that is extraneous and heterogeneous, with the conclusion of Cavendish, in which there is nothing hypothetical or superfluous, we may confidently assent to the decision which has been pronounced by one* of our own time in favour of Cavendish. And we may with pleasure recognise, in this enlightened umpire, a due appreciation of the value of the maxim on which we are now insisting. "Cavendish," says Mr. Vernon Harcourt, "pared off from the hypotheses their theories of combustion, and *affinities of imponderable for ponderable matter*, as complicating chemical with physical considerations."

6. *Relation of Heat to Chemistry.*—But while we thus condemn the attempts to explain the thermotical phenomena of chemical processes by means of chemical considerations, it may be asked if we are altogether to renounce the hope of understanding such phenomena? It is plain, it may be said, that heat generated in chemical changes is always a very important circumstance, and can sometimes be measured, and perhaps reduced to laws; are we prohibited from speculating concerning the causes of such circumstances and

* The Rev. W. Vernon Harcourt, Address to the British Association, 1839.

such laws? And to this we reply, that we may properly attempt to connect chemical with thermotical processes, *so far as* we have obtained a clear and probable view of the nature of the thermotical processes. When our theory of thermotics is tolerably complete and certain, we may with propriety undertake to connect it with our theory of chemistry. But at present we are not far enough advanced in our knowledge of heat to make this attempt with any hope of success. We can hardly expect to understand the part which heat plays in the union of two bodies, when we cannot as yet comprehend in what manner it produces the liquefaction or vaporization of one body. We cannot look to account for Gay Lussac and Dalton's Law, that all gases expand equally by heat, till we learn how heat causes a gas to expand. We cannot hope to see the grounds of Dulong and Petit's Law, that the specific heat of all atoms is the same, till we know much more, not only about atoms, but about specific heat. We have as yet no thermotical theory which even professes to account for all the prominent facts of the subject*: and the theories which have been proposed are of the most diverse kind. Laplace assumes particles of bodies surrounded by atmospheres of caloric†; Cauchy makes heat consist in longitudinal vibrations of the ether of which transverse vibrations produce light: in Ampère's theory†, heat consists in the vibrations of the particles of bodies. And so long as we have nothing more certain in our conceptions of heat than the alternative of these and other precarious hypotheses, how can we expect to arrive at any real knowledge, by connecting the results of such hypotheses with the speculations of chemistry, of which science the theory is at least equally obscure?

* *Hist. Ind. Sci.*, ii., 530.   † *Ib.*, ii., 531.
        ‡ *Ib.*, ii., 529.

The largest attempts at chemical theory have been made in the form of the Atomic Theory, to which I have just had occasion to allude. I must, therefore, before quitting the subject, say a few words respecting this theory.

CHAPTER V.

THE ATOMIC THEORY.

1. *The Atomic Theory considered on Chemical Grounds.*—We have already seen that the combinations which result from chemical affinity are definite, a certain quantity of one ingredient uniting, not with an uncertain, but with a certain quantity of another ingredient. But it was found, in addition to this principle, that one ingredient would often unite with another in different proportions, and that, in such cases, these proportions are multiples one of another. In the three salts formed by potassa with oxalic acid, the quantities of acid which combine with the same quantity of alkali are exactly in the proportion of the numbers 1, 2, 4. And the same rule of the existence of multiple proportions is found to obtain in other cases.

It is obvious that such results will be accounted for, if we suppose the base and the acid to consist each of definite equal particles, and that the formation of the salts above mentioned consists in the combination of one particle of the base with one particle of acid, with two particles of acid, and with four particles of acid, respectively. But further; as we have already stated, chemical affinity is not only definite, but reciprocal. The proportions of potassa and soda which form neutral salts are 590 and 391 in one case, and therefore in all. These numbers represent the *proportions* of weight in which the two bases, potassa and soda, enter into analogous combinations; 590 of potassa is *equivalent* to 391 of soda. These facts with regard to combination are still expressed by the above supposition of equal particles, assuming that the weights of a particle of potassa and of soda are in the proportion of 590 to 391.

But we pursue our analysis further. We find that potassa is a compound of a metallic base, potassium, and of oxygen, in the proportion of 490 to 100; we suppose, then, that the particle of potassa consists of a particle of potassium and a particle of oxygen, and these latter particles, since we see no present need to suppose them divided, potassium and oxygen being simple bodies, we may call *atoms*, and assume to be indivisible. And by supposing all simple bodies to consist of such atoms, and compounds to be formed by the union of two, or three, or more of such atoms, we explain the occurrence of definite and multiple proportions, and we construct the Atomic Theory.

2. *Hypothesis of Atoms.*—So far as the assumption of such atoms as we have spoken of serves to express those laws of chemical composition which we have referred to, it is a clear and useful generalization. But if the Atomic Theory be put forwards (and its author, Dr. Dalton, appears to have put it forwards with such an intention,) as asserting that chemical elements are really composed of atoms, that is, of such particles not further divisible, we cannot avoid remarking, that for such a conclusion, chemical research has not afforded, nor can afford, any satisfactory evidence whatever. The smallest observable quantities of ingredients, as well as the largest, combine according to the laws of proportions and equivalence which have been cited above. How are we to deduce from such facts any

inference with regard to the existence of certain smallest possible particles? The Theory, when dogmatically taught as a physical truth, asserts that all observable quantities of elements are composed of proportional numbers of particles which can no further be subdivided; but all which observation teaches us is, that *if* there be such particles, they are smaller than the smallest observable quantities. In chemical experiment, at least, there is not the slightest positive evidence for the existence of such atoms. The assumption of indivisible particles, smaller than the smallest observable, which combine, particle with particle, will explain the phenomena; but the assumption of particles bearing this proportion, but *not* possessing the property of indivisibility, will explain the phenomena at least equally well. The decision of the question, therefore, whether the Atomic Hypothesis be the proper way of conceiving the chemical combinations of substances, must depend, not upon chemical facts, but upon our conception of substance. In this sense the question is an ancient and curious controversy, and we shall hereafter have to make some remarks upon it.

3. *Chemical Difficulties of the Hypothesis.* — But before doing this, we may observe that there is no small difficulty in reconciling this hypothesis with the facts of chemistry. According to the theory, all salts, compounded of an acid and a base, are analogous in their atomic constitution; and the number of atoms in one such compound being known or assumed, the number of atoms in other salts may be determined. But when we proceed in this course of reasoning to other bodies, as metals, we find ourselves involved in difficulties. The protoxide of iron is a base which, according to all analogy, must consist of one atom of iron and one of oxygen : but the peroxide of iron is also a base, and it appears by the analysis of this substance that it must consist of *two-*

thirds of an atom of iron and one atom of oxygen. Here, then, our indivisible atoms must be divisible, even upon chemical grounds. And if we attempt to evade this difficulty by making the peroxide of iron consist of two atoms of iron and three of oxygen, we have to make a corresponding alteration in the theoretical constitution of all bodies analogous to the protoxide; and thus we overturn the very foundation of the theory. Chemical facts, therefore, not only do not prove the Atomic Theory as a physical truth, but they are not, according to any modification yet devised of the theory, reconcilable with its scheme.

Nearly the same conclusions result from the attempts to employ the Atomic Hypothesis in expressing another important chemical law;—the law of the combinations of gases according to definite proportions of their volumes, experimentally established by Guy Lussac [*]. In order to account for this law, it has been very plausibly suggested that all gases, under the same pressure, contain an equal number of atoms in the same space; and that when they combine, they unite atom to atom. Thus one volume of chlorine unites with one volume of hydrogen, and form hydrochloric acid [†]. But then this hydrochloric acid occupies the space of the two volumes; and therefore the proper number of particles cannot be supplied, and the uniform distribution of atoms in all gases maintained, without dividing into two each of the compound particles, constituted of an atom of chlorine and an atom of hydrogen. And thus in this case, also, the Atomic Theory becomes untenable if it be understood to imply the indivisibility of the atoms.

In all these attempts to obtain a distinct physical conception of chemical union by the aid of the Atomic Hypothesis, the atoms are conceived to be associated by

---

[*] *Hist. Ind. Sc.*, iii., 153,   [†] DUMAS, *Phil. Chim.* 263.

certain forces of the nature of mechanical attractions. But we have already seen* that no such mode of conception can at all explain or express the facts of chemical combination; and therefore it is not wonderful that when the Atomic Theory attempts to give an account of chemical relations by contemplating them under such an aspect, the facts on which it grounds itself should be found not to authorise its positive doctrines; and that when these doctrines are tried upon the general range of chemical observation, they should prove incapable of even expressing, without self-contradiction, the laws of phenomena.

4. *Grounds of the Atomic Doctrine.*—Yet the doctrine of atoms, or of substance as composed of indivisible particles, has in all ages had great hold upon the minds of physical speculators; nor would this doctrine ever have suggested itself so readily, or have been maintained so tenaciously, as the true mode of conceiving chemical combinations, if it had not been already familiar to the minds of those who endeavour to obtain a general view of the constitution of nature. The grounds of the assumption of the atomic structure of substance are to be found rather in the idea of substance itself, than in the experimental laws of chemical affinity. And the question of the existence of atoms, thus depending upon an idea which has been the subject of contemplation from the very infancy of philosophy, has been discussed in all ages with interest and ingenuity. On this very account it is unlikely that the question, so far as it bears upon chemistry, should admit of any clear and final solution. Still it will be instructive to look back at some of the opinions which have been delivered respecting this doctrine.

5.. *Ancient Prevalence of the Atomic Doctrine.*—The doctrine that matter consists of minute, simple, indivisible,

* See Chapter I. of this Book.

indestructible particles as its ultimate elements, has been current in all ages and countries, whenever the tendency of man to wide and subtle speculations has been active. I need not attempt to trace the history of this opinion in the schools of Greece and Italy. It was the leading feature in the physical tenets of the Epicureans, and was adopted by their Roman disciples, as the poem of Lucretius copiously shows us. The same tenet had been held at still earlier periods, in forms more or less definite, by other philosophers. It is ascribed to Democritus, and is said to have been by him derived from Leucippus. But this doctrine is found also, we are told*, among the speculations of another intellectual and acute race, the Hindoos. According to some of their philosophical writers, the ultimate elements of matter are atoms, of which it is proved by certain reasonings, that they are each one-sixth of one of the motes that float in the sunbeam.

This early prevalence of controversies of the widest and deepest kind, which even in our day remain undecided, has in it nothing which need surprise us; or, at least, it has in it nothing which is not in conformity with the general course of the history of philosophy. As soon as any ideas are clearly possessed by the human mind, its activity and acuteness in reasoning upon them are such, that the fundamental antitheses and ultimate difficulties which belong to them are soon brought into view. The Greek and Indian philosophers had mastered completely the Idea of Space, and possessed the Idea of Substance in tolerable distinctness. They were, therefore, quite ready, with their lively and subtle minds, to discuss the question of the finite and infinite divisibility of matter, so far as it involved only the ideas of space and of sub-

* By Mr. Colebrook. *Asiatic Res.* 1824.

stance, and this accordingly they did with great ingenuity and perseverance.

But the ideas of Space and of Substance are far from being sufficient to enable men to form a complete general view of the constitution of matter. We must add to these ideas, that of mechanical Force with its antagonist Resistance, and that of the Affinity of one kind of matter for another. Now the former of these ideas the ancients possessed in a very obscure and confused manner; and of the latter they had no apprehension whatever. They made vague assumptions respecting the impact and pressure of atoms on each other; but of their mutual attraction and repulsion they never had any conception, except of the most dim and wavering kind; and of an affinity different from mere local union they did not even dream. Their speculations concerning atoms, therefore, can have no value for us, except as a part of the history of science. If their doctrines appear to us to approach near to the conclusions of our modern philosophy, it must be because our modern philosophy has not fully profited by the additional light which the experiments and meditations of later times have thrown upon the constitution of matter.

6. *Bacon.*—Still, when modern philosophers look upon the Atomic Theory of the ancients in a general point of view merely, without considering the special conditions which such a theory must fulfil, in order to represent the discoveries of modern times, they are disposed to regard it with admiration. Accordingly we find Francis Bacon strongly expressing such a feeling. The Atomic Theory is selected and dwelt upon by him as the chain which connects the best parts of the physical philosophy of the ancient and the modern world. Among his works is a remarkable dissertation *On the Philosophy of Democritus, Parmenides, and Telesius:* the last mentioned of whom was one of the revivers of physical science in modern times. In

this work he speaks of the atomic doctrine of Democritus as a favourable example of the exertions of the undisciplined intellect. "Hæc ipsa placita, quamvis paulo emendatiora, talia sunt qualia esse possunt ille quæ ab intellectu sibi permisso, nec continenter et gradatim sublevato, profecta videntur."—"Accordingly," he adds, "the doctrine of Atoms, from its going a step beyond the period in which it was advanced, was ridiculed by the vulgar, and severely handled in the disputations of the learned, notwithstanding the profound acquaintance with physical science by which its author was allowed to be distinguished, and from which he acquired the character of a magician."

"However," he continues, "neither the hostility of Aristotle, with all his skill and vigour in disputation, (though, like the Ottoman sultans, he laboured to destroy all his brother philosophers that he might rest undisputed master of the throne of science,) nor the majestic and lofty authority of Plato, could effect the subversion of the doctrine of Democritus. And while the opinions of Plato and Aristotle were rehearsed with loud declamation and professorial pomp in the schools, this of Democritus was always held in high honour by those of a deeper wisdom, who followed in silence a severer path of contemplation. In the days of Roman speculation it kept its ground and its favour; Cicero everywhere speaks of its author with the greatest praise; and Juvenal, who, like poets in general, probably expressed the prevailing judgment of his time, proclaims his merit as a noble exception to the general stupidity of his countrymen.

> . . . . Cujus prudentia monstrat
> Magnos posse viros et magna exempla daturos
> Verveccum in patriâ crassoque sub aere nasci.

"The destruction of this philosophy was not effected by Aristotle and Plato, but by Genseric and Attila, and their barbarians. For then, when human knowledge had suffered shipwreck, those fragments of the Aristotelian and Platonic philosophy floated on the surface like things of some lighter and emptier sort, and so were preserved; while more solid matters went to the bottom, and were almost lost in oblivion."

7. *Modern Prevalence of the Atomic Doctrine.*—It is our business here to consider the doctrine of Atoms only in its bearing upon existing physical sciences, and I must therefore abstain from tracing the various manifestations of it in the schemes of hypothetical cosmologists;—its place among the *vortices* of Descartes, its exhibition in the *monads* of Leibnitz. I will, however, quote a passage from Newton to show the hold it had upon his mind.

At the close of his *Opticks* he says, "All these things being considered, it seems probable to me that God, in the beginning, formed matter in solid, massy, hard, impenetrable, moveable particles, of such sizes and figures, and with such other properties, and in such proportions to space, as most conduced to the end for which He formed them; and that these primitive particles, being solids, are incomparably harder than any porous bodies compounded of them, even so very hard as never to wear or break in pieces; no ordinary power being able to divide what God had made one in the first creation. While the particles continue entire, they may compose bodies of one and the same nature and texture in all ages: but should they wear away or break in pieces, the nature of things depending on them would be changed. Water and earth composed of old worn particles and fragments of particles would not be of the same nature and texture now with water and earth composed of entire particles in the beginning. And therefore that nature may be lasting, the changes of corporeal things are to be placed only in the various separations and new associations and motions of these permanent particles; compounded bodies being apt to break, not in the midst of solid particles, but where those particles are laid together and only touch in a few points."

We shall hereafter see how extensively the atomic doctrine has prevailed among still more recent philosophers. Not only have the chemists assumed it as the fittest form for exhibiting the principles of multiple proportions; but the physical mathematicians, as Laplace and Poisson, have made it the basis of their theories of heat, electricity, capillary action; and the crystallographers have been supposed to have established both the existence and the arrangement of such ultimate molecules.

In the way in which it has been employed by such writers, the hypothesis of ultimate particles has been of great use, and is undoubtedly permissible. But when we would assert this theory, not as a convenient hypothesis for the expression or calculation of the laws of nature, but as a philosophical truth respecting the constitution of the universe, we find ourselves checked by difficulties of reasoning which we cannot overcome, as well as by conflicting phenomena which we cannot reconcile. I will attempt to state briefly the opposing arguments on this question.

8. *Arguments for and against Atoms.*—The leading arguments on the two sides of the question, in their most general form, may be stated as follows:—

For the Atomic Doctrine.—The appearances which nature presents are compounded of many parts, but if we go on resolving the larger parts into smaller, and so on successively, we must at last come to something simple. For that which is compound can be so no otherwise than by composition of what is simple; and if we suppose all composition to be removed, which hypothetically we may do, there can remain nothing but a number of simple substances, capable of composition, but themselves not compounded. That is, matter being dissolved, resolves itself into atoms.

Against the Atomic Doctrine.—Space is divisible without limit, as may be proved by geometry; and matter occupies space, therefore matter is divisible without limit, and no portion of matter is *indivisible,* or an *atom.*

And to the argument on the other side just stated, it is replied that we cannot even hypothetically divest a body of composition, if by composition we mean the relation of point to point in space. However small be a particle, it is compounded of parts having relation in space.

The Atomists urge again, that if matter be infinitely divisible, a finite body consists of an infinite number of parts, which is a contradiction. To this it is replied, that the finite body consists of an infinite number of parts in the same sense in which the parts are infinitely small, which is no contradiction.

But the opponents of the Atomists not only rebut, but retort this argument drawn from the notion of infinity. Your atoms, they say, are indivisible by any finite force; therefore they are infinitely hard; and thus your finite particles possess infinite properties. To this the Atomists are wont to reply, that they do not mean the hardness of their particles to be infinite, but only so great as to resist all usual natural forces. But here it is plain that their position becomes untenable; for, in the first place, their assumption of this precise degree of hardness in the particles is altogether gratuitous; and in the next place, if it were granted, such particles are not atoms, since in the next moment the forces of nature may be augmented so as to divide the particle, though hitherto undivided.

Such are the arguments for and against the Atomic Theory in its original form. But when these atoms are conceived, as they have been by Newton, and commonly by his followers, to be solid, hard particles exerting attractive and repulsive forces, a new set of arguments come into play. Of these, the principal one may be thus stated: According to the Atomic Theory thus modified, the properties of bodies depend upon the attractions and repulsions of the particles. Therefore, among other properties of bodies, their hardness depends upon such forces. But if the hardness *of the bodies* depends upon the forces, the repulsion, for instance, of the particles, upon what does the hardness *of the particles* depend? what progress do we make in explaining the properties of bodies, when we assume the same properties in our explanation? and to what purpose do we assume that the particles *are* hard?

9. *Transition to Boscovich's Theory.*—To this difficulty it does not appear easy to offer any reply. But if the hardness and solidity of the particles be given up as an incongruous and untenable appendage to the Newtonian view of the Atomic Theory, we are led to the theory of Boscovich, according to which matter consists not of solid particles, but of mere mathematical centres of force. According to this theory, each body is composed of a number of geometrical points from which emanate forces, following certain mathematical laws in virtue of which they become, at certain small distances attractive, at certain other distances repulsive, and at greater distances attractive again. From these forces of the points arise the cohesion of the parts of the same body, the resistance which it exerts against the pressure of another body, and finally the attraction of gravitation which it exerts upon bodies at a distance.

This theory is at least a homogeneous and consistent mechanical theory, and it is probable that it may be used as an instrument for investigating and expressing true laws of nature; although, as we have already said, the attempt to identify the forces by which the particles of bodies are bound together with mechanical attraction appears to be a confusion of two separate ideas.

10. *Use of the Molecular Hypothesis.*—In this form, representing matter as a collection of molecules or centres of force, the Atomic Theory has been abundantly employed in modern times as an hypothesis on which calculations respecting the elementary forces of bodies might be conducted. When thus employed it is to be considered as expressing the principle that the properties of bodies depend upon forces emanating from immovable points of their mass. This view of the way in which the properties of bodies are to be treated by the mechanical philosopher was introduced by Newton, and was a natural sequel to the success which he had obtained by reasoning concerning central forces on a large scale. I have already quoted his Preface to the *Principia*, in which he says, "Many things induce me to believe that the rest of the phenomena of nature, as well as those of astronomy, may depend upon certain forces by which the particles of bodies, in virtue of causes not yet known, are urged towards each other and cohere in regular figures, or are mutually repelled and recede ; and philosophers, knowing nothing of these forces, have hitherto failed in their examination of nature." Since the time of Newton, this line of speculation has been followed with great assiduity, and by some mathematicians with great success. In particular Laplace has shown that it may, in many instances, be made a much closer representation of nature, if we suppose the forces exerted by the particles to deciease so rapidly with the increasing distance from them, that the force is finite only at distances imperceptible to our senses, and vanishes at all remoter points. He has taught the method of expressing and calculating such forces, and he and other mathematicians of his

school have applied this method to many of the most important questions of physics; as capillary action, the elasticity of solids, the conduction and radiation of heat. The explanation of many apparently unconnected and curious observed facts by these mathematical theories gives us a strong assurance that its essential principles are true. But it must be observed that the actual constitution of bodies as composed of distinct and separate particles is by no means proved by these coincidences. The assumption, in the reasoning, of certain centres of force acting at a distance, is to be considered as nothing more than a method of reducing to calculation that view of the constitution of bodies, which supposes that they exert force at *every* point. It is a mathematical artifice of the same kind as the hypothetical division of a body into infinitesimal parts, in order to find its centre of gravity; and no more implies a physical reality than that hypothesis does.

11. *Poisson's Inference.*—When, therefore, M. Poisson, in his views of Capillary Action, treats this hypothetical distribution of centres of force as if it were a physical fact, and blames Laplace for not taking account of their different distribution at the surface of the fluid and below it*, he appears to push the claims of the molecular hypothesis too far. The only ground for the assumption of separate centres, is that we can thus explain the action of the whole mass. The intervals between the centres nowhere enter into this explanation: and therefore we can have no reason for assuming these intervals different in one part of the fluid and in the other. M Poisson asserts that the density of the fluid diminishes when we approach very near the surface; but he allows that this diminution is not detected by experiment, and that the formulæ on his supposition, so far as the results

* Poisson, *Theorie de l'Action Capillaire.*

go, are identical with those of Laplace. It is clear, then, that his doctrine consists merely in the assertion of the necessary truth of a part of the hypothesis which cannot be put to the test of experiment. It is true, that so long as we have before us the hypothesis of separate centres, the particles very near the surface are not in a condition symmetrical with that of the others: but it is also true that this hypothesis is only a step of calculation. There results, at one period of the process of deduction, a stratum of smaller density at the surface of the fluid; but at a succeeding point of the reasoning the thickness of this stratum vanishes; it has no physical existence.

Thus the *molecular* hypothesis, as used in such cases, does not differ from the doctrine of forces acting at *every point* of the mass; and this principle, which is common to both the opposite views, is the true part of each.

12. *Wollaston's Argument.* — An attempt has been made in another case, but depending on nearly the same arguments, to bring the doctrine of ultimate atoms to the test of observation. In the case of the air, we know that there *is* a diminution of density in approaching the upper surface of the atmosphere, if it have a surface: but it is held by some that except we allow the doctrine of ultenate molecules, it will not be bounded by any surface, but will extend to an infinite distance. This is the reasoning of Wollaston*. "If air consists of any ultimate particles no longer divisible, then must the expansion of the medium composed of them cease at that distance where the force of gravity downwards is equal to the resistance arising from the repulsive force of the medium." But if there be no such ultimate particles, every stratum will require a stratum beyond it to prevent by its weight a further expansion, and thus the atmosphere must extend to an infinite distance. And Wollaston con-

* *Phil. Trans.*, 1822, p. 89.

ceived that he could learn from observation whether the atmosphere was thus diffused through all space; for if so, it must, he argued, be accumulated about the larger bodies of the system, as Jupiter and the Sun, by the law of universal gravitation; and the existence of an atmosphere about these bodies, might, he remarked, be detected by its effects in producing refraction. His result is, that "all the phenomena accord entirely with the supposition that the earth's atmosphere is of finite extent, limited by the weight of ultimate atoms. of definite magnitude, no longer divisible by repulsion of their parts."

A very little reflection will show us that such a line of reasoning cannot lead to any result. For we know nothing of the law which connects the density with the compressing force, in air so extremely rare as we must suppose it to be near the boundary of the atmosphere. Now there are possible laws of dependence of the density upon the compressing force such that the atmosphere would terminate in virtue of the law without any assumption of atoms. This may be proved by mathematical reasoning. If we suppose the density of air to be as the square root of the compressing force, it will follow that at the very limits of the atmosphere, the strata of equal thickness may observe in their densities such a law of proportion as is expressed by the numbers 7, 5, 3, 1*.

If it be asked how, on this hypothesis, the density of the highest stratum can be as 1, since there is nothing to

---

* For the compressing force on each being as the whole weight beyond it, will be for the four highest strata, 16, 9, 4 and 1, of which the square roots are as 4, 3, 2, 1, or, as 8, 6, 4, 2; and though these numbers are not exactly as the densities 7, 5, 3, 1, those who are a little acquainted with mathematical reasoning, will see that the difference arises from taking so small a number of strata. If we were to make the strata indefinitely thin, as to avoid error we ought to do, the coincidence would be exact; and thus, according to this law, the series of strata terminates as we ascend, without any consideration of atoms.

compress it, we answer that the upper part of the highest stratum compresses the lower, and that the density diminishes continually to the surface, so that the need of compression and the compressing weight vanish together.

The fallacy of concluding that because the height of the atmosphere is finite, the weight of the highest stratum must be finite, is just the same as the fallacy of those who conclude that when we project a body vertically upwards, because it occupies only a finite time in ascending to the highest point, the velocity at the last instant of the ascent must be finite. For it might be said, if the last velocity of ascent be not finite, how can the body describe the last particle of space in a finite time? and the answer is, that there is no last finite particle of space, and there fore no last finite velocity.

13. *Permanence of Properties of Bodies.*—We nave already seen that, in explaining the properties of mattei as we find them in nature, the assumption of solid, hard, indestructible particles is of no use or value. But we may remark, before quitting the subject, that Newton appears to have had another reason for assuming such particles, and one well worthy of notice. He wished to express, by means of this hypothesis, the doctrine that the laws of nature do not alter with the course of time. This we have already seen in the quotation from Newton. " The ultimate particles of matter are indestructible, unalterable, impenetrable ; for if they could break or wear, the structure of material bodies now would be different from that which it was when the particles were new." No philosopher will deny the truth which is thus conveyed by the assertion of atoms; but it is obviously equally easy for a person who rejects the atomic view, to state this truth by saying that the forces which matter exerts do not vary with time; but however modified by the new modifications of its form, are always unimpaired

in quantity, and capable of being restored to their former mode of action.

We now proceed to speculations in which the fundamental conceptions may, perhaps, be expressed, at least in some cases, by means of the arrangement of atoms; but in which the philosophy of the subject appears to require a reference to a new Fundamental Idea.

BOOK VII.

THE PHILOSOPHY OF MORPHOLOGY, INCLUDING CRYSTALLOGRAPHY.

CHAPTER I.

EXPLICATION OF THE IDEA OF SYMMETRY.

1. WE have seen in the History of the Sciences, that a principle which I have there termed* the principle of developed and metamorphosed Symmetry, has been extensively applied in botany and physiology, and has given rise to a province of science termed Morphology. In order to understand clearly this principle, it is necessary to obtain a clear idea of the Symmetry of which we thus speak. But this Idea of Symmetry is applicable in the inorganic, as well as in the organic kingdoms of nature; it is presented to our eyes in the forms of minerals, as well as of flowers and animals; we must, therefore, take it under our consideration here, in order that we may complete our view of mineralogy, which, as I have repeatedly said, is an essential part of chemical science. I shall accordingly endeavour to unfold the Idea of Symmetry with which we here have to do.

It will of course be understood that by the term *Symmetry* I here intend, not that more indefinite attribute of form which belongs to the domain of the fine arts, as when we speak of the symmetry of an edifice or of a

* *Hist. Ind. Sci.,* iii., 433,

sculptured figure, but a certain definite relation or property, no less rigorous and precise than other relations of number and position, which is thus one of the sure guides of the scientific faculty, and one of the bases of our exact science.

2. In order to explain what Symmetry is in this sense, let the reader recollect that the bodies of animals consist of *two* equal and similar sets of members, the right and the left side ;—that some flowers consist of three or of five equal sets of organs, similarly and regularly disposed, as the iris has *three* straight petals, and three reflexed ones, alternately disposed, the rose has *five* equal and similar sepals of the calyx, and alternate with these as many petals of the corolla. This orderly and exactly similar distribution of two, or three, or five, or any other number of parts, is Symmetry; and according to its various modifications, the forms thus determined are said to be *symmetrical* with various numbers of members. The classification of these different kinds of symmetry has been most attended to in Crystallography, in which science it is the highest and most general principle by which the classes of forms are governed. Without entering far into the technicalities of the subject, we may point out some of the features of such classes.

 The first of the figures (1) in the margin may represent the summit of a crystal as it appears to an eye looking directly down upon it; the centre of the figure represents the summit of a pyramid, and the spaces of various forms which diverge from this point represent sloping sides of the pyramid. Now it will be observed that the figure consists of three portions exactly similar to one another, and that each part or member is repeated in each of these portions. The faces, or pairs of faces, are repeated in

threes, with exactly similar forms and angles. This figure is said to be *three-membered,* or to have *triangular* symmetry. The same kind of symmetry may exist in a flower, as presented in the accompanying figure, and does, in fact, occur in a large class of flowers, as for example, all the lily tribe. The next pair of figures (2) have four equal and similar portions, and have their members or pairs of members four times repeated. Such figures are termed *four-membered,* and are said to have *square* or *tetragonal* symmetry. The *pentagonal* symmetry, formed by *five* similar *members,* is represented in the next figures (3). It occurs abundantly in the vegetable world, but never among crystals; for the pentagonal figures which crystals sometimes assume, are never exactly regular. But there is still another kind of symmetry (4) in which the opposite ends are exactly similar to each other and also the opposite sides; this is *oblong,* or *two-and-two-membered* symmetry. And finally, we have the case of *simple* symmetry (5) in which the two sides of the object are exactly alike (in opposite positions) without any further repetition.

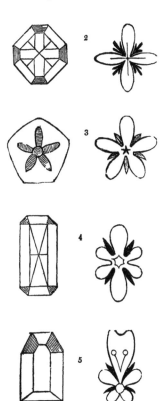

3. These different kinds of symmetry occur in various ways in the animal, vegetable, and mineral kingdom; thus vertebrate animals have a right and a left side exactly alike and thus possess *simple* symmetry. The same kind of symmetry (simple symmetry) occurs very largely in the forms of vegetables, as in most leaves, in *papilionaceous,*

personate, and *labiate* flowers.   Among minerals, crystals
which possess this symmetry are called *oblique-prismatic,*
and are of very frequent occurrence.   The *oblong,* or
two-and-two membered symmetry belongs to *right-prismatic*
crystals; and may be seen in *cruciferous* flowers, for
though these are cross-shaped, the cross has two longer
and two shorter arms, or pairs of arms.   The *square* or
tetragonal symmetry occurs in crystals abundantly; to
the vegetable world it appears to be less congenial;
for though there are flowers with four exactly similar
and regularly-disposed petals, as the herb Paris (*Paris
quadrifolia*), these flowers appear, from various circumstances, to be deviations from the usual type of vegetable forms.   The *trigonal,* or *three-membered* symmetry is
found abundantly both in plants and in crystals, while the
pentagonal symmetry, on the other hand, though by far
the most common among flowers, nowhere occurs in
minerals, and does not appear to be a possible form of
crystals.   This pentagonal form further occurs in the animal kingdom, which the oblong, triangular, and square
forms do not.   Many of Cuvier's *radiate* animals appear
in this pentagonal form, as *echini* and *pentacrinites,* which
latter have hence their name.

4. The regular, or as they may be called, the *normal*
types of the vegetable world appear to be the forms which
possess triangular and pentagonal symmetry; from these
the others may be conceived to be derived, by transformations resulting from the expansion of one or more parts.
Thus it is manifest that if in a three-membered or five-
membered flower, one of the petals be expanded more
than the other, it is immediately reduced from pentagonal
or trigonal, to simple symmetry.   And the oblong or two-
and-two-membered symmetry of the flowers of cruciferous plants, (in which the stamens are four large and two
small ones, arranged in regular opposition,) is held by

botanists to result from a normal form with ten stamens; Meinecke explaining this by adhesion, and Sprengel by the metamorphosis of the stamens into petals*.

It is easy to see that these various kinds of symmetry include relations both of form and of number, but more especially of the latter kind; and as this symmetry is often an important character in various classes of natural objects, such classes have often curious numerical properties. One of the most remarkable and extensive of these is the distinction which prevails between monocotyledonous and dicotyledonous plants; the number *three* being the ground of the symmetry of the former, and the number *five*, of the latter. Thus liliaceous and bulbous plants, and the like, have flowers of three or six petals, and the other organs follow the same numbers: while the vast majority of plants are pentandrous, and with their five stamens have also their other parts in fives. This great numerical distinction corresponding to a leading difference of physiological structure cannot but be considered as a highly curious fact in phytology. Such properties of numbers, thus connected in an incomprehensible manner with fundamental and extensive laws of nature, give to numbers an appearance of mysterious importance and efficacy. We learn from history how strongly the study of such properties, as they are exhibited by the phenomena of the heavens, took possession of the mind of Kepler; perhaps it was this, which, at an earlier period, contributed in no small degree to the numerical mysticism of the Pythagoreans in antiquity, and of the Arabians and others in the middle ages. In crystallography, numbers are the primary characters in which the properties of substances are expressed;—they appear, first, in that classification of forms which depends on the degree of symmetry, that is, upon the number of

* SPRENGEL, *Gesch. d. Bot.*, ii., 304.

correspondencies; and next, in the laws of derivation, which, for the most part, appear to be common in their occurrence in proportion to the numerical simplicity of their expression. But the manifestation of a governing numerical relation in the organic world strikes us as more unexpected; and the selection of the number *five* as the index of the symmetry of dicotyledonous plants and radiated animals, (a number which is nowhere symmetrically produced in inorganic bodies,) makes this a new and remarkable illustration of the constancy of numerical relations. We may observe, however, that the moment one of these radiate animals has one of its five members expanded, or in any way peculiarly modified, (as happens among the echini) it is reduced to the common type of animals simply symmetrical, with a right and left side.

5. It is not necessary to attempt to enumerate all the kinds of Symmetry, since our object is only to explain what Symmetry is, and for this purpose enough has probably been said already. It will be seen, as soon as the notion of Symmetry in general is well apprehended, that it is or includes a peculiar Fundamental Idea, not capable of being resolved into any of the ideas hitherto examined. It may be said, perhaps, that the Idea of Symmetry is a modification or derivative of our ideas of space and number;—that a symmetrical shape is one which consists of parts exactly similar, repeated a certain number of times, and placed so as to correspond with each other. But on further reflection it will be seen that this repetition and correspondence of parts in symmetrical figures are something peculiar; for it is not *any* repetition or any correspondence of parts to which we should give the name of symmetry, in the manner in which we are now using the term. Symmetrical arrangements may no doubt be concerned with space and position, time and number; but there appears to be implied

in them a Fundamental Idea of regularity, of completeness, of complex simplicity, which is not a mere modification of other ideas.

6. It is, however, not necessary, in this and in similar
cases to determine whether the idea which we have
before us be a peculiar and independent Fundamental
Idea or a modification of other ideas, provided we clearly
perceive the evidence of those Axioms by means of which
the Idea is applied in scientific reasonings. Now in the
application of the Idea of Symmetry to crystallography,
phytology and zoology, we must have this idea embodied
in some principle which asserts more than a mere geometrical or numerical accordance of members. We must
have it involved in some vital or productive action, in
order that it may connect and explain the facts of the
organic world. Nor is it difficult to enunciate such a
principle. We may state it in this manner. *All the
symmetrical members of a natural product are, under like
circumstances, alike affected.* The parts which we have
termed symmetrical, resemble each other, not only in
their form and position, but also in the manner in which
they are produced and modified by natural causes. And
this principle we assume to be necessarily true, however
unknown and inconceivable may be the causes which
determine the phenomena. Thus it has not yet been
found possible to discover or represent to ourselves, in
any intelligible manner, the forces by which the various
faces of a crystal are consequent upon its primary form;
but the whole of crystallography rests upon this principle,
that if one of the primary planes or axes be modified in
any manner, all the symmetrical planes and axes must be
modified in the same manner. And though accidental
mechanical or other causes may interfere with the actual
exhibition of such faces, we do not the less assume their
crystallographical reality, as inevitably implied in the

law of symmetry of the crystal*. And we apply similar considerations to organized beings. We assume that in a regular flower, each of the similar members has the same organization and similar powers of developement; and hence if among these similar parts some are much less developed than others, we consider them as *abortive;* and if we wish to remove doubts as to what are symmetrical members in such a case, we make the inquiry by tracing the anatomy of these members, or by following them in their earlier states of developement, or in cases where their capabilities are magnified by monstrosity or otherwise. The power of developement may be modified by external causes, and thus we may pass from one kind of symmetry to another; as we have already remarked. Thus a regular flower with pentagonal symmetry, growing on a lateral branch, has one petal nearest to the axis of the plant: if this petal be more or less expanded than the others, the pentagonal symmetry is interfered with, and the flower may change to a symmetry of another kind. But it is easy to see that all such conceptions of expansion, abortion, and any other kind of metamorphosis go, upon the supposition of identical faculties and tendencies in each similar member, in so far as such tendencies have any relation to the symmetry. And thus the principle we have stated above is the basis of that which, in the History, we termed the Principle of Developed and Metamorphosed Symmetry.

We shall not at present pursue the other applications of this Idea of Symmetry, but we shall consider some of the results of its introduction into Crystallography.

* Some crystalline forms, instead of being *holohedral* (provided with their whole number of faces), are *hemihedral* (provided with only half their number of faces). But in these hemihedral forms, the half of the faces are still *symmetrically* suppressed.

CHAPTER II.

APPLICATION OF THE IDEA OF SYMMETRY
TO CRYSTALS.

1. MINERALS and other bodies of definite chemical
composition often exhibit that marked regularity of form
and structure which we designate by terming them
Crystals; and in such crystals, when we duly study them,
we perceive the various kinds of symmetry of which we
have spoken in the previous chapter. And the different
kinds of symmetry which we have there described are
now usually distinguished from each other, by writers on
crystallography. Indeed it is mainly to such writers that
we are indebted for a sound and consistent classification
of the kinds and degrees of symmetry of which forms are
capable. But this classification was by no means invented
as soon as mineralogists applied themselves to the study
of crystals. These first attempts to arrange crystalline
forms were very imperfect; those, for example, of Linnæus, Werner, Romé de Lisle, and Haüy. The essays of
these writers implied a classification at once defective
and superfluous. They reduced all crystals to one or
other of certain *fundamental forms;* and this procedure
might have been a perfectly good method of dividing
crystalline forms into classes, if the fundamental forms
had been selected so as to exemplify the different kinds
of symmetry. But this was not the case. Haüy's fundamental or "primitive" forms, were, for instance, the
following: the parallelepiped, the octahedron, the tetrahedron, the regular hexagonal prism, the rhombic dodecahedron, and the double hexagonal pyramid. Of these,
the octahedron, the tetrahedron, the rhombic dodecahedron, all belong to the same kind of symmetry (the

tessular systems); also the hexagonal prism and the hexagonal pyramid both belong to the rhombic system; while the parallelepiped is so employed as to include all kinds of symmetry.

It is, however, to be recollected that Haüy, in his selection of primitive forms, not only had an eye to the external form of the crystal and to its degree and kind of regularity, but also made his classification with an especial reference to the cleavage of the mineral, which he considered as a primary element in crystalline analysis. There can ·be no doubt that the cleavage of a crystal is one of its most important characters: it is a relation of form belonging to the interior, which is to be attended to no less than the form of the exterior. But still the cleavage is to be regarded only as determining the degree of geometrical symmetry of the body, and not as defining a special geometrical figure to which the body *must* be referred. To have looked upon it in the latter light was a mistake of the earlier crystallographic speculators, on which we shall shortly have to remark.

2. I have said that the reference of crystals to primitive forms might have been well employed as a mode of expressing a just classification of them. This follows as a consequence from the application of the principle stated in the last chapter, that *all symmetrical members are alike affected.* Thus we may take an upright triangular prism as the representative of the rhombic system, and if we then suppose one of the upper edges to be cut off, or *truncated*, we must, by the principle of symmetry, suppose the other two upper edges to be truncated in precisely the same manner. By this truncation we may obtain the upper part of a rhombohedron; and by truncations of the same kind, symmetrically affecting all the analogous parts of the figure, we may obtain any other form possessing three-membered symmetry. And the same is true of any of

the other kinds of symmetry, provided we make a proper selection of a fundamental form. And this was really the method employed by Demeste, Werner, and Romé de Lisle. They assumed a primitive form, and then conceived other forms, such as they found in nature, to be derived from the primitive form by truncation of the edges, acumination of the corners, and the like processes. This mode of conception was a perfectly just and legitimate expression of the general idea of symmetry.

3. The true view of the degrees of symmetry was, as I have already said, impeded by the attempts which Haüy and others made to arrive at primitive forms by the light which cleavage was supposed to throw upon the structure of minerals. At last, however, in Germany, as I have narrated in the History of Mineralogy*, Weiss and Mohs introduced a classification of forms implying a more philosophical principle, dividing the forms into *Systems*; which, employing the terms of the latter writer, we shall call the *tessular*, the *pyramidal* or *square pyramidal*, the *prismatic* or *oblong*, and the *rhombohedral* systems.

Of these forms, the three latter may be at once referred to those kinds of symmetry of which we have spoken in the last chapter. The rhombohedral system has triangular symmetry, or is three-membered: the pyramidal has square symmetry, or is four-membered: the prismatic has oblong symmetry, and is two-and-two-membered. But the kinds of symmetry which were spoken of in the former chapter, do not exhaust the idea when applied to minerals. For the symmetry which was there explained was such only as can be exhibited on a surface, whereas the forms of crystals are solid. Not only have the right and left parts of the upper surface of a crystal relations to each other; but the upper surface

* *Hist. Ind. Sci.*, iii. 209.

and the lateral faces of the crystal have also their relations; they may be different, or they may be alike.
If we take a cube, and hold it so that four of its faces
are vertical, not only are all these four sides exactly similar, so as to give square symmetry; but also we may turn
the cube, so that any one of these four sides shall become
the top, and still the four sides which are thus made
vertical, though not the same which were vertical before,
are still perfectly symmetrical. Thus this cubical figure
possesses more than square symmetry. It possesses
square symmetry in a vertical as well as in a horizontal
sense. It possesses a symmetry which has the same
relation to a *cube* which four-membered symmetry has to
a *square*. And this kind of symmetry is termed the
cubical or *tessular* symmetry. All the other kinds of
symmetry have reference to an axis, about which the corresponding parts are disposed; but in tessular symmetry
the horizontal and vertical axes are also symmetrical, or
interchangeable; and thus the figure may be said to have
no axis at all.

4. It has already been repeatedly stated that, by the
very idea of symmetry, all the incidents of form must
affect alike all the corresponding parts. Now in crystals
we have, among these incidents, not only external figure,
but *cleavage*, which may be considered as internal figure.
Cleavage, then, must conform to the degree of symmetry
of the figure. Accordingly cleavage, no less than form, is
to be attended to in determining to what system a mineral
belongs. If a crystal were to occur as a square prism or
pyramid, it would not on that account necessarily belong
to the square pyramidal system. If it were found that
it was cleavable parallel to one side of the prism, but not
in the transverse direction, it has only oblong symmetry;
and the equality of the sides which makes it square is
only accidental.

Thus no cleavage is admissible in any system of crystallization which does not agree with the degree of symmetry of the system. On the other hand, *any* cleavage which *is* consistent with the symmetry of the system, is (hypothetically at least) allowable. Thus in the oblong prismatic system we may have a cleavage parallel to one side only of the prism; or parallel to both, but of different distinctness; or parallel to the two diagonals of the prism but of the same distinctness; or we may have both these cleavages together. In the rhombohedral system, the cleavage may be parallel to the sides of the rhombohedron, as in Calc Spar: or, in the same system, the cleavage, instead of being thus oblique to the axis, may be along the axis in those directions which make equal angles with each other: this cleavage easily gives either a triangular or a hexagonal prism. Again, in the tessular system, the cleavage may be parallel to the surface of the cube, which is thus readily separable into other cubes, as in Galena; or the cleavage may be such as to cut off the solid angle of the cube, and since there are eight of these, such cleavage gives us an octahedron, which, however, may be reduced to a tetrahedron, by rejecting all parallel faces, as being mere repetitions of the same cleavage; this is the case with Fluor Spar: or the cube of the tessular system may be cleavable in planes which truncate all the edges of the cube; and as these are twelve, we thus obtain the dodecahedron with rhombic faces: this occurs in Zinc Blende. And thus we see the origin of Haüy's various primitive forms, the tetrahedron, octahedron, and rhombic dodecahedron, all belonging to the tessular system:—they are, in fact, different cleavage forms of that system.

5. I do not dwell upon other incidents of crystals which have reference to form, nor upon the lustre, smoothness, and striation of the surfaces. To all such inci dents

the general principle applies, that similar parts are similarly affected; and hence if any parts are found to be constantly and definitely different from other parts of the same sort, they are not similar parts; and the symmetry is to be interpreted with reference to this difference.

We have now to consider the inferences which have been drawn from these incidents of crystallization, with regard to the intimate structure of bodies.

CHAPTER III.

SPECULATIONS FOUNDED UPON THE SYMMETRY OF CRYSTALS.

1. When a crystal, as, for instance, a crystal of galena, (sulphuret of lead,) is readily divisible into smaller cubes, and these into smaller ones, and so on without limit, it is very natural to represent to ourselves the original cube as really consisting of small cubical elements; and to imagine that it is a philosophical account of the physical structure of such a substance to say that it is made up of cubical molecules. And when the galena crystal has externally the form of a cube, there is no difficulty in such a conception; for the surface of the crystal is also conceived as made up of the surfaces of its cubical molecules. We conceive the crystal so constituted, as we conceive a wall built of bricks.

But if, as often happens, the galena crystal be an octahedron, a further consideration is requisite in order to understand its structure, pursuing still the same hypothesis. The mineral is still, as in the other case, readily cleavable into small cubes, having their corners turned to the faces of the octahedron. Therefore these faces can no longer be conceived as made up of the faces of cubical elements of which the whole is constituted. If we suppose a pile of such small cubes to be closely built together, but with decreasing width above, so as to form a pyramid, the face of such a pyramid will no longer be plane; it will consist of a great number of the corners or edges of the small elementary cubes. It would appear at first sight, therefore, that such a face cannot represent the smooth polished surface of a crystal.

But when we come to look more closely, this difficulty disappears. For how large are these elementary cubes? We cannot tell, even supposing they really have any size. But we know that they must be, at any rate, very small; so small as to be inappreciable by our senses, for our senses find no limit to the divisibility of minerals by cleavage. Hence the surface of the pyramid above described would not consist of visible corners or edges, but would be roughened by specks of imperceptible size; or rather, by supposing these specks to become still smaller, the roughness becomes smoothness. And thus we may have a crystal with a smooth surface, made up of small cubes in such a manner that their surfaces are all oblique to the surface of the crystal.

Haüy, struck by some instances in which the supposition of such a structure of crystals appeared to account happily for several of their relations and properties, adopted and propounded it as a general theory. The small elements, of which he supposed crystals to be thus built up, he termed *integrant molecules*. The form of these molecules might or might not be the same as the *primitive form* with which his construction was supposed to begin; but there was, at any rate, a close connexion between these forms, since both of them were founded on the cleavage of the mineral. The tenet that crystals are constituted in the manner which I have been describing, I shall call the *Theory of Integrant Molecules,* and I have now to make some remarks on the grounds of this theory.

2. In the case of which I have spoken, the mineral used as the example, galena, readily splits into cubes, and cubes are easily placed together so as to fit each other, and fill the space which they occupy. The same is the case in the mineral which suggested to Haüy his theory, namely, calc spar. The crystals of this substance are readily divisible into rhombohedrons, a form like a brick with oblique angles; and such bricks can be built together so as to produce crystals of all the immense varieties of form which calc spar presents. This kind of masonry is equally possible in many other minerals; but as we go through the mineral kingdom in our survey, we soon find cases which offer difficulties. Some minerals cleave only in two directions, some in one only; in such cases we cannot by cleavage obtain an integrant molecule of definite form; one of its dimensions, at least, must remain indeterminate and arbitrary. Again, in some instances, we have more than three different planes of cleavage, as in fluor spar, where we have four. The solid, bounded by four planes, is a tetrahedron; or if we take four pairs of parallel faces, an octahedron. But if we attempt to take either of these forms for our integrant molecule, we are met by this difficulty : that a collection of such forms will not fill space. Perhaps this difficulty will be more readily conceived by the general reader if it be contemplated with reference to plane figures. It will readily be seen that a number of equal squares may be put together so as to fill the space which they occupy; but if we take a number of equal regular octagons, we may easily convince ourselves that no possible arrangement can make them cover a flat space without leaving blank spots between. In like manner octahedrons or tetrahedrons cannot be arranged in solid space

so as to fill it. They necessarily leave vacancies. Hence the structure of fluor spar, and similar crystals, was a serious obstacle in the way of the theory of integrant molecules. That theory had been adopted in the first instance because portions of the crystal, obtained by cleavage, could be built up into a solid mass; but this ground of the theory failed altogether in such instances as I have described, and hence the theory, even upon the representations of its adherents, had no longer any claim to assent.

The doctrine of Integral Molecules, however, was by no means given up at once, even in such instances. In this and in other subjects, we may observe that a theory, once constructed and carried into detail, has such a hold upon the minds of those who have been in the habit of applying it, that they will attempt to uphold it by introducing suppositions inconsistent with the original foundations of the theory. Thus those who assert the atomic theory, reconcile it with facts by taking the halves of atoms; and thus the theory of integrant molecules was maintained for fluor spar, by representing the elementary octahedrons of which crystals are built up, as touching each other only by the edges. The contact of surface with surface amongst integrant molecules had been the first basis of the theory; but this supposition being here inapplicable, was replaced by one which made the theory no longer a representation of the facts (the cleavages) but a mere geometrical construction. Although, however, the inapplicability of the theory to such cases was thus, in some degree, disguised to the disciples of Haüy, it was plain that, in the face of such difficulties, the Theory of Integrant Molecules could not hold its place as a philosophical truth. But it still answered the purpose (a very valuable one, and one to which crystallography is much indebted,) of an instrument for calculating the geometrical relations of the parts of crystals to each other: for the integrant molecules were supposed to be placed layer above layer, each layer as we ascend, *decreasing* by a certain number of molécules and rows of molecules; and the calculation of these *laws of decrement* was, in fact, the best mode then known of determining the positions of the faces. The Theory of Decrements served to express and to determine, in a great number of the most obvious cases, *the laws of phenomena* in crystalline forms, though the Theory of Integrant Molecules could not be maintained as a just view of the structure of crystals.

3. The Theory of Integrant Molecules, however, involved this just and important principle: that a true view of the intimate structure of crystals must include and explain the facts of crystallization, that is, crystalline form and cleavage; and that it must take these into account, according to their degree of symmetry. So far all theories concerning the elements of crystals must agree. And it was soon seen that this was, in reality, all that had been established by the investigations of Haüy and his school. I have already, in the History, quoted Weiss's reflections on making this step. "When in 1809," he says*, "I published my Dissertation, I shared the common opinion as to the necessity of the assumption, and the reality of the existence of a primitive form, at least in a sense not very different from the usual sense of the expression." He then proceeds to relate that he sought a ground for such an opinion, independent of the doctrine of atoms, which he, in common with a great number of philosophers of that time in his own country, was disposed to reject, inclining to believe that the properties of bodies were determined by *forces* which acted in them, and not by *molecules* of which they were composed. He adds, that in pursuing this train of thought,

* Acad. Berlin. 1816. p. 307.

he found, " that out of his primitive forms there was gradually unfolded to his hands that which really governs them, and is not affected by their casual fluctuations; namely, the fundamental relations of their Dimensions," or as we now may call them, Axes of Symmetry. With reference to these axes, he found, as he goes on to say, that "a multiplicity of internal oppositions, necessarily and mutually interdependent, are developed in the crystalline mass, each relation having its own polarity; so that the crystalline character is co-extensive with these polarities." The character of these polarities, whether manifested in crystalline faces, cleavage, or any other incidents of crystallization, is necessarily displayed in the degree and kind of symmetry which the crystal possesses: and thus this symmetry, in all our speculations concerning the structure of crystals, necessarily takes the place of that enumeration of primitive forms which were rejected as inconsistent with observed facts, and destitute of sound scientific principle.

I may just notice here what I have stated in the History of Mineralogy*, that the distinction of systems of crystallization, as introduced by Weiss and Mohs, was strikingly confirmed by Sir David Brewster's discoveries respecting the optical properties of minerals. The splendid phenomena which were produced by passing polarized light through crystals, were found to vary according as the crystals were of the rhombohedral, square pyramidal, oblong prismatic, or tessular system. The optical exactly corresponded with the geometrical symmetry. In the two former systems were crystals *uniaxal* in respect of their optical properties; the oblong prismatic was *biaxal;* while in the tessular, the want of a predominant axis prevented the phenomena here spoken of from occurring at all. The optical experiments must have led to a classifi-

* *Hist. Ind. Sci.*, iii. 217. *

cation of crystals into the above systems or something
nearly equivalent, even had they not been already so
arranged by attention to their forms.

4. While in Germany Weiss and Mohs with their
disciples, were gradually rejecting what was superfluous
in the previous crystallographical hypotheses, philosophers
in England were also trying to represent to themselves
the constitution of crystals in a manner which should be
free from the obviously arbitrary and untenable fictions
of the Haüyian school.   These attempts, however, were
not crowned with much success.   One mode of representing the structure of crystals which suggested itself,
was to reject the polyhedral forms which Haüy gave to
his integrant molecules, and to conceive the elements of
crystals as spheres, the properties of the crystal being
determined not by the *surfaces*, but by the *position* of
the elements.   This was done by Wollaston, in the *Philosophical Transactions* for 1813.   He applied this view to
the tessular system, in which, indeed, the application is
not difficult; and he showed that octahedral and tetrahedral figures may be deduced from symmetrical arrangements of equal spherules.   But though in doing this, he
manifested a perception of the conditions of the problem,
he appeared to lose his hold on the real question when he
tried to pass on to other systems of crystallization.   For
he accounted for the rhombohedral system by supposing
the spheres changed into spheroids.   Such a procedure
involved him in a gratuitous and useless hypothesis: for
to what purpose do we introduce the arrangement of
atoms (instead of their figure,) as a mode of explaining
the symmetry of the crystallization, when at the next
step we ascribe to the atom, by an arbitrary fiction, a
symmetry of figure of the same kind as that which we
have to explain?   It is just as easy, and as allowable, to
assume an elementary rhombohedron, as to assume ele-

mentary spheroids, of which the rhombohedrons are constructed.

5. Many hypotheses of the same kind might be adduced, devised both by mineralogists and chemists. But almost all such speculations have been pursued with a most surprising neglect of the principle which obviously is the only sound basis on which they can proceed. The principle is this:—that *all hypotheses concerning the arrangement of the elementary atoms of bodies in space must be constructed with reference to the general facts of crystallization.* The truth and importance of this principle can admit of no doubt. For if we make any hypothesis concerning the mode of connexion of the elementary particles of bodies, this must be done with the view of representing to ourselves the forces which connect them, and the results of these forces as manifested in the properties of the bodies. Now the forces which connect the particles of bodies so as to make them crystalline, are manifestly chemical forces. It is only definite chemical compounds which crystallize; and in crystals the force of cohesion by which the particles are held together cannot in any way be distinguished or separated from the chemical force by which their elements are combined. The elements are understood to be combined, precisely because the result is a definite, apparently homogeneous substance. The properties of the compound bodies depend upon the elements and their mode of combination; for, in fact, these include everything on which they *can* depend. There are no other circumstances than these which can affect the properties of a body. Therefore all those properties which have reference to space, namely, the crystalline properties, cannot depend upon anything else than the arrangement of the elementary molecules in space. These properties are the facts which any hypothesis of the arrangement of

molecules must explain, or at least render conceivable; and all such hypotheses, all constructions of bodies by supposed arrangements of molecules, can have no other philosophical object than to account for facts of this kind. If they do not do this, they are mere arbitrary geometrical fictions, which cannot be in any degree confirmed or authorized by an examination of nature, and are therefore not deserving of any regard.

6. Those philosophers who have endeavoured to represent the mode in which bodies are constructed by the combination of their chemical atoms, have often undertaken to show, not only that the atoms are combined, but also in what positions and configurations they are combined. And it is truly remarkable, as I have already said, that they have done this, almost in every instance, without any consideration of the crystalline character of the resulting combinations; from which alone we receive any light as to the relation of their elements in space. Thus Dr. Dalton, in his *Elements of Chemistry*, in which he gave to the world the Atomic Theory as a representation of the doctrine of definite and multiple proportions, also published a large collection of diagrams, exhibiting what he conceived to be the configuration of the atoms in a great number of the most common combinations of chemical elements. Now these hypothetical diagrams do not in any way correspond, as to the nature of their symmetry, with the compounds, as we find them displaying their symmetry when they occur crystallized. Carbonate of lime has in reality a triangular symmetry, since it belongs to the rhombohedral system; Dr. Dalton's carbonate of lime would be an oblique rhombic prism or pyramid. Sulphate of baryta is really two-and-two membered; Dr. Dalton's diagram makes it two-and-one membered. Alum is really octahedral or tessular; but according to the diagram it could not be so, since the

two ends of the atom are not symmetrical. And the same want of correspondence between the facts and the hypothesis runs through the whole system. It need not surprise us that the theoretical arrangement of atoms does not explain the facts of crystallization; for to produce such an explanation would be a second step in science quite as great as the first, the discovery of the atomic theory in its chemical sense. But we may allow ourselves to be surprised that an utter discrepance between all the facts of crystallization and the figures assumed in the theory, did not suggest any doubt as to the soundness of the mode of philosophizing by which this part of the theory was constructed.

7. Some little accordance between the hypothetical arrangements of chemical atoms and the facts of crystallization, does appear to have been arrived at by some of the theorists to whom we here refer, although by no means enough to show a due conviction of the importance of the principle stated above. Thus Wollaston, in the Essay above noticed, after showing that a symmetrical arrangement of equal spherules would give rise to octahedral and other tessular figures, remarks, very properly, that the metals, which are simple bodies, crystallize in such forms. M. Ampère* also, in 1814, published a brief account of an hypothesis of a somewhat similar nature, and stated himself to have developed this speculation in a Memoir which has not yet, so far as I am aware, been published. In this notice he conceives bodies to be compounded of *molecules*, which, arranged in a polyhedral form, constitute *particles*. These *representative forms* of the particles depend on chemical laws. Thus the particles of oxygen, of hydrogen, and of azote, are composed each of four molecules. Hence it is collected that the particles of nitrous gas are composed of two

* *Ann. de Chimie*, tom. xc. p. 43.

molecules of oxygen and two of azote; and similar conclusions are drawn respecting other substances. These
conclusions, though expressed by means of the polyhedrons thus introduced, are supported by chemical, rather
than by crystallographical comparisons. The author does,
indeed, appeal to the crystallization of sal ammoniac as
an argument*; but as all the forms which he introduces
appear to belong to the tessular system of crystallization,
there is, in his reasonings, nothing distinctive; and
therefore nothing, crystallographically speaking, of any
weight on the side of this theory.

8. Any hypothesis which should introduce any principle of chemical order among the actual forms of minerals, would well deserve attention. At first sight, nothing
can appear more anomalous than the forms which occur.
We have, indeed, one broad fact, which has an encou-
raging aspect, the tessular forms in which the pure metals
crystallize. The highest degree of chemical and of geometrical simplicity coincide: irregularity disappears precisely where it is excluded by the consideration above
stated, that the symmetry of chemical composition must
determine the symmetry of crystalline form.

But if we go on to any other class of crystalline
forms, we soon find ourselves lost in our attempts to
follow any thread of order. We have indeed many large
groups connected by obvious analogies; as the rhombohedral carbonates of lime, magnesia, iron, manganese;—
the prismatic carbonates and sulphates of lime, baryta,
strontia, lead. But even in these, we cannot form any
plausible hypothesis of the arrangement of the elements;
and in other cases to which we naturally turn, we can
find nothing but confusion. For instance, if we examine
the oxides of metals:—those of iron are rhombohedral
and tessular; those of copper, tessular; those of tin, of

* *Ann. de Chimie*, tom. xc. p. 83.

titanium, of manganese, square pyramidal; those of antimony, prismatic ; and we have other forms for other substances.

It may be added, that if we take account of the optical properties which, as we have already stated, have constant relations to the crystalline forms, the confusion is still further increased; for the optical dimensions vary in amount, though not in symmetry, where chemistry can trace no difference of composition.

9. We will not quit the subject, however, without noticing the much more promising aspect which it has assumed by the detection of such groups as are referred to in the last article; or in other words, by Mitscherlich's discovery of *Isomorphism*. According to that discovery, there are various elements which may take the place of each other in crystalline bodies, either without any alteration of the crystalline form, or at most with only a slight alteration of its dimensions. Such a group of elements we have in the earths lime and magnesia, the protoxides of iron and manganese: for the carbonates of all these bases occur crystallized in forms of the rhombohedral system, the characteristic angle being nearly the same in all. Now lime and magnesia, by the discoveries of modern chemistry, are really oxides of metals; and therefore all these carbonates have a similar chemical constitution, while they have also a similar crystalline form. Whether or no we can devise any arrangement of molecules by which this connexion of the chemical and the geometrical property can be represented, we cannot help considering the connexion as an extremely important fact in the constitution of bodies; and such facts are more likely than any other to give us some intelligible view of the relations of the ultimate parts of bodies. The same may be said of all the other

isomorphous or plesiomorphous groups*. For instance, we have a number of minerals which belong to the same system of crystallization, but in which the chemical composition appears at first sight to be very various: namely, spinelle, pleonaste, gahnite, franklinite, chromic iron oxide, magnetic iron oxide: but Abich has shown that all these may be reduced to a common chemical formula; —they are bioxides of one set of bases, combined with trioxides of another set. Perhaps some mathematician may be able to devise some geometrical arrangement of such a group of elements which may possess the properties of the tessular system. Hypothetical arrangements of atoms, thus expressing both the chemical and the crystalline symmetry which we know to belong to the substance, would be valuable steps in analytical science; and when they had been duly verified, the hypotheses might easily be divested of their atomic character.

Thus, as we have already said, mineralogy, understood in its wider sense, as the counterpart of chemistry, has for one of its main objects to discover those relations of the elements of bodies which have reference to space. In this research, the foundation of all sound speculation is the kind and degree of symmetry of form which we find in definite chemical compounds: and the problem at present before the inquirer is, to devise such arrangements of molecules as shall answer the conditions alike of chemistry and of crystallography.

We now proceed to the Classificatory Sciences, of which mineralogy is one, though hitherto by far the least successful.

* See *Hist. Ind. Sci.*, iii. 222.

BOOK VIII.

THE PHILOSOPHY OF THE CLASSIFICATORY SCIENCES.

CHAPTER I.

THE IDEA OF LIKENESS AS GOVERNING THE USE OF COMMON NAMES.

1. *Object of the Chapter.*—Not only the Classificatory Sciences, but the application of names to things in the rudest and most unscientific manner, depends upon our apprehending them as *like* each other. We must therefore endeavour to trace the influence and operation of the Idea of Likeness in the common use of language, before we speak of the conditions under which it acquires its utmost exactness and efficacy.

It will be my object to show in this, as in previous cases, that the impressions of sense are apprehended by acts of the mind; and that these mental acts necessarily imply certain relations which may be made the subjects of speculative reasoning. We shall have, if we can, to seize and bring into clear view the principles which the relation of *like* and *unlike* involves, and the mode in which these principles have been developed.

2. *Unity of the Individual.*—But before we can attend to several things as like or unlike, we must be able to apprehend each of these by itself as *one thing*. It may at first sight perhaps appear that this apprehension results immediately from the impressions on our senses, without

any act of our thoughts. A very little attention, however, enables us to see that thus to single out special
objects requires a mental operation as well as a sensation.
How, for example, without an exertion of mental activity,
can we see one tree, in a forest where there are many? We
have, spread before us, a collection of colours and forms,
green and brown, dark and light, irregular and straight:
this is all that sensation gives or can give. But we associate one brown trunk with one portion of the green mass,
excluding the rest, although the neighbouring leaves are
both nearer in contiguity and more similar in appearance
than is the stem. We thus have before us one tree; but
this unity is given by the mind itself. We see the green
and the brown, but we must *make* the tree before we can
see *it*.

That this composition of our sensations so as to form
one thing implies an act of our own, will perhaps be more
readily allowed, if we once more turn our attention to
the manner in which we sometimes attempt to imitate
and record the objects of sight, by drawing. When we do
this, as we have already observed, we mark this unity of each
object, by drawing a line to separate the parts which we
include from those which we exclude;—an *Outline*. This
line corresponds to nothing which we see; the beginner
in drawing has great difficulty in discerning it; he has in
fact to make it. It is, as has been said by a painter of
our own time*, a fiction: but it is a fiction employed to
mark a real act of the mind; to designate the singleness
of the object in our conception. As we have said elsewhere, we see lines, but especially outlines, by mentally
drawing them ourselves.

The same act of conception which the outline thus
represents and commemorates in visible objects,—the same
combination of sensible impressions into a unit,—is exer-

---

* PHILLIPS *on Painting*,—Design.

cised also with regard to the objects of all our senses: and
the singleness thus given to each object, is a necessary
preliminary to its being named or represented in any
other way.

But it may be said, Is it then by an arbitrary act of
our own that we put together the branches of the same
tree, or the limbs of the same animal? Have we equally
the power and the right to make the branch of the fir a
part of the neighbouring oak? Can we include in the
outline of a man any object with which he happens to be
in contact?

Such suppositions are manifestly absurd. And the
answer is, that though we give unity to objects by an
act of thought, it is not by an *arbitrary* act; but by a
process subject to certain conditions: to conditions which
exclude such incongruous combinations as have just been
spoken of.

What are these conditions which regulate our apprehension of an object as one? which determine what portion of our impressions does, and what portion does not
belong to the same thing?

2. *Condition of Unity.*—I reply, that the primary and
fundamental condition is, that we must be able to make
intelligible assertions respecting the object, and to entertain that belief of which assertions are the exposition. A
tree *grows, sheds* its leaves in autumn, and *buds* again in
the spring, *waves* in the wind, or *falls* before the storm.
And to the tree belong all those parts which must be
included in order that such declarations, and the thoughts
which they convey, shall have a coherent and permanent
meaning. Those are *its* branches which wave and fall with
its trunk; those are *its* leaves which grow on *its* branches.
The permanent connexions which we observe,—permanent, among unconnected changes which affect the surrounding appearances,—are what we bind together as

belonging to one object. This permanence is the condition of our conceiving the object *as* one. The connected changes may always be described by means of assertions; and the connexion is seen in the identity of the subject of successive predications; in the possibility of applying many verbs to one substantive. We may therefore express the condition of the unity of an object to be this: that *assertions concerning the object shall be possible:* or rather we should say, that the acts of belief which such assertions enunciate shall be possible.

It may seem to be superfluous to put in a form so abstract and remote, the grounds of a process apparently so simple as our conceiving an object to be one. But the same condition to which we have thus been led, as the essential principle of the unity of objects, namely, that propositions shall be possible, will repeatedly occur in the present chapter; and it may serve to illustrate our views, to show that this condition pervades even the simplest cases.

4. *Kinds.*—The mental synthesis of which we have thus spoken, gives us our knowledge of *individual* things; it enables me to apprehend that particular tree or man which I now see, or, by the help of memory, the tree or the man I saw yesterday. But the knowledge with which we have mainly here to do is not a knowledge of individuals but of kinds; of such classes as are indicated by common names. We have to make assertions concerning a tree or a man in general, without regarding what is peculiar to this man or that tree.

Now it is clear that certain individual objects are all called *man,* or all called *tree,* in virtue of some resemblance which they have. If we had not the power of perceiving in the appearances around us, likeness and unlikeness, we could not consider objects as distributed into kinds at all. The impressions of sense would throng upon us, but being uncompared with each other, they would flow away like the waves of the sea, and each vanish from our contemplation when the sensation faded. That we do apprehend surrounding objects as belonging to permanent kinds, as being men and horses, oaks and roses, arises from our having the idea of likeness, and from our applying it habitually, and so far as such a classification requires.

Not only can we employ the idea of likeness in this manner, but we apply it incessantly and universally to the whole mass and train of our sensations. For we have no external sensations to which we cannot apply some language or other, and all language necessarily implies recognition of resemblances. We cannot call an object *green* or *round* without comparing in our thoughts its colour or its shape, with a shape and a colour seen in other objects. All our sensations, therefore, without any exception of kind or time, are subject to this constant process of classification; and the idea of likeness is perpetually operating to distribute them into kinds, at least so far as the use of language requires.

We come then again to the question, Upon what principle, under what conditions, is the idea of likeness thus operative? What are the limits of the classes thus formed? Where does that similarity end, which induces and entitles us to call a thing a tree? What universal rule is there for the application of common names, so that we may not apply them wrongly?

5. *Not made by Definitions.*—Perhaps some one might expect in answer to these inquiries a definition or a series of definitions;—might imagine that some description of a tree might be given which might show when the term was applicable and when it was not; and that we might construct a body of rules to which such descriptions must conform. But on consideration it will be clear that the real solution of our difficulty cannot be obtained in such

a manner. For *first;* such descriptions must be given in words, and therefore suppose that we have already satisfied ourselves how words are to be used. If we define a tree to be a living thing without the power of voluntary motion, we shall be called upon to define a living thing; and it is manifest that this renewal of the demand for definition might be repeated indefinitely; and, therefore, we cannot in this way come to a final principle. And in the *next* place, most of those who use language, even with great precision and consistency, would find it difficult or impossible to give good definitions even of a few of the general names which they use; and therefore their practice cannot be regulated by any tacit reference to such definitions. That definitions of terms are of great use and importance in their right place, we shall soon see; but their place is not to regulate the use of common language.

What then, once more, is this regulative principle? What rules do men follow in the use of words, so as commonly to avoid confusion and ambiguity? How do they come to understand each other so well as they ordinarily do, respecting the limits of classes never defined, and which they cannot define? What is the common convention, or condition to which they conform?

6. *Condition of the Use of Terms.*—To this we reply, that the condition which regulates the use of language, is that it shall be capable of being used;—that is, that general assertions shall be possible. The term tree is applicable as far as it is useful in expressing our knowledge concerning trees:—thus we know that trees are fixed in the ground, have a solid stem, branches, leaves, and many other properties. With regard to all the objects which surround us, we have an immense store of knowledge of such properties, and we employ the names of the objects in such a manner as enables us to express these properties.

But the connexion of such properties is variable and indefinite. Some properties are constantly combined, others occasionally only. The leaves of different oaks resemble each other, the branches resemble far less, and may differ very widely. The term *oak* does not enable us to say that all oaks have straight branches or all crooked. Terms can only express properties as far as they are constant. Not only, therefore, the accumulation of a vast mass of knowledge of the properties and attributes of objects, but also an observation of the habitual *connexion* of such properties is needed, to direct us to the consistent application of terms:—to enable us to apply them so as to express truths. But here again we are largely provided with the requisite knowledge and observation by the common course of our existence. The unintermitting stream of experience supplies us with an incalculable amount of such observed connexions. All men have observed that the associations of the same form of leaves are more constant than of the same form of branches;—that though persons walk in different attitudes none go on all fours; and thus the term *oak* is so applied as to include those cases in which the leaves are alike in form though the branches be unlike; and though we should refuse to apply the term *man* to a class of creatures which habitually and without compulsion used four legs, we make no scruple of affixing it to persons of very different figures. The whole of human experience being composed of such observed connexions, we have thus materials even for the immense multiplicity of names which human language contains; all which names are, as we have said, regulated in their application by the condition of expressing such experience.

Thus amid the countless combinations of properties and divisions of classes which the structure of language implies, scarcely any are arbitrary or capricious. A word which expressed a mere wanton collection of unconnected attributes could hardly be called a word; for of such a collection of properties no truth could be asserted, and the word would disappear, for want of some occasion on which it could be used. Though much of the fabric of language appears, not unnaturally, fantastical and purely conventional, it is in fact otherwise. The associations and distinctions of phraseology are not more fanciful than is requisite to make them correspond to the apparent caprices of nature or of thought; and though much in language may be called conventional, the conventions exist for the sake of expressing some truth or opinion, and not for their own sake. The principle, that *the condition of the use of terms is the possibility of general, intelligible, consistent assertions,* is true in the most complete and extensive sense.

7. *Terms may have different Uses.*—The terms with which we are here most concerned are names of classes of natural objects; and when we say that the principle and the limit of such names are their use in expressing propositions concerning the classes, it is clear that much will depend on the kind of propositions which we mainly have to express: and that the same name may have different limits, according to the purpose we have in view. For example, is the whale properly included in the general term *fish?* When men are concerned in catching marine animals, the main features of the process are the same however the animals may differ; hence whales are classed with fishes, and we speak of the *whale-fishery.* But if we look at the analogies of organization, we find that, according to these, the whale is clearly not a fish, but a *beast,* (confining this term, for the sake of distinctness, to suckling beasts or *mammals*). In Natural History, therefore, the whale is not included among fish. The indefinite and miscellaneous propositions which language is

employed to enunciate in the course of common practical
life, are replaced by a more coherent and systematic
collection of properties, when we come to aim at scientific
knowledge. But we shall hereafter consider the principle
of the classifications of Natural History; our present
subject is the application of the Idea of Likeness in
common practice and common language.

8. *Gradation of Kinds.*—Common names, then, include many individuals associated in virtue of resemblances, and of permanently connected properties; and
such names are applicable as far as they serve to express
such properties. These collections of individuals are
termed kinds, sorts, classes.

But this association of particulars is capable of degrees.
As individuals by their resemblances form kinds, so kinds
of things, though different, may resemble each other so as
to be again associated in a higher class; and there may
be several successive steps of such classification. *Man,
horse, tree, stone,* are each a name of a kind; but *animal*
includes the two first and excludes the others; *living
thing* is a term which includes animal and tree but not
stone; *body* includes all the four. And such a subordination of kinds may be traced very widely in the arrangements of language.

The condition of the use of the wider is the same as
that of the narrower names of classes;—they are good as
far as they serve to express true propositions. In common language, though such an order of generality may in
a variety of instances be easily discerned, it is not systematically and extensively referred to; but this subordination and graduated comprehensiveness is the essence of
the methods and nomenclatures of Natural History, as we
shall soon have to show.

But such subordination is not without its use, even in
common cases, and when it is expressed in the terms of

common language. Thus *organized body* is a term which includes plants and animals; *animal* includes beasts, birds, fishes; *beast* includes horses and dogs; *dogs*, again, are greyhounds, spaniels, terriers.

9. *Characters of Kinds.*—Now when we have such a series of names and classes, we find that we take for granted irresistibly that each class has some *character* which distinguishes it from other classes included in the superior division. We ask what kind of beast a dog is; what kind of animal a beast is; and we assume that such questions admit of answer;—that each kind has some mark or marks by which it may be described. And such descriptions may be given: an animal is an organized body having sensation and volition; man is a reasonable animal. Whether or no we assent to the exactness of these definitions, we allow the propriety of their form. If we maintain these to be wrong, we must believe some others to be right, however difficult it may be to hit upon them. We entertain a conviction that there must be, among things so classed and named, a possibility of defining each.

Now what is the foundation of this postulate? What is the ground of this assumption, that there must exist a definition which we have never seen, and which perhaps no one has seen in a satisfactory form? The knowledge of this definition is by no means necessary to our using the word with propriety; for any one can make true assertions about dogs, but who can define a dog? And yet if the definition be not necessary to enable us to use the word, why is it necessary at all? We allow that we possess an indestructible conviction that there must be such a character of each kind as will supply a definition; but we ask, on what this conviction rests.

I reply, that our persuasion that there must needs be characteristic marks by which things can be defined in words, is founded on the assumption of *the necessary possibility of reasoning.*

The reference of any object or conception to its class without definition, may give us a persuasion that it shares the properties of its class, but does not enable us to reason upon those properties. When we consider man as an animal, we ascribe to him in thought the appetites, desires, affections, which we habitually include in our notion of animal: but except we have expressed these in some definition or acknowledged description of the term animal, we can make no use of the persuasion in ratiocination. But if we have described animals as "beings impelled to action by appetites and passions," we can not only think, but say, "man is an animal, and therefore he is impelled to act by appetites and passions." And if we add a further definition, that "man is a reasonable animal," and if it appear that "reason implies conformity to a rule of action," we can then further infer that man's nature is to conform the results of animal appetite and passion to a rule of action.

The possibility of pursuing any such train of reasoning as this, depends on the definitions, of *animal* and of *man*, which we have introduced; and the possibility of reasoning concerning the objects around us being inevitably assumed by us from the constitution of our nature, we assume consequently the possibility of such definitions as may thus form part of our deduction, and the existence of such defining characters.

10. *Difficulty of Definitions.*—But though men are, on such grounds, led to make constant and importunate *demands* for definitions of the terms which they employ in their speculations, they are, in fact, far from being able to carry into complete effect the postulate on which they proceed, that they must be able to find definitions which by logical consequence shall lead to the truths

they seek. The postulate overlooks the process by which our classes of things are formed and our names applied. This process consisting, as we have already said, in observing permanent connexions of properties, and in fixing them by the attribution of names, is of the nature of the process of induction, of which we shall afterwards have to speak. And the postulate is so far true, that this process of induction being once performed, its result may usually be expressed by means of a few definitions, and may thus lead by a deduction to a train of real truths.

But in the subjects where we principally find such a subordination of classes as we have spoken of, this process of deduction is rarely of much prominence: for example, in the branches of natural history. Yet it is in these subjects that the existence and importance of these characteristic marks, which we have spoken of, principally comes into view. In treating of these marks, however, we enter upon methods which are technical and scientific, not popular and common. And before we make this transition, we have a remark to make on the manner in which writers, without reference to physics or natural history, have spoken of kinds, their subordination, and their marks.

11. *" The Five Words."*—These things,—the nature and relations of classes,—were, in fact, the subjects of minute and technical treatment by the logicians of the school of Aristotle. Porphyry wrote an Introduction to the *Categories* of that philosoper, which is entitled *On the Five Words.* The " Five Words" are *genus, species, difference, property, accident.* Genus and species are superior and inferior classes, and are stated* to be capable of repeated subordination. The "most general genus" is the widest class, the " most special species" the narrowest. Between these are intermediate classes, which

* PORPHYR. *Isagog.* c. 23.

are genera with regard to those below, and species with regard to those above them. Thus Being is the most general genus; under this is Body; under Body is Living Body; under this again Animal; under Animal is Rational Animal, or Man; under Man are Socrates and Plato, and other individual men.

The Difference is that which is added to the genus to make the species; thus Rational is the Difference by which the genus Animal is made the species Man; the Difference in this Technical sense is the "Specific," or species-making Difference*. It forms the Definition for the purposes of logic, and corresponds to the "Character" (specific or generic) of the Natural Historians. Indeed several of them, as, for instance, Linnæus, in his *Philosophia Botanica*, always call these Characters the Difference, by a traditional application of the Peripatetic terms of art.

Of the other two words, the Property is that which though not employed in defining the class, belongs to every part of it†: it is, "What happens to all the class, to it alone, and at all times; as to be capable of laughing is a property of a man."

The Accident is that which may be present and absent without the destruction of the subject, as to sleep is an Accident (a thing which happens) to man.

I need not dwell further on this system of technicalities. The most remarkable points in it are those which I have already noticed; the doctrine of the successive subordination of genera, and the fixing attention upon the specific difference. These doctrines, though invented in order to make reasoning more systematic, and at a period anterior to the existence of any classificatory science, have, by a curious contrast with the intentions of their founders, been of scarcely any use in sciences of reasoning, but have been amply applied and developed

---

* εἰδοποιός.   † *Isagog*. c. 4.

in the Natural History which arose in later times. We
must now treat of the principles on which this science
proceeds, and explain what peculiar and technical processes it employs in addition to those of common thought
and common language.

CHAPTER II.

THE METHODS OF NATURAL HISTORY, AS REGU-
LATED BY THE IDEA OF LIKENESS.

1. *Idea of Likeness in Natural History.*—The various
branches of Natural History, in so far as they are classificatory sciences merely, and do not depend upon physiological views, rest upon the same Idea of Likeness which
is the ground of the application of the names, more or
less general, of common language. But the nature of
science requires that for her purposes this idea should be
applied in a more exact and rigorous manner than in its
common and popular employment; just as occurs with
regard to the other Ideas on which science is founded;—
for instance, as the idea of space gives rise, in popular use,
to the relations implied in the prepositions and adjectives
which refer to position and form, and in its scientific
developement gives rise to the more precise relations of
geometry.

The way in which the Idea of Likeness has been
applied, so as to lead to the construction of a science, is
best seen in Botany: for, in the Classification of Animals,
we are inevitably guided by a consideration of the *function*
of parts; that is, by an idea of *purpose*, and not of likeness merely: and in Mineralogy the attempts at classification on the principles of Natural History have been
hitherto very imperfectly successful. But in Botany we
have an example of a branch of knowledge in which sys-

tematic classification has been effected with great beauty and advantage; and in which the peculiarities and principles on which such classification must depend have been carefully studied. Many of the principal botanists, as Linnæus, Adanson, Decandolle, have not only practically applied, but have theoretically enunciated, what they held to be the sound maxims of classificatory science: and have thus enabled us to place before the reader with confidence the philosophy of this kind of science.

2. *Condition of its Use.*—We may begin by remarking that the Idea of Likeness, in its systematic employment, is governed by the same principle which we have already spoken of as regulating the distribution of things into kinds, and the assignment of names in unsystematic thought and speech; namely, the condition that *general propositions shall be possible.* But as in this case the propositions are to be of a scientific form and exactness, the likeness must be treated with a corresponding precision; and its consequences traced by steady and distinct processes. Naturalists must, for their purposes, employ the resemblances of objects in a technical manner. This technical process may be considered as consisting of three steps;—The fixation of the resemblances; The use of them in making a classification; The means of applying the classification. These three steps may be spoken of as the *Terminology,* the *Plan of the System,* and the *Scheme of the Characters.*

3. (I.) *Terminology* *.—Terminology signifies the collection of *terms*, or technical words, which belong to the science. But in fixing the meaning of the terms, at

---

* Decandolle and others use the term Glossology instead of Terminology, to avoid the blemish of a word compounded of two parts taken from different languages. The convenience of treating the termination *ology* (and a few other parts of compounds) as not restricted to Greek combinations, is so great, that I shall venture, in these cases, to disregard this philological scruple.

least of the descriptive terms, we necessarily fix, at the
same time, the perceptions and notions which the terms
are to convey; and thus the terminology of a classificatory science exhibits the elements of its substance as
well as of its language. A large but indispensable part
of the study of botany (and of mineralogy and zoology
also,) consists in the acquisition of the peculiar vocabulary of the science.

The meaning of technical terms can be fixed in the
first instance only by convention, and can be made intelligible only by presenting to the senses that which the
terms are to signify. The knowledge of a colour by its
name can only be taught through the eye. No description can convey to a hearer what we mean by *apple green*
or *French grey*. It might, perhaps, be supposed that, in
the first example, the term *apple*, referring to so familiar
an object, sufficiently suggests the colour intended. But
it may easily be seen that this is not true; for apples are
of many different hues of green, and it is only by a conventional selection that we can appropriate the term to
one special shade. When this appropriation is once made,
the term refers to the sensation, and not to the parts of
the term; for these enter into the compound merely as
a help to the memory, whether the suggestion be a
natural connexion as in "apple green," or a casual one as in
"French grey." In order to derive due advantage from
technical terms of this kind, they must be associated
immediately with the perception to which they belong;
and not connected with it through the vague usages of
common language. The memory must retain the sensation; and the technical word must be understood as
directly as the most familiar word, and more distinctly.
When we find such terms as *tin-white* or *pinchbeck-
brown*, the metallic colour so denoted ought to start up
in our memory without delay or search.

This, which it is most important to recollect with respect to the simpler properties of bodies, as colour and form, is no less true with respect to more compound notions. In all cases the term is fixed to a peculiar meaning by convention; and the student, in order to use the word, must be completely familiar with the convention, so that he has no need to frame conjectures from the word itself. Such conjectures would always be insecure, and often erroneous. Thus the term *papilionaceous* applied to a flower is employed to indicate, not only a resemblance to a butterfly, but a resemblance arising from five petals of a certain peculiar shape and arrangement; and even if the resemblance were much stronger than it is in such cases, yet if it were produced in a different way, as, for example, by one petal, or two only, instead of a "standard," two "wings," and a "keel" consisting of two parts more or less united into one, we should no longer be justified in speaking of it as a "papilionaceous" flower.

The formation of an exact and extensive descriptive language for botany has been executed with a degree of skill and felicity, which, before it was attained, could hardly have been dreamt of as attainable. Every part of a plant has been named; and the form of every part, even the most minute, has had a large assemblage of descriptive terms appropriated to it, by means of which the botanist can convey and receive knowledge of form and structure, as exactly as if each minute part were presented to him vastly magnified. This acquisition was part of the Linnæan reform, of which we have spoken in the History. "Tournefort," says Decandolle*, "appears to have been the first who really perceived the utility of fixing the sense of terms in such a way as always to employ the same word in the same sense, and always to express the same idea by the same word; but it was Linnæus who

* *Theor. Elem.*, p. 327.

really created and fixed this ·botanical language, and
this is his fairest claim to glory, for by this fixation of
language he has shed clearness and precision over all
parts of the science."

It is not necessary here to give any detailed account
of the terms of botany. The fundamental ones have been
gradually introduced, as the parts of plants were more
carefully and minutely examined. Thus the flower was
successively distinguished into the *calyx*, the *corolla*, the
stamens, and the *pistils*: the sections of the corolla were
termed *petals* by Columna; those of the calyx were
called *sepals* by Necker*. Sometimes terms of greater
generality were devised; as *perianth* to include the calyx
and corolla, whether one or both of these were present†;
pericarp for the part inclosing the grain, of whatever kind
it be, fruit, nut, pod, &c. And it may easily be imagined
that descriptive terms may, by definition and combination,
become very numerous and distinct. Thus leaves may be
called *pinnatifid*‡, *pinnatipartite*, *pinnatisect*, *pinnatilobate*,
palmatifid, *palmatipartite*, &c., and each of these words
designates different combinations of the modes and extent
of the divisions of the leaf with the divisions of its outline.
In some cases arbitrary numerical relations are introduced
into the definition: thus a leaf is called *bilobate*§ when it
is divided into two parts by a notch; but if the notch go
to the middle of its length, it is *bifid*; if it go near the
base of the leaf, it is *bipartite*; if to the base, it is *bisect*.
Thus, too, a pod of a cruciferous plant is a *silica*‖ if it be
four times as long as it is broad, but if it be shorter than
this it is a *silicula*. Such terms being established, the
form of the very complex leaf or frond of a fern is exactly
conveyed by the following phrase: "fronds rigid pinnate,

---

* Dec. 329.

† For this Erhart and Decandolle use *Perigone*.

‡ Dec. 318.   § *Ib*. 493.   ‖ *Ib*. 422.

pinnæ recurved subunilateral pinnatifid, the segments linear undivided or bifid spinuloso-serrate*."

Other characters, as well as form, are conveyed with the like precision: Colour by means of a classified scale of colours, as we have seen in speaking of the measures of secondary qualities; to which, however, we must add, that the naturalist employs arbitrary names, (such as we have already quoted,) and not mere numerical exponents, to indicate a certain number of selected colours. This was done with most precision by Werner, and his scale of colours is still the most usual standard of naturalists. Werner also introduced a more exact terminology with regard to other characters which are important in mineralogy, as lustre, hardness. But Mohs improved upon this step by giving a numerical scale of hardness, in which talc is 1, gypsum 2, calc spar 3, and so on, as we have already explained in the History of Mineralogy. Some properties, as specific gravity, by their definition give at once a numerical measure; and others, as crystalline form, require a very considerable array of mathematical calculation and reasoning, to point out their relations and gradations. In all cases the features of likeness in the objects must be rightly apprehended, in order to their being expressed by a distinct terminology. Thus no terms could describe crystals for any purpose of natural history, till it was discovered that in a class of minerals the proportion of the faces might vary, while the angle remained the same. Nor could crystals be described so as to distinguish species, till it was found that the derived and primitive forms are connected by very simple relations of space and number. The discovery of the mode in which characters must be apprehended so that they may be considered as *fixed* for a class, is an important

---

* HOOKER, *Brit. Flo.*, p. 450. *Hymenophyllum Wilsoni*, Scottish filmy-fern, abundant in the Highlands of Scotland and about Killarney.　　　　　　　　　　　　2 H 2

step in the progress of each branch of Natural History; and hence we have had, in the History of Mineralogy and Botany, to distinguish as important and eminent persons those who made such discoveries, Romé de Lisle and Haüy, Cesalpinus and Gesner.

By the continued progress of that knowledge of minerals, plants, and other natural objects, in which such persons made the most distinct and marked steps, but which has been constantly advancing in a more gradual and imperceptible manner, the most important and essential features of similarity and dissimilarity in such objects have been selected, arranged, and fitted with names; and we have thus in such departments, systems of terminology which fix our attention upon the resemblances which it is proper to consider, and enable us to convey them in words. We have now to speak of the mode in which such resemblances have been employed in the construction of a systematic classification.

4. (II.) *The Plan of the System.*—The collection of sound views and maxims by which the resemblances of natural objects are applied so as to form a scientific classification, is a department of the philosophy of natural history which has been termed by some writers (as Decandolle,) *Taxonomy*, as containing the *Laws* of the *Taxis*, (arrangement). By some Germans this has been denominated *Systematik*; if we could now form a new substantive after the analogy of the words Logic, Rhetoric, and the like, we might call it *Systematick*. But though our English writers commonly use the expression *Systematical Botany* for the Botany of Classification, they appear to prefer the term *Diataxis* for the method of constructing the classification. The rules of such a branch of science are curious and instructive.

In framing a classification of objects we must attend to their resemblances and differences. But here the question occurs, to *what* resemblances and differences? for

a different selection of the points of resemblance would
give different results : a plant frequently agrees in leaves
with one group of plants, in flowers with another.   Which
set of characters are we to take as our guide?

The view already given of the regulative principle of
all classification, namely, that it must enable us to assert
true and general propositions, will obviously occur as
applicable here.   The object of a scientific classification
is to enable us to enunciate scientific truths: we must
therefore classify according to those resemblances of
objects (plants or any others,) which bring to light such
truths.

But this reply to the inquiry, On what characters of
resemblance we are to found our system, is still too general and vague to be satisfactory.   It carries us, however, as
far as this ; that since the truths we are to attend to are
scientific truths, governed by precise and homogeneous
relations, we must not found our scientific classification on
casual, indefinite, and unconnected considerations.   We
must not, for instance, be satisfied with dividing plants,
as Dioscorides does, into *aromatic, esculent, medicinal,*
and *vinous ;* or even with the long prevalent distribution
into *trees, shrubs,* and *herbs ;* since in these subdivisions
there is no consistent principle.

5. *Latent Reference to Natural Affinity.*—But there
may be several kinds of truths, all exact and coherent,
which may be discovered concerning plants or any other
natural objects ; and if this should be the case, our rule
leaves us still at a loss in what manner our classification
is to be constructed.   And, historically speaking, a much
more serious inconvenience has been this ;—that the task
of classification of plants was necessarily performed when
the general laws of their form and nature were very little
known ; or rather, when the existence of such laws was
only just beginning to be discerned.   Even up to the
present day, the general propositions which botanists are

able to assert concerning the structure and properties of plants, are extremely imperfect and obscure.

We are thus led to this conclusion :—that the idea of likeness could not be applied so as to give rise to a scientific classification of plants, till considerable progress was made in studying the general relations of vegetable form and life ; and that the selection of the resemblances which should be taken into account, must depend upon the nature of the relations which were then brought into view.

But this amounts to saying that, in the consideration of the classification of vegetables, other Ideas must be called into action as well as the Idea of Likeness.   The new general views to which the more intimate study of plants leads, must depend, like all general truths, upon some regulating Idea which gives unity to scattered facts : no progress could be made in botanical knowledge without the operation of such principles : and such additional Ideas must be employed, besides those of mere likeness and unlikeness, in order to point out that classification which has a real scientific value.

Accordingly in the classificatory sciences Ideas other than Likeness do make their appearance.   Such Ideas in botany have influenced the progress of the science, even before they have been clearly brought into view. We have especially the Idea of Affinity, which is the basis of all Natural Systems of Classification, and which we shall consider in a succeeding chapter.   The assumption that there *is* a Natural System, an assumption made by all philosophical botanists, implies a belief in the existence of Natural Affinity, and is carried into effect by means of principles which are involved in that Idea. But as the formation of all systems of classification must involve, in a great degree, the Idea of Resemblance and Difference, I shall first consider the effect of that Idea, before I treat specially of Natural Affinity.

6. *Natural Classes.*—Many attempts were made to

classify vegetables before the rules which govern a natural system were clearly apprehended. Botanists agreed in esteeming some characters as of more value than others, before they had agreed upon any general rules or principles for estimating the relative importance of the characters. They were convinced of the necessity of adding other considerations to that of resemblance, without seeing clearly what these ought to be. They aimed at a Natural Classification, without knowing distinctly in what manner it was to be Natural.

The attempts to form *Natural Classes*, therefore, in the first part of their history, belong to the Idea of Likeness, though obscurely modified, even from an early period, by the Ideas of Affinity, and even of Function and of Developement. Hence Natural Classes may, to a certain extent be treated of in this place.

Natural Classes are opposed to Artificial Classes, which are understood to be regulated by an assumed character. Yet no classes can be so absolutely Artificial in this sense, as to be framed upon characters arbitrarily assumed; for instance, no one would speak of a class of shrubs defined by the circumstance of each having a hundred leaves: for of such a class no assertion could be made, and therefore the class could never come under our notice. In what sense then are Artificial Classes to be understood, as opposed to Natural ?

7. *Artificial Classes.*—To this question the following is the answer. When Natural Classes of a certain small extent have been formed, a system may be devised which shall be.regulated by a few selected characters, and which shall not dissever these small Natural Classes, but conform to them as far as they go. If these selected characters be made absolute and imperative, and if we abandon all attempt to obtain Natural Classes of any higher order and wider extent, we form an Artificial System.

Thus in the Linnæan System of Botanical Classification, it is assumed that certain natural groups, namely, species and genera, are established; it is conceived, moreover, that the division of classes according to the number of stamens and of pistils does not violate the natural connexions of species and genera. This arrangement, according to the number of stamens and pistils, (further modified in certain cases by other considerations,) is then made the ground of all the higher divisions of plants, and thus we have an Artificial System.

It has been objected to this view, that the Linnæan Artificial System does not in all cases respect the boundaries of genera, but would, if rigorously applied, distribute the species of the same genus into different artificial classes; it would divide, for instance, the genera Valeriana, Geranium*, &c. To this we must reply, that so far as the Linnæan System does this, it is an imperfect Artificial System. Its great merit is in its making such a disjunction in comparatively so few cases; and in the artificial characters being, for the most part, obvious and easily applied.

8. *Are Genera Natural?*—It has been objected also that Genera are not Natural groups. Linnæus asserts in the most positive manner that they are†. On which Adanson observes‡, "I know not how any Botanist can maintain such a thesis: that which is certain is, that up to the present time no one has been able to prove it, nor to give an exact definition of a natural genus, but only of an artificial." He then brings several arguments to confirm this view.

But we are to observe, in answer to this, that Adanson improperly confounds the recognition of the existence of a natural group with the invention of a technical mark or definition of it. Genera are groups of species

---

* Decand. *Th. El.*, p. 45.

† *Phil. Bot.* Art. 165.   ‡ *Famille de Ph.*, Pref. cv.

associated in virtue of natural affinity, of general resemblance, of real propinquity: of such groups, certain selected characters, one or few, may usually be discovered, by which the species may be referred to their groups. These Artificial characters do not constitute, but indicate the genus: they are the *Diagnosis*, not the basis of the *Diataxis*: and they are always subject to be rejected, and to have others substituted for them, when they violate the natural connexion of species which a minute and enlarged study discovers.

It is, therefore, no proof that Genera are not Natural, to say that their artificial characters are different in different systems. Such characters are only different attempts to confine the variety of nature within the limits of definition. Nor is it sufficient to say that these groups themselves are different in different writers; that some botanists make genera what others make only species; as *Pedicularis, Rhinanthus, Euphrasia, Antirrhinun**. This discrepancy shows only that the natural arrangement is not yet completely known, even in the smaller groups; a conclusion to which we need not refuse our assent. But in opposition to these negatives, the manner in which Genera have been established proves that they are regulated by the principle of being natural and that alone. For they are not formed according to any *à priori* rule. The Botanist does not take any selected or arbitrary part or parts of the plants, and marshal his genera according to the differences of this part. On the contrary, the divisions of genera are sometimes made by means of the flower; sometimes by means of the fruit; the anthers, the stamens, the seeds, the pericarp, and the most varied features of these parts, are used in the most miscellaneous and unsystematic manner. Linnæus has indeed laid down a maxim that the characteristic differences of

* ADANSON, p. cvi.

genera must reside in the fructification* : but Adanson has justly remarked†, that an arbitrary restriction like this makes the groups artificial : and that in some families other characters are more essential than those of the fructification; as the leaves in the families of *Aparineæ* and *Leguminosæ,* and the disposition of the flowers in *Labiatæ.* And Naturalists are so far from thinking it sufficient to distribute species into genera by *arbitrary* marks, that we find them in many cases lamenting the absence of good *natural* marks : as in the families of Umbelliferæ, where Linnæus declared that any one who could find good characters of genera would deserve great admiration, and where it is only of late that good characters have been discovered and the arrangement settled‡ by means principally of the ribs of the fruit§.

It is thus clear that genera are not established on any assumed or preconceived basis. What, then, is the principle which regulates botanists when they try to fix genera? What is the arrangement which they thus wish for, without being able to hit upon it? What is the tendency which thus drives them from the corolla to the anthers, from the flower to the fruit, from the fructification to the leaves? It is plain that they seek something, not of their own devising and creating;—not anything merely conventional and systematic; but something which they conceive to exist in the relations of the plants themselves;—something which is without the mind, not within;—in nature, not in art;—in short, a natural order.

Thus the regulative principle of a genus, or of any other natural group is, that it is, or is supposed to be, natural. And by reference to this principle as our guide,

---

* *Phil. Bot.* Art. 162.   † ADANSON, Pref., p. cxx.
‡ LINDLEY, *Nat. Syst.*, p. 5.
§ In like manner we find Cuvier saying of Rondelet that he has " un *sentiment* très vrai des genres." *Hist. Ichth.*, p. 39.

we shall be able to understand the meaning of that indefiniteness and indecision which we frequently find in the
descriptions of such groups, and which must appear so
strange and inconsistent to any one who does not suppose
these descriptions to assume any deeper ground of connexion than an arbitrary choice of the botanist. Thus
in the family of the Rose-tree, we are told that the
ovules are *very rarely* erect*, the *stigmata usually* simple.
Of what use, it might be asked, can such loose accounts
be? To which the answer is, that they are not inserted
in order to distinguish the species, but in order to
describe the family, and the total relations of the ovules
and of the stigmata of the family are better known by
this general statement. A similar observation may be
made with regard to the Anomalies of each group, which
occur so commonly, that Mr. Lindley, in his *Introduction
to the Natural System of Botany*, makes the "Anomalies"
an article in each family. Thus, part of the character of
the Rosaceæ is that they have alternate *stipulate* leaves,
and that the *albumen* is *obliterated :* but yet in *Lowea*, one
of the genera of this family, the stipulæ are *absent ;* and
the albumen is *present* in another, *Neillia*. This implies,
as we have already seen, that the artificial character (or
diagnosis as Mr. Lindley calls it) is imperfect. It is,
though very nearly, yet not exactly, commensurate with
the natural group : and hence in certain cases this character is made to yield to the general weight of natural
affinities.

9. *Difference of Natural History and Mathematics.*—
These views,—of classes determined by characters which
cannot be expressed in words,—of propositions which
state, not what happens in all cases, but only usually,—
of particulars which are included in a class though they
transgress the definition of it, may very probably surprise

* Lindley, *Nat. Syst.*, p. 81.

the reader. They are so contrary to many of the received opinions respecting the use of definitions and the nature of scientific propositions, that they will probably appear to many persons highly illogical and unphilosophical. But a disposition to such a judgment arises in a great measure from this ;—that the mathematical and mathematico-physical sciences have, in a great degree, determined men's views of the general nature and form of scientific truth ; while Natural History has not yet had time or opportunity to exert its due influence upon the current habits of philosophizing. The apparent indefiniteness and inconsistency of the classifications and definitions of Natural History belongs, in a far higher degree, to all other except mathematical speculations : and the modes in which approximations to exact distinctions and general truths have been made in Natural History, may be worthy our attention, even for the light they throw upon the best modes of pursuing truth of all kinds.

10. *Natural Groups given by Type not by Definition.*— The further developement of this suggestion must be considered hereafter. But we may here observe, that though in a Natural group of objects a definition can no longer be of any use as a regulative principle, classes are not, therefore, left quite loose, without any certain standard or guide. The class is steadily fixed, though not precisely limited; it is given, though not circumscribed; it is determined, not by a boundary line without, but by a central point within ; not by what it strictly excludes, but by what it eminently includes; by an example, not by a precept; in short, instead of Definition we have a *Type* for our director.

A Type is an example of any class, for instance, a species of a genus, which is considered as eminently possessing the characters of the class. All the species which have a greater affinity with this type-species than with any others, form the genus, and are ranged about it, deviating from it in various directions and different degrees. Thus a genus may consist of several species which approach very near the type, and of which the claim to a place with it is obvious; while there may be other species which straggle further from this central knot, and which yet are clearly more connected with it than with any other. And even if there should be some species of which the place is dubious, and which appear to be equally bound to two generic types, it is easily seen that this would not destroy the reality of the generic groups, any more than the scattered trees of the intervening plain prevent our speaking intelligibly of the distinct forests of two separate hills.

The type-species of every genus, the type-genus of every family, is, then, one which possesses all the characters and properties of the genus in a marked and prominent manner. The type of the Rose family has alternate stipulate leaves, wants the albumen, has the ovules not erect, has the stigmata simple, and besides these features, which distinguish it from the exceptions or varieties of its class, it has the features which make it prominent in its class. It is one of those which possess clearly several leading attributes; and thus, though we cannot say of any one genus that it *must* be the type of the family, or of any one species that it *must* be the type of the genus, we are still not wholly to seek: the type must be connected by many affinities with most of the others of its group; it must be near the centre of the crowd, and not one of the stragglers.

11. It has already been repeatedly stated, as the great rule of all classification, that the classification must serve to assert general propositions. It may be asked what propositions we are able to enunciate by means of such classifications as we are now treating of. And the

answer is, that the collected knowledge of the characters, habits, properties, organization, and functions of these groups and families, as it is found in the best botanical works, and as it exists in the minds of the best botanists, exhibits to us the propositions which constitute the science, and to the expression of which the classification is to serve.   All that is not strictly definition, that is, all that is not artificial character, in the descriptions of such classes, is a statement of truths, more or less general, more or less precise, but making up, together, the positive knowledge which constitutes the science.   As we have said, the consideration of the properties of plants in order to form a system of classification, has been termed Taxonomy, or the Systematick of Botany; all the parts of the descriptions, which, taking the system for granted, convey additional information, are termed the *Physiography* of the science; and the same terms may be applied in the other branches of Natural History.

12. *Artificial and Natural Systems.*—If I have succeeded in making it apparent that an artificial system of characters necessarily implies natural classes which are not severed by the artificial marks, we shall now be able to compare the nature and objects of the Artificial and Natural Systems; points on which much has been written in recent times.

The Artificial System is one which is, or professes to be, entirely founded upon marks selected according to the condition which has been stated, of not violating certain narrow natural groups; namely, in the Linnæan system, the natural genera of plants.   The marks which form the basis of the system are applied rigorously and universally without any further regard to any other characters or indications of affinity.   Thus in the Linnæan system, which depends mainly on the number of male organs or stamens, and on the number of female organs or styles, the largest divisions, or the *Classes*, are arranged according to the number of the stamens, and are *monandria, diandria, triandria, tetrandria, pentandria, hexandria,* and so on: the names being formed of the Greek numerical words, and of the word which implies *male*. And the Orders of each of these Classes are distinguished by the number of styles, and are called *monogynia, digynia, trigynia,* and so on, the termination of these words meaning *female*. And so far as this numerical division and subdivision go on, the system is a rigorous system, and strictly artificial.

But the condition that the artificial system shall leave certain natural affinities untouched, makes it impossible to go through the vegetable kingdom by a method of mere numeration of stamens and styles. The distinction of flowers with twenty and with thirty stamens is not a fixed distinction: flowers of one and the same kind, as roses, have, some fewer than the former, some more than the latter number. The Artificial System, therefore, must be modified. And there are various relations of connexion and proportion among the stamina which are more permanent and important than their mere number. Thus flowers with two longer and two shorter stamens are not placed in the class tetrandria, but are made a separate class *didynamia;* those with four longer and two shorter are in like manner *tetradynamia,* not hexandria; those in which the filaments are bound into two bundles are *diadelphia.* All these and other classes are deviations from the plan of the earlier classes, and are so far defects of the artificial system; but they are requisite in order that it may leave a basis of natural groups, without which it would not be a system of vegetables. And as the division is still founded on some properties of the stamens, it combines not ill with that part of the system which depends on the number. The classes framed in virtue of these various considerations make up an artificial system which is tolerably coherent.

But since the Artificial System thus regards natural groups, in what does it differ from a Natural System? It differs in this :—That though it allows certain subordinate natural groups, it merely allows these, and does not endeavour to ascend to any wider natural groups. It takes all the higher divisions of its scheme from its artificial characters, its stamens and pistils, without looking to any natural affinities. It accepts natural *genera*, but it does not seek natural *families*, or orders, or classes. It assumes natural groups, but does not investigate any; it forms wider and higher groups, but professes to 'frame them arbitrarily.

But then, on the other hand, the question occurs, this being the case, what can be the use of the Artificial System? If its characters, in the higher stages of classification, be arbitrary, how can it lead us to the natural relations of plants? And the answer is, that it does so in virtue of the original condition, that there shall be certain natural relations which the artificial system shall not transgress; and that its use arises from the facility with which we can follow the artificial arrangement as far as it goes. We can count the stamens and pistils, and thus we know the Class and Order of our plant; and we have then to discover its Genus and Species by means less symmetrical but more natural. The Artificial System, though arbitrary in a certain degree, brings us to a Class in which the whole of each genus is contained, and there we can find the proper Genus by a suitable method of seeking. No Artificial System can conduct us into the extreme of detail, but it can place us in a situation where the detail is within our reach. We cannot find the house of a foreign friend by its latitude and longitude; but we may be enabled, by a knowledge of the latitude and longitude, to find the city in which he dwells, or at least the island; and we then can reach his abode by following the road or exploring the locality. The Artificial System is such a method of travelling by latitude and longitude ; the Natural System is that which is guided by a knowledge of the country.

The Natural System, then, is that which endeavours to arrange by the natural affinities of objects ; and more especially, which attempts to ascend from the lower natural groups to the higher; as for example from genera to natural families, orders, and classes. But as we have already hinted, these expressions of natural affinities, natural groups, and the like, when considered in reference to the idea of resemblance alone, without studying analogy or function, are very vague and obscure. We must notice some of the attempts which were made under the operation of this imperfect view of the subject.

13. *Modes of framing Natural Systems.*—Decandolle* distinguishes the attempts at Natural Classifications into three sorts : those of *blind trial*, (tâtonnement,) those of *general comparison*, and those of *subordination of characters.* The two former do not depend distinctly upon any principle, except resemblance ; the third refers us to other views, and must be considered in a future chapter.

Method of Blind Trial.—The notion of the existence of natural classes dependent on the general resemblance of plants,—of an affinity showing itself in different parts and various ways,—though necessarily somewhat vague and obscure, was acted upon at an early period, as we have seen in the formation of genera; and was enunciated in general terms soon after. Thus Magnolius† says that he discerns in plants an affinity, by means of which they may be arranged in families. " Yet it is impossible to

---

* *Th. El.*, art. 41.

† Dec. *Th. El.*, art. 42. Petri Magnoli, *Prodromus Hist. Gen. Plant.*, 1689.

obtain from the fructification alone the Characters of these families; and I have therefore chosen those parts of plants in which the principal characteristic marks are found, as the root, the stem, the flower, the seed. In some plants there is even a certain resemblance; an affinity which does not consist in the parts considered separately, but in their totality; an affinity which may be felt but not expressed; as we see in the families of agrimonies and cinquefoils, which every botanist will judge to be related, though they differ by their roots, their leaves, their flowers, and their seeds."

This obscure feeling of a resemblance on the whole, an affinity of an indefinite kind, appears fifty years later in Linnæus's attempts. "In the Natural Classification," he says*, "no *à priori* rule can be admitted, no part of the fructification can be taken exclusively into consideration; but only the simple symmetry of all its parts." Hence though he proposed natural families, and even stated the formation of such families to be the first and last object of all methods, he never gave the characters of those groups, or connected them by any method. He even declared it to be impossible to lay down such a system of characters. This persuasion was the result of his having refused to admit into his mind any idea more profound than that notion of resemblance of which he had made so much and such successful use; he would not attempt to unravel the ideas of symmetry and of function on which the clear establishment of natural relations must depend. He even despised the study of the inner organization of plants; and reckoned† the *Anatomici*, who studied the anatomy and physiology of plants and the laws of vegetation, among the *Botanophili*, the mere amateurs of his science.

The same notion of general resemblance and affinity,

* Dec., *Th. El.*, art. 42.   † *Phil. Bot.*, s. 44.

accompanied with the same vagueness, is to be found in the writer who least participated in the general admiration of Linnæus, Buffon. Though it was in a great measure his love of higher views which made him dislike what he considered the pedantry of the Swedish school, he does not seem to have obtained a clearer sight of the principle of the natural method than his rival, except that he did not restrict his Characters to the fructification. Things must be arranged by their resemblances and differences, (he says in 1750*,) " but the resemblances and differences must be taken not from one part but from the whole; and we must attend to the form, the size, the habit, the number and position of the parts, even the substance of the part ; and we must make use of these elements in greater or smaller number, as we have need."

14. *Method of General Comparison.*—A countryman of Buffon, who shared with him his depreciating estimate of the Linnæan system, and his wish to found a natural system upon a broader basis, was Adanson ; and he invented an ingenious method of apparently avoiding the vagueness of the practice of following the general feeling of resemblance. This method consisted in making many artificial systems, in each of which plants were arranged by some one part ; and then collecting those plants which came near each other in the greatest number of those artificial systems, as plants naturally the most related. Adanson gives an account† of the manner in which this system arose in his mind. He had gone to Senegal, animated by an intense zeal for natural history; and there, amid the luxuriant vegetation of the torrid zone, he found that the methods of Linnæus and Tournefort failed him altogether as means of arranging his new botanical treasures. He was driven to seek a new

---

* ADANSON, p. clvi. BUFFON, *Hist. Nat.*, t. i., p. 21.
† Pref., p. clvii.

system. "For this purpose," he says, "I examined plants in all their parts, without omitting any, from the roots to the embryo, the folding of the leaves in the bud, their mode of sheathing*, the situation and folding of the embryo and of its radicle in the seed, relatively to the fruit; in short, a number of particulars which few botanists notice. I made in the first place a complete description of each plant, putting each of its parts in separate articles, in all its details; when new species occurred I put down the points in which they differed, omitting those in which they agreed. By means of the aggregate of these comparative descriptions, I perceived that plants arranged themselves into classes or families which could not be artificial or arbitrary, not being founded upon one or two parts, which might change at certain limits, but on all the parts; so that the disproportion of one of these parts was corrected and balanced by the introduction of another." Thus the principle of resemblance was to suffice for the general arrangement, not by means of a new principle, as symmetry or organization, which should regulate its application, but by a numeration of the peculiarities in which the resemblance consisted.

The labour which Adanson underwent in the execution of this thought was immense. By taking each organ, and considering its situation, figure, number, &c., he framed sixty-five artificial systems; and collected his natural families by a numerical combination of these. For example, his sixty-fifth artificial system† is that which depends upon the situation of the ovary with regard to the flower; according to this system he frames ten artificial classes, including ninety-three sections: and of these sections the resulting natural arrangement retains thirty-five, above one-third: the same estimate is applied in other cases.

---

* "Leur manière de s'engainer."   † ADANSON, Pref., p. cccxii.

But this attempt to make number supply the defects which the vague notion of resemblance introduces, however ingenious, must end in failure. For, as Decandolle observes *, it supposes that we know, not only all the organs of plants, but all the points of view in which it is possible to consider them; and even if this assumption were true, which it is, and long must be, very far from being, the principle is altogether vicious; for it supposes that all these points of view, and all the resulting artificial systems are of equal importance: a supposition manifestly erroneous. We are thus led back to the consideration of the *relative importance* of organs and their qualities, as a basis for the classification of plants, which no artificial method can supersede; and thus we find the necessity of attending to something besides mere external and detached resemblance. The method of general comparison cannot, any more than the method of blind trial, lead us, with any certainty or clearness, to the natural method. Adanson's families are held by the best botanists to be, for the greater part natural; but his hypotheses are unfounded; and his success is probably more due to the dim feeling of affinity, by which he was unconsciously guided, than to the help he derived from his numerical processes.

15. In a succeeding chapter I shall treat of that Natural Affinity on which a Natural System must really be founded. But before proceeding to this higher subject, we must say a few words on some of the other parts of the philosophy of Natural History,—the Gradation of Groups, the Nomenclature, the Diagnosis, and the application of the methods to other subjects.

Gradation of Groups.—It has been already noticed (last chapter,) that even that vague application of the idea of resemblance. which gives rise to the terms of

* DEC., *Th. El.*, p. 67.

common language, introduces a subordination of classes, as man, animal, body, substance. Such a subordination appears in a more precise form when we employ this idea in a scientific manner as we do in Natural History. We have then a series of divisions, each inclusive of the lower ones, which are expressed by various metaphors in different writers. Thus some have gone as far as eight terms of the series*, and have taken, for the most part, military names for them; as *Hosts, Legions, Phalanxes, Centuries, Cohorts, Sections, Genera, Species.* But the most received series is *Classes, Orders, Genera,* and *Species;* in which, however, we often have other terms interpolated, as *Sub-genera,* or Sections of genera. The expressions *Family* and *Tribe,* are commonly appropriated to natural groups; and we speak of the Vegetable, Animal, Mineral *Kingdom;* but the other metaphors of Provinces, Districts, &c., which this suggests, have not been commonly used.

It will of course be understood that each ascending step of classification is deduced by the same process from the one below. A genus is a collection of species which resemble each other more than they resemble other species; an order is a collection of genera having, in like manner, the first degree of resemblance, and so on. What the degrees of resemblance are, much depend upon the nature of the objects compared, and cannot possibly be prescribed before-hand. Hence the same term, Class and Order for instance, may imply in different provinces of nature very different degrees of resemblance. The Classes of Animals are Insects, Birds, Fish, Beasts, &c, The Orders of Beasts are *Ruminants, Tardigrades, Plantigrades,* &c. The two Classes of Plants (according to the Natural Order†,) are *Vascular* and *Cellular,* the latter having neither sexes, flowers, nor spiral vessels.

---

* ADANSON, p. cvi.   † LINDLEY.

The Vascular Plants are divided into Orders, as *Umbelli-feræ, Ranunculaceæ*, &c., but between this Class and its Orders are interposed two other steps: two Sub-classes, *Dicotyledonous* and *Monocotyledonous*, and two Tribes of each: *Angiospermiæ, Gymnospermiæ* of the first; and *Petaloideæ, Glumaciæ* of the second. Such interpolations are modifications of the general formula of subordination for the purpose of accommodating it to the most prominent natural affinities.

16. *Species.*—As we have already seen in tracing the principles of the natural method, when by the intimate study of plants we seek to give fixity and definiteness to the notion of resemblance and affinity on which all these divisions depend, we are led to the study of organization and analogy. But we make a reference to physio-logical conditions even from the first, with regard to the lowest step of our arrangement, the *species;* for we consider it a proof of the impropriety of separating two species, if it be shown that they can by any course of propagation, culture, and treatment, the one pass into the other. It is in this way, for example, that it has been supposed to be established that the common prim-rose, oxlip, polyanthus, and cowslip, are all the same species. Plants which thus, in virtue of external cir-cumstances, as soil, exposure, climate, exhibit differences which may disappear by changing the circumstances, are called *varieties* of the species. And thus we cannot say that a species is a collection of individuals which possess the first degree of resemblance; for it is clear that a primrose resembles another primrose more than it does a cowslip; but this resemblance only constitutes a variety. And we find that we must necessarily include in our conception of species, the notion of propagation from the same stock. And thus a species has been well de-fined*. " The collection of the individuals descended from

---

* Cuv., *Règne Animal*, p. 19.

one another, or from common parents, and of those which resemble these as much as these resemble each other." And thus the sexual doctrine of plants, or rather the consideration of them as things which propagate their kind, (whether by seed, shoot, or in any other way,) is at the basis of our classifications.

17. The first degree of resemblance among organized beings is thus that which depends on this relation of generation, and we might expect that the groups which are connected by this relation would derive their names from the notion of generation. It is curious that both in Greek and Latin languages and in our own, the words which have this origin (γένος, *genus*, *kind*) do not, in the phraseology of science at least, denote the nearest degree of relationship, but have other terms subordinate to them, which appear etymologically to indicate a mere resemblance of appearance, (εἶδος, *species*, *sort*,) and which are appropriated to the groups resulting from propagation. Probably the reason of this is, that the former terms had been applied so widely and loosely before the scientific fixation of terms, that to confine them to what we call species would have been to restrict them in a manner too unusual to be convenient.

18. *Varieties. Races.*—The Species, as we have said, is the collection of individuals which resemble each other as much as do the offspring of a common stock. But within the limits of this boundary, there are often observable differences permanent enough to attract our notice, though capable of being obliterated by mixture in the course of generation. Such different groups are called *Varieties*. Thus the primrose and cowslip, as has been stated above, are found to be varieties of the same plant; the poodle and the greyhound are well marked varieties of the species *dog*. Such differences are hereditary, and as we have seen, it may be long doubtful whether such here-

ditary differences are varieties only, or different species. In such cases the term *Race* has been applied.

19. (III.) *Nomenclature.*—The Nomenclature of any branch of Natural History is the collection of names of all its species; which, when they become extremely numerous, requires some artifice to make it possible to recollect or apply them. The known species of plants, for example, were 10,000 at the time of Linnæus, and are now probably 60,000. It would be useless to endeavour to frame and employ separate names for each of these species.

The division of the objects into a subordinated system of classification enables us to introduce a Nomenclature which does not require this enormous number of names. The artifice employed to avoid this inconvenience is to name a species by means of two (or it might be more) steps of the successive division. Thus in Botany each of the genera has its name, and the species are marked by the addition of some epithet to the name of the genus. In this manner about 1,700 generic names, with a moderate number of specific names, were found by Linnæus sufficient to designate with precision all the species of vegetables known at his time. And this *Binary Method* of Nomenclature has been found so convenient that it has been universally adopted in every other department of the Natural History of organized beings.

Many other modes of Nomenclature have been tried, but no other has at all taken root. Linnæus himself appears at first to have intended marking each species by the generic name accompanied by a characteristic descriptive phrase; and to have proposed the employment of a *trivial* specific name, as he termed it, only as a method of occasional convenience. The use of these trivial names, has, however, become universal, as we have said, and is by many persons considered the greatest improvement introduced at the Linnæan reform.

Both Linnæus and other writers (as Adanson) have given many maxims with a view of regulating the selection of generic and specific names. The maxims of Linnæus were intended as much as possible to exclude barbarism and confusion, and have, upon the whole, been generally adopted; though many of them were objected to by his contemporaries (Adanson and others*), as capricious or unnecessary innovations. Many of the names, introduced by Linnæus, certainly appear fanciful enough: thus he gives the name of *Bauhinia* to a plant with leaves in pairs, because the Bauhins were a pair of brothers; *Banisteria* is the name of a climbing plant, in honour of Banister, who travelled among mountains. But such names, once established by adequate authority, lose all their inconvenience and easily become permanent; and hence the reasonableness of the Linnæan rule†, that as such a perpetuation of the names of persons by the names of plants is the only honour botanists have to bestow, it ought to be used with care and caution.

The generic name must, as Linnæus says, be fixed‡ before we attempt to form a specific name; "the latter without the former is like the clapper without the bell." The name of the genus being established, the species may be marked by adding to it "a single word taken at will from any quarter;" that is, not involving a description or any essential property of the plant, but a casual or arbitrary appellation. Thus the various species of Hieracium‖ are *Hieracium Alpinum, H. Halleri, H. Pilosella, H. dubium, H. murorum,* &c., where we see how different may be the kind of origin of the words.

Attempts have been made at various times to form the names of species from those of genera in some more

---

* Pp. cxxix, clxxii.   † *Phil. Bot.*, Sec. 239.
‡ *Ib.*, Sec. 222.   § *Ib.*, Sec. 260.
‖ HOOKER, *Fl. Scot.*, 228.

symmetrical manner. Thus some have numbered the species of genus 1, 2, 3, &c., but this method is liable to the inconveniences, first, that it offers nothing for the memory to take hold of; and second, that if a new species intermediate between 1 and 2, 2 and 3, &c., be discovered, it cannot be put in its place It has also been proposed to mark the species by altering the termination of the genus. Thus Adanson*, denoting a genus by the name *Fonna* (*Lychnidea*), conceived he might mark five of its species by altering the last vowel, *Fonna, Fonna-e, Fonna-i, Fonna-o, Fonna-u;* then others by *Fonna-ba, Fonna-ka,* and so on. This course would be liable to the same evils which have been noticed as belonging to the numerical method.

The names of plants (and the same is true of animals) have in common practice been binary only, consisting of a generic and a specific name. The Class and Order have not been admitted to form part of the appellation of the species.   Indeed it is easy to see that a name which must be identical in so many instances as that of an order would be, would be felt as superfluous and burdensome.   Accordingly, Linnæus makes it a precept†, that the name of the Class and the Order must not be expressed but understood : and hence, he says, Royen, who took *Lilium* for the name of a class, rightly rejected it as a generic name and substituted *Lirium*, with the Greek termination.

Yet we must not too peremptorily assume such maxims as these to be universal for all classificatory sciences.   It is very possible that it may be found advisable to use *three* terms, that of order, genus and species, in designating minerals, as is done in Mohs's nomenclature; for example, *Rhombohedral Calc Haloide, Paratomous Hal Baryte.*   It is possible also that it may

* Pref., clxxvi.   † *Phil. Bot.*, Sec. 215.

be found useful in the same science to mark some of the steps of classification by the termination.

Thus it has been proposed to confine the termination *ite* to the Order *Silicides* of Naumann, as Apophyll*ite*, Stilb*ite*, Leuc*ite*, &c., and to use names of different form in other orders, as Talc *Spar* for Brennerite, Pyramidal Titanium *Oxide* for Octahedrite. Some such method appears to be the most likely to give us a tolerable mineralogical nomenclature.

20. (IV.) *Diagnosis.*—German Naturalists speak of a part of the general method which they call the *Characteristik* of Natural History, and which is distinguished from the *Systematik* of the science. The *Systematick* arranges the objects by means of all their resemblances, the *Characteristick* enables us to detect their place in the arrangement by means of a few of their characters. What these characters are to be, must be discovered by observation of the groups and divisions of the system when they are formed. To construct a collection of such as shall be clear and fixed, is a useful, and generally a difficult task'; for there is usually no apparent connexion between the marks which are used in discriminating the groups, and the nature of the groups themselves. They are assumed only because the Naturalist, extensively and exactly acquainted with the groups and the properties of the objects which compose them, sees, by a survey of the field, that these marks divide it properly.

The Characteristick has been termed by some English Botanists the *Diagnosis* of plants; a word which we may conveniently adopt. The Diagnosis of any genus or species is different according to the system we follow. Thus in the Linnæan system the Diagnosis of the Rose is in the first place given by its Class and Order: it is Icosandrous, and Polygynous; and then the generic distinction is that the calyx is five-cleft, the tube urceolate, including many hairy achenia, the receptacle villous*. In the Natural System the Rose-Tribe are distinguished as being† "Polypetalous dicotyledons, with lateral styles, superior simple ovaria, regular perigynous stamens, exalbuminous definite seeds, and alternate stipulate leaves." And the true Roses are further distinguished by having "Nuts, numerous, hairy, terminated by the persistent lateral style and inclosed within the fleshy tube of the calyx," &c.

It will be observed that in a rigorous artificial system the *Systematick* coincides with the *Characteristick;* the *Diataxis* with the *Diagnosis;* the reason why a plant is put in a division is identical with the mode by which it is known to be in the division. The Rose is in the class *icosandria*, because it has many stamens inserted in the calyx ; and when we see such a set of stamens we immediately know the class. But this is not the case with the Diagnosis of natural families. Thus the genera *Lamium* and *Galeopsis* (Dead Nettle and Hemp Nettle), are each formed into a separate group in virtue of their general resemblances and differences, and not because the former has one tooth on each side of the lower lip, and the latter a notch in its upper lip, though they are distinguished by these marks.

Thus, so far as our Systems are natural, (which, as we have shown, all systems to a certain extent must be), the Characteristick is distinct both from a Natural and an Artificial System; and is, in fact, an Artificial key to a Natural System. As being Artificial, it takes as few characters as possible; as being Natural, its characters are not selected by any general or prescribed rule, but follow the natural affinities. The Botanists who have made any steps› in the formation of a natural method of plants since Linnæus, have all attempted to give a Diagnosis corresponding to the Diataxis of their method.

---

* LINDLEY, *Nat. Syst.*, p. 149.   † *Ib.*, p. 81. 3.

CHAPTER III.

APPLICATION OF THE NATURAL HISTORY METHOD TO MINERALOGY.

1. THE philosophy of the Sciences of Classification has had great light thrown upon it by discussions concerning the methods which are used in Botany: for that science is one of the most complete examples which can be conceived of the consistent and successful application of the principles and ideas of Classification; and this application has been made in general without giving rise to any very startling paradoxes, or disclosing any insurmountable difficulties. But the discussions concerning methods of Mineralogical Classification have been instructive for quite a different reason: they have brought into view the boundaries and the difficulties of the process of Classification; and have presented examples in which every possible mode of classifying appeared to involve inextricable contradictions. I will notice some of the points of this kind which demand our attention, referring to the works published recently by several mineralogists.

In the History of Mineralogy we noticed the attempt made by Mohs and other Germans to apply to minerals a method of arrangement similar to that which has been so successfully employed for plants. The survey which we have now taken of the grounds of that method will point out some of the reasons of the very imperfect success of this attempt. We have already said that the *Terminology* of Mineralogy was materially reformed by Werner, and including in this branch of the subject (as we must do) the Crystallography of later writers, it may be considered as to a great extent complete. Of the attempts at a Natural arrangement, that of Mohs appears

to proceed by the method of *blind trial*, the undefinable
perception of relationship by which the earliest attempts
at a Natural Arrangement of plants were made. Breithaupt, however, has made (though I do not know that he
has published) an essay in a mode which corresponds very
nearly to Adanson's process of *multiplied comparisons*.
Having ascertained the specific gravity and hardness of
all the species of minerals, he arranged them in a table,
representing by two lines at right angles to each other
these two numerical quantities. Thus all minerals were
distributed according to two co-ordinates representing
specific gravity and hardness. He conceived that the
groups which were thus brought together were natural
groups. On both these methods, and on all similar ones,
we might observe, that in minerals as in plants, the mere
general notion of likeness cannot lead us to a real arrangement: it requires to have precision and aim given it by
some other relation;—the relation of chemical composition in minerals, as the relation of organic function in
vegetables. The physical and crystallographical properties
of minerals must be studied with reference to their constitution; and they must be arranged into groups which
have some common chemical character, before we can
consider any advance as made towards a natural arrangement.

In reality, it happens in Mineralogy as it happened in
Botany, that those speculators are regulated by an obscure
perception of this ulterior relation, who do not profess to
be regulated by it. Several of the Orders of Mohs have
really great unity of chemical character, and thus have
good evidence of their being really Natural Orders.

2. Supposing the Diataxis of minerals thus obtained,
Mohs attempted the Diagnosis; and his *Characteristick of
the Mineral Kingdom*, published at Dresden, in 1820, was
the first public indication of his having constructed a

system. From the nature of a Characteristick, it is necessarily brief, and without any ostensible principle ; but its
importance was duly appreciated by the author's countrymen. Since that time, many attempts have been made
at improved arrangements of minerals, but none, I think,
(except perhaps that of Breithaupt,) professing to proceed rigorously on the principles of Natural History ;—to
arrange by means of external characters, neglecting altogether, or rather postponing, the consideration of chemical
properties. By relaxing from this rigour, however, and
by combining physical and chemical considerations,
arrangements have been obtained (for example, that of
Naumann,) which appear more likely than the one of Mohs
to be approximations to an ultimate really natural system.
Naumann's Classes are *Hydrolytes, Haloides, Silicides,
Metal Oxides, Metals, Sulphurides, Anthracides*, with subdivisions of Orders, as *Anhydrous unmetallic Silicides*.
It may be remarked that the designations of these are
mostly chemical. As we have observed already, chemistry, and mineralogy in its largest sense, are each the
necessary supplement of the other. If chemistry furnish
the nomenclature, mineralogy must supply the physiography : if the arrangement be founded on **external**
characters, and the names independent of chemistry, the
chemical composition of each species is an important
scientific truth respecting it.

3. The inquiry may actually occur, whether any subordination of groups in the mineral kingdom has really
been made out. The ancient chemical arrangements,
for instance, that of Haüy, though professing to distribute
minerals according to Classes, Orders, Genera, and Species,
were not only arbitrary, but inapplicable ; for the first
postulate of any method, that the species should have
constant characters of unity and difference, was not
satisfied. It was not ascertained that carbonate of lime

was really distinguishable in all cases from carbonate of
magnesia, or of iron; yet these species were placed in remote parts of the system: and the above carbonates made
just so many species, although, if distinct from one another
at all, they were further distinguishable into additional species. Even now, we may, perhaps, say that the limits of
mineralogical species, and their laws of fixity, are not yet
clearly seen. For the discovery of the isomorphous relations and optical properties of minerals have rather shown
us in what direction the object lies, than led us to the goal.
It is clear that, in the mineral kingdom, the Definition of
Species, borrowed from the laws of the continuation of
the kind, which holds throughout the organic world, fails
us altogether, and must be replaced by some other condition: nor is it difficult to see that the definite atomic
relations of the chemical constituents, and the definite
crystalline angle, must supply the principles of the *specific*
identity for minerals. Yet the exact limits for the definiteness in both these cases (when we admit the effect of
mechanical mixtures, &c.) have not yet been completely
disentangled. It is clear that any *arbitrary* assumption
(as the allowance of a certain per centage of mixture, or
a certain small deviation in the angle,) is altogether contrary to the philosophy of the natural system, and can
lead to no stable views. It is only by laborious, extensive, and minute research, that we can hope to attain to
any solid basis of arrangement.

4. Still, though there are many doubts respecting
mineralogical species, a large number of such species are
so far fixed that they may be supposed capable of being
united under the higher divisions of a system with approximate truth. Of these higher divisions, those which have
been termed *Orders* appear to tend to something like a
fixed chemical character. Thus the *Haloids* of Naumann,
and mostly those of Mohs, are combinations of an oxide with

an acid, and thus resemble Salts, whence their name. The
Silicides contain most of Mohs's *Spaths:* and the Orders
Pyrites, Glance, and *Blende,* are common to Naumann
and Mohs; being established by the latter on a difference
of external character, which difference is, indeed, very
manifest; and being included by the former in one chemical *Class, Sulphurides.* The distinctions of *Hydrous*
and *Anhydrous, Metallic* and *Unmetallic,* are, of course,
chemical distinctions, but occur as the differences of
Orders in Naumann's mixed system.

We may observe that some French writers, following
Haüy's last edition, use, instead of *metallic* and *unmetallic,*
autopside metallic and *heteropside metallic;* meaning by this
phraseology to acknowledge the discovery that earths, &c.,
are metallic, though they do not *appear* to be so, while
metals both are and appear metallic. But this seems to
be a refinement not only useless but absurd. For what is
gained by adding the word *metallic,* which is common to
all, and therefore makes no distinction? If certain metals
are distinguished by their *appearing* to be metals, this
appearance is a reason for giving them the peculiar name,
metals. Nothing is gained by first bringing earths and
metals together, and then immediately separating them
again by new and inconvenient names. No proposition
can be expressed better by calling *earths heteropside metallic substances,* and therefore such nomenclature is to be
rejected.

Granting, then, that the Orders of the best recent mineralogical systems approximate to natural groups, we are led
to ask whether the same can be said of the Genera of the
Natural History systems, such as those of Mohs and Breithaupt. And here I must confess that I see no principle
in these genera, and have failed to apprehend the conceptions by the application of which they have been constructed: I shall therefore not pass any further judgment

upon them. The subordination of Mineralogical Species to Orders is a manifest gain to science: in the interposition of Genera I see nothing but a source of confusion.

5. In Mineralogy, as in other branches of natural history, a reformed arrangement ought to give rise to a reformed Nomenclature; and for this, there is more occasion at present in Mineralogy than there was in Botany at the worst period, at least as far as the extent of the subject allows. The characters of minerals are much more dimly and unfrequently developed than those of plants; hence arbitrary chemical arrangements, which could not lead to any natural groups, and therefore not to any good names, prevailed till recently; and this state of things produced an anarchy in which every man did what seemed right in his own eyes,—proposed species without any ascertained distinction, and without a thought of subordination, and gave them arbitrary names; and thus with only about two or three hundred known species, we have thousands upon thousands of names, of anomalous form and uncertain application.

Mohs has attempted to reform the Nomenclature of the subject in a mode consistent with his attempt to reform the System. In doing this, he has fatally transgressed a rule always insisted upon by the legislators of Botany, of altering usual names as little as possible; and his names are both so novel and so cumbrous, that they appear to have little chance of permanent currency. They are, perhaps, more unwieldy than they need to be, by referring, as we have said, to *three* of the steps of his classification, the Species, Genus, and Order. We may, however, assert confidently, from the whole analogy of natural history, that no good names can be found which do not refer to at least *two* terms of the arrangement. This rule has been practically adopted to a great extent by Naumann, who gives to most of his Haloids the name

Spar, as Calc spar, Iron spar, &c.; to all his Oxides the terminal word *Erz* (*Ore*); and to the species of the orders *Kies* (*Pyrites*), *Glance*, and *Blende*, these names. It has also been theoretically assented to by Beudant, who proposes that we should say *silicate stilbite, silicate chabasie; carbonate calcaire, carbonate witherite; sulphate couperose,* &c. One great difficulty in this case would arise from the great number of *silicides;* it is not likely that any names would obtain a footing which tacked the term *silicide* to another word for each of these species. The artifice which I have proposed, in order to obviate this difficulty, is that we should make the names of the silicides, and those alone, end in *ite* or *lite,* which a large proportion of them do already.

By this and a few similar contrivances, we might, I conceive, without any inconvenient change, introduce into mineralogy a systematic nomenclature.

6. I shall now proceed to make a few remarks on a work on mineralogy more recent than those which I have above noticed, and written with express reference to such difficulties as I have been discussing. I allude to the treatise of M. Necker, *Le Règne Mineral ramené aux Methodes d'Histoire Naturelle**, which also contains various dissertations on the philosophy of classification in general, and its application to mineralogy in particular.

M. Necker remarks very justly, that mineralogy, as it has hitherto been treated, differs from all other branches of Natural History in this:—that while it is invested with all the forms of the sciences of classification,— Classes, Divisions, Genera, and the like,—the properties of those bodies to which the mineralogical student's attention is directed have no bearing whatever on the classification. A person, he remarks†, might be perfectly well acquainted with all the characters of minerals which

---

* Paris, 1835.   † *Règne Mineral,* p. 3.

Werner or Haüy examined so carefully, and might yet be quite unable to assign to any mineral its place in the divisions of their methods. There is* a complete separation between the study of mineralogical characters and the recognition of the name and systematic place of a mineral. Those who know *mineralogy* well, may know *minerals* ill, or hardly at all; the systematist may be in such knowledge vastly inferior to the mineral-dealer or the miner. In this respect there is a complete contrast between this science and other classificatory sciences.

Again, in the best-known systems of mineralogy, (as those of Werner and Haüy,) the bodies which are grouped together as belonging to the same division, have not, as they have in other classificatory sciences, any resemblance. The different members of the larger classes are united by the common possession of some abstract property,—as, that they all contain iron. This is a property to which no common circumstance in the bodies themselves corresponds. What is there common to the minerals named oxidulous iron, sulphuret of iron, carbonate of iron, sulphate of iron, except that they all contain iron? And when we have classed these bodies together, what general assertion can we make concerning them, except that which is the ground of our classification, that they contain iron? They have nothing in common with iron or with each other in any other way.

Again, as these classes have no general properties, all the properties are particular to the species; and the descriptions of these necessarily become both tediously long, and inconveniently insulated.

7. These inconveniences arise from making chemical composition the basis of mineralogical classification without giving chemical analysis the first place among mineral properties. Shall we, then, correct this omission, so far

* *Règne Mineral*, p. 8.

as it has affected mineralogical systems? Shall we teach
the student the chemical analysis of minerals, and then
direct him to classify them according to the results of his
analysis*?

But why should we do this? To what purpose, or on
what ground, do we arrange the results of chemical analysis according to the forms and subordination of natural
history? Is not chemistry a science distinct from natural
history? Are not the sciences opposed? Is not natural
history confined to organic bodies? Can mere chemical
elements and their combinations be, with any propriety
or consistency, arranged into species, genera, and families? What is the principle on which genera and species
depend? Do not species imply individuals? What is
an individual in the case of a chemical substance?

8. We thus find some of the widest and deepest
questions of the philosophy of classification brought under
our consideration when we would provide a method for
the classification of minerals. The answers to these questions are given by M. Necker; and I shall state some of
his opinions; taking the liberty of adding such remarks
as are suggested by referring the subject to those principles which have already been established in this work.

M. Necker asserts† that the distinctions of different
sciences depend, not on the objects they consider, but on
the different and independent points of view on which
they proceed. Each science has its logic, that is, its
mode of applying the general rules of human reason to
its own special case. It has been said by some‡, that in
minerals, natural history and chemistry contemplate common objects, and thus form a single science. But do
chemistry and natural history consider minerals in the
same point of view?

* *Règne Mineral*, p. 18.   † *Ib.*, p. 23.
‡ *Ib.*, p. 27.

The answer is, that they do not. Physics and chemistry consider the properties of bodies in an abstract manner; as, their composition, their elements, their mutual actions, with the laws of these; their forces, as attraction, affinity; all which objects are abstract ideas. In these cases we have nothing to do with bodies themselves, but as the vehicles of the powers and properties which we contemplate.

Natural history, on the other hand, has to do with natural bodies: their properties are not considered abstractedly, but only as characters. If the properties are abstracted, it is but for a moment. Natural history has to describe and class bodies as they are. All which cannot be perceived by the senses, belongs not to its domain, as molecules, atoms, elements.

Natural history* may have recourse to physics or chemistry in order to recognise those properties of bodies which serve as characters; but natural history is not, on that account, physics or chemistry. Classification is the essential business of the natural historian†, to which task chemistry and physics are only instrumental, and the further account of properties only complementary.

It has been said, in support of the doctrine that chemistry and mineralogy are identical, that chemistry does not neglect external characters. "The chemist in describing sulphur, mentions its colour, taste, odour, hardness, transparence, crystalline form, specific gravity; how does he then differ from the mineralogist?" But to this it is replied, that these notices of the external characters of this or any substance are introduced in chemistry merely as convenient marks of recognition; whereas they are essential in mineralogy. If we had taken the account given of several substances instead of one, we should have seen that the chemist and the naturalist consider

* *Règne Mineral*, p. 37.   † *Ib.*, p. 41.

them in ways altogether different. The chemist will make it his business to discover the mutual action of the substances; he will combine them, form new products, determine the proportions of the elements. The mineralogist will divide the substances into groups according to their properties, and then subdivide these groups, till he refers each substance to its species. Exterior and physical characters are merely accessory and subordinate for the chemist; chemistry is merely instrumental for the mineralogist.

This view agrees with that to which we have been led by our previous reasonings; and may, according to our principles, be expressed briefly by saying, that the Idea which chemistry has to apply is the idea of Elementary Composition, while natural history applies the Idea of graduated Resemblances, and thus performs the task of classification.

9. The question occurs*, whether Natural History can be applied to Inorganic Substances? And the answer to this question is, that it can be applied, if there are such things as inorganic *individuals,* since the resemblances and differences with which natural history has to do are the resemblances and differences of individuals.

What is an Individual? It certainly is not that which is so simple that it cannot be divided. Individual animals are composed of many parts. But if we examine, we shall find that our idea of an individual is, that it is a whole composed of parts, which are not similar to the whole, and have not an independent existence, while the whole has an independent existence and a definite form †.

What then is the Mineralogical Individual? At first, while minerals were studied for their use, the most precious of the substances which they contained was looked upon as the characteristic of the mineral. The smallest

* *Règne Mineral,* p. 46.   † *Ib.,* p. 52.

trace of silver made a mineral an *ore of silver*. Thus forms and properties were disregarded, and *substance* was considered as identical with *mineral*. And hence * Daubenton refused to recognise *species* in the mineral kingdom, because he recognised no individuals. He proposed to call *sorts* what we call species. In this way of considering minerals, there are no individuals.

10. But still this is not satisfactory: for if we take a well formed and distinct crystal, this clearly *is* an individual†.

It may be objected, that the crystal is divisible (according to the theory of crystallography) into smaller solids; that these small solids are really the simple objects; and that actual crystals are formed by combinations of these molecules according to certain laws.

But, as we have already said, an individual is such, not because it cannot be divided, but because it cannot be divided into parts similar to the whole. As to the division of the form into its component *laws*, this is an abstract proceeding, foreign to natural history‡. Therefore there is so far nothing to prevent a crystal from being an individual.

11. We cannot (M. Necker goes on to remark) consider the *Integrant Molecules* as individuals. These are useful abstractions, but abstractions only, which we must not deal with as real objects. Haüy himself warns us § that his doctrine of increments is a purely abstract conception, and that nature, in fact, follows a different process. Accordingly, Weiss and Mohs express laws identical with those of Haüy, without even speaking of molecules; and Wollaston and Davy have deemed it probable that the molecules are not polyhedrons, but spheres or spheroids. Such mere creations of the mind can never be treated as individuals. If the maxim of

* *Règne Mineral*, p. 54.   † *Ib.*, p. 56.   ‡ *Ib.*, p. 58.
§ *Ib.*, p. 61.

natural history, that the species is a collection of individuals, be applied so as to make those individuals mere abstractions; or if, instead of individuals, we take such an abstraction as substance or matter, the course of natural history is altogether violated. And yet this error has hitherto generally prevailed; and mineralogists have classified, not things, but abstract ideas*.

12. But it may be said †, will not the small solids obtained by Cleavage better answer the idea of individuals? To this it is replied, that these small solids have no independent existence. They are only the result of a mode of division. They are never found separate and independent. The secondary forms which they compose are determined by various circumstances (the nature of the solution, &c.), and the cleavage which produces these small solids is only one result among many from the crystalline forces‡.

Thus neither integrant molecules, nor solids obtained by cleavage, can be such mineralogical individuals as the spirit of natural history requires. Hence it appears that we must take the real crystals for individuals §.

13. We must, however, reject crystals (generally large ones) which are obviously formed of several smaller ones of a similar form (as occurs so often in quartz and calc spar). We must also distinguish cases in which a large regular form is composed of smaller but different regular forms (as octahedrons of fluor spar made up of cubes). Here the small component forms are the individuals. Also we must notice the cases‖ in which we have a natural crystal, similar to the primary form. Here the face will show whether the body is a result obtained by cleavage or a natural individual.

14. It will be objected ¶, that the crystalline form ought

---

* *Règne Mineral*, p. 67.　　† *Ib.*, p. 69.　　‡ *Ib.*, p. 71.
§ *Ib.*, p. 73.　　‖ *Ib.*, p. 75.　　¶ *Ib.*, p. 79.

not to be made the dominant character in mineralogy, since it rarely occurs perfect. To this it is replied, that even if the application of the principle be difficult, still it has been shown to be the only true principle, and therefore we have no alternative. But further*, it is not true that amorphous substances are more numerous than crystals. In LEONHARD's *Manual of Oryctognosy*, there are 377 mineral substances. Of these, 281 have a crystalline structure, and 96 only have not been found in a regular form.

Again, the 281 crystalline forms have each its varieties, some of which are crystalline, and some are not so. Now the crystalline varieties amount to 1453, and the uncrystalline to 186 only. Thus mineralogy, according to the view of it here presented, has a sufficiently wide field†.

15. It will be objected‡, that according to this mode of proceeding, we must reject from our system all non-crystalline minerals. But we reply, that if the mass be composed of crystals, the size of the crystals makes no difference. Now lamellar and other compact masses are very generally groups of crystals in various positions. Individuals mutilated and mixed together are not the less individuals; and therefore such masses may be treated as objects of natural history.

If we cannot refer all rocks to crystalline species, those which elude our method may appear as an appendix, corresponding to those which botanists call *genera incertæ sedis*§.

But these genera and species will often be afterwards removed into the crystalline part of the system, by being identified with crystalline species. Thus *pyrope*, &c., have been referred to *garnet*, and *basalt*, *wacke*, &c., to

* *Règne Mineral*, p. 82.   † *Ib.*, p. 85.   ‡ *Ib.*, p. 86.
§ *Ib.*, p. 91.

compound rocks. Thus veins of *dolerite*, visibly composed of two or three elements, pass to an apparently
simple state by becoming fine-grained*.

16. Finally†, we have to ask, are artificial crystals to
enter into our classification? M. Necker answers, No;
because they are the result of art, like mules, mestizos,
hybrids, and the like.

17. Upon these opinions, we may observe, that they
appear to be, in the main, consistent with the soundest
philosophy. That each natural crystal is an individual,
is a doctrine which is the only basis of mineralogy as a
Natural Historical science; yet the imperfections and
confused unions of crystals make this principle difficult
to apply. Perhaps it may be expressed in a more precise
manner by referring to the crystalline forces, and to the
axes by which their operation is determined, rather than
to the external form. *That* portion of a mineral substance is a mineralogical *individual* which is determined
by crystalline forces acting to the *same axes*. In this
way we avoid the difficulty arising from the absence of
faces, and enable ourselves to use either cleavage, or optical
properties, or any others, as indications of the identity of
the individual. The individual extends so far as the polar
forces extend by which crystalline form is determined,
whether or not those forces produce their full effect, a
perfectly circumscribed polyhedron.

18. There is only one material point on which our
principles lead us to differ from M. Necker;—the propriety of including artificial crystals in our mineralogical
classification. To exclude them, as he does, is a conclusion so entirely at variance with the whole course of his
own reasonings, that it is difficult to conceive that he would
persist in his conclusion, if his attention were drawn to
the question more steadily. For, as he justly says‡, each

* *Règne Mineral*, p. 93.　† *Ib.*, p. 95.　‡ *Ib.*, p. 23.

science has its appropriate domain, determined by its peculiar point of view. Now artificial and natural crystals are considered in the same point of view, (namely, with reference to crystalline, physical, and optical properties, as subservient to classification,) and ought, therefore, to belong to the same science. Again, he says*, that chemistry would reject as useless all notice of the physical properties and external characters of substances, if a special science were to take charge of the description and classification of these products. But such a special science must be mineralogy; for we cannot well make one science of classification of natural, and another of artificial substances: or if we do, the two sciences will be identical in method and principles, and will extend over each other's boundaries, so that it will be neither useful nor possible to distinguish them. Again, M. Necker's own reasonings on the selection of the individual in mineralogy are supported by well chosen examples†; but these examples are taken from artificial salts; as, for instance, common salt crystallizing in different mixtures. Again, the analogy of mules and mestizos, as products of art, with chemical compounds, is not just. Chemical compounds correspond rather to natural species, propagated by man under the most natural circumstances, in order that he may study the laws of their production‡.

19. But the decisive argument against the separation of natural and artificial crystals in our schemes of classification is, that we *cannot* make such a separation. Substances which were long known only as the products of the laboratory, are often discovered, after a time, in natural deposits. Are the crystals which are found in a forgotten retort or solution to be considered as belonging

---

* *Règne Minéral*, p. 36.   † *Ib.*, p. 71.

‡ We may remark that M. Necker, in his own arrangement of minerals, inserts among his species iron and lead, which do not occur native.

to a different science from those which occur in a deserted mine? And are the crystals which are produced where man has turned a stream of water or air out of its course, to be separated from natural crystals, when the composition, growth, and properties, are exactly the same in both? And again: How many natural crystals can we already produce by synthesis! How many more may we hope to imitate hereafter! M. Necker himself states[*], that Mitscherlich found, in the scoriæ of the mines of Sweden and Germany, artificial minerals having the same composition and the same crystalline form with natural minerals: as silicates of iron, lime, and magnesia agreeing with peridot; bisilicate of iron, lime, and magnesia agreeing with pyroxene; red oxide of copper; oxide of zinc; protoxide of iron (fer oxydulé); sulphurets of iron, zinc, lead; arseniuret of nickel; black mica. These were accidental results of fusion. But M. Berthier, by bringing together the elements in proper quantities, has succeeded in composing similar minerals, and has thus obtained artificial silicates, with the same forms and the same characters as natural silicates. Other chemists (M. Haldat, M. Becquerel) have, in like manner, obtained, by artificial processes, other crystals, known previously as occurring naturally. How are these crystals, thus identical with natural minerals, to be removed out of the domain of mineralogy, and transferred to a science which shall classify artificial crystals only? If this be done, the mineralogist will not be able to classify any specimen till he has human testimony whether it was found naturally occurring or produced by chemical art. Or is the other alternative to be taken, and are these crystals to be given up to mineralogy because they occur naturally also? But what can be more unphilosophical than to refer to separate sciences the results of chemical processes closely

[*] *Règne Minéral*, p. 151.

allied, and all but identical? The chemist constructs
bisilicates, and these are classified by the mineralogist:
but if he constructs a trisilicate, it belongs to another
science. All these intolerable incongruities are avoided
by acknowledging that artificial, as well as natural,
crystals belong to the domain of mineralogy. It is, in
fact, the *name* only of mineralogy which appears to discover any inconsistency in this mode of proceeding.
Mineralogy is the representative of a science which has
a wider office than mineralogists first contemplated; but
which must exist, in order that the body of science may
be complete. There must, as we have already said, be a
Science, the object of which is to classify bodies by their
physical characters, in order that we may have some
means of asserting chemical truths concerning bodies;
some language in which we may express the propositions
which chemical analysis discovers. And this Science will
have its object prescribed, not by any accidental or arbitrary
difference of the story belonging to each specimen;—not
by knowing whether the specimen was found in the
mine or in the laboratory; produced by attempting to
imitate nature, or to do violence to her:—but will have
its course determined by its own character. The range
and boundaries of this Science will be regulated by the
ideas with which it deals. Like all other sciences, it
must extend to everything to which its principles apply.
The limits of the province which it includes are fixed
by the consideration that it must be a connected whole.
No previous definition, no historical accident, no casual
phrase, can at all stand in the way of philosophical consistency;—can make this Science exclude what that
includes, or oblige it to admit what that rejects. And thus,
whatever we call our Science;—whether we term it
External Chemistry, Mineralogy, the Natural History of
Inorganic Bodies;—since it can be nothing but the

Science of the Classification of Inorganic Bodies of definite forms and properties, it must classify all such bodies, whether or not they be minerals, and whether or not they be natural.

20. In the application of the principles of classification to minerals, the question occurs, What are to be considered as mineral *Species?* By Species we are to understand, according to the usage of other parts of natural history, the lowest step of our subordinate divisions;—the most limited of the groups which have definite distinctions. What definite distinctions of groups of objects of any kind really occur in nature, is to be learnt from an examination of nature : and the result of our inquiries will be some general principle which connects the members of each group, and distinguishes the members of groups which, though contiguous, are different. In the classification of organized bodies, the rule which thus presides over the formation of Species is the principle of *reproduction.* Those animals and those plants are of the same Species which are produced from a common stock, or which resemble each other as much as the progeny of a common stock. Accordingly in practice, if any questions arise whether two varieties of form be of the same or different species, it is settled by reference to the fact of reproduction; and when it is ascertained that the two forms come within the habitual and regular limits of a common circle of reproduction, they are held to be of the same species. Now in crystals, this principle of reproduction disappears altogether, and the basis of the formation of species must be sought elsewhere. We must have some other principle to replace the reproduction which belongs only to organic life. This principle will be, we may expect, one which secures the permanence and regularity of mineral forms, as the reproductive power does of animal and vegetable.

Such a principle is the *Power of Crystallization.* The forces of which solidity, cohesion, and crystallization are the result, are those which give to minerals their permanent existence and their physical properties; and ever since the discovery of the distinctions of crystalline forms and crystalline systems, it is certain that this force distinguishes groups of crystals in the most precise and definite manner. The rhombohedral carbonates of lime and of iron, for instance, are distinguished exactly by the angles of their rhombohedrons. And if, in the case of any proposed crystal, we should doubt to which kind the specimen belongs, the measurement of the angles of cleavage would at once decide the question. The principle of crystallization therefore appears, from analogy, to be exactly fitted to take the place of the principle of animal generation. The forces which make the individual permanent and its properties definite, here stand in the place of the forces which preserve the race, while individuals are generated and die.

21. According to this view, the different modifications of the same crystalline form would be *Varieties* only of the same species. All the various solids, for example, which are produced by the different laws of derivation of rhombohedral carbonate of lime, would fall within the same Species. And this appears to be required by the general analogy of natural history. For these differences of form, produced by the laws of crystalline derivation, are not *definite.* The faces which are added to one form in order to produce another, may be of any size, small or large, and thus the crystal which represents one modification passes by insensible degrees to another. The forms of calc spar, which we call *dog-tooth spar, cannon spar, nail-head spar,* and the like, appear at first, no doubt, distinct enough; but so do the races of dogs. And we find, in the mineral as in the animal, that the

distinction is obliterated by taking such intermediate steps as really occur. And if a fragment of any of these crystals is given us, we can determine that it is rhombohedral carbonate of lime; but it is not possible, in general, to determine to which of the kinds of crystal it has belonged.

22. Notwithstanding these considerations, M. Necker has taken for his basis of mineral species* the *Secondary* Modifications, and not the Primary Forms. Thus *cubical galena, octahedral galena,* and *triform galena,* are, with him, three *species* of crystals.

On this I have to observe, as I have already done, that on this principle we have no *definite* distinction of species; for these forms may and do pass into each other: among cubo-octahedrons of galena occur cubes and octahedrons, as one face or another vanishes, and the transition is insensible. We shall, on this principle, find almost always three or four species in the same tuft of crystals; for almost every individual in such assemblages may exhibit a different combination of secondary faces. Again, in cases where the secondary laws are numerous, it would be impracticable to enumerate all their combinations, and impossible therefore to give a list of species. Accordingly M. Necker† gives seventy-one Species of *spath calcaire,* and then says, " Nous n'avons pas enuméré la dixieme partie des espèces connues de ce genre, qui se montent à plus de huit cents." Again, in many substances, of which few crystals are found, every new specimen would be a new species; if indeed it were perfect enough to be referred to a species at all. But from a specimen without perfect external form however perfect in crystalline character, although everything else might be known,—angles, optical pro-

---

* *Règne Mineral,* p. 396.   † *Ib.* ii. 634.

perties, physical properties, and chemical constitution,— the species could not be determined. Thus Necker says[*] of the micas, "Quant aux espèces propre à chaque genre, la lacune sera presque complète; car jusqu' ici les cristaux entiers de Mica et de Talc n'ont pas été fort communs."

These inconveniencies arise from neglecting the leading rule of natural history, that the *predominant principle* of the existence of an object must determine the Species; whether this principle be reproduction operating for development, or crystallization operating for permanence of form. We may add to the above statement of inconveniencies this;—that if M. Necker's view of mineralogical species be adopted, the distinction of species is vague and indefinite, while that of genera is perfectly precise and rigorous;—an aspect of the system entirely at variance with other parts of natural history; for in all these the species is a more definite group than the genus.

This result follows, as has already been said, from M. Necker's wish to have individuals marked by external form. If, instead of this, we are contented to take for an individual that portion of a mass, of whatever form, which is connected by the continuous influence of the same crystalline forces, by whatever incidents these forces may be manifested, (as cleavage, physical and optical properties,) our mode of proceeding avoids all the above inconveniencies, applies alike to the most perfect and most imperfect specimens, and gives a result agreeable to the general analogy of natural history, and the rules of its methods[†].

---

[*] *Règne Mineral*, ii. 414.

[†] I will not again enter into the subject of Nomenclature; but I may remark that M. Necker has adopted (i. 415) the Nomenclature of Beudant, latinising the names, and thus converting each into a single word. He has also introduced, besides the names of Genera, names of

I now quit the subject of mere Resemblance, and proceed to treat of that natural affinity which Natural Systems of Classification for organic bodies must involve.

CHAPTER IV.

OF THE IDEA OF NATURAL AFFINITY.

1. IN the Second Chapter of this Book it was shown that although the Classificatory Sciences proceed ostensibly upon the Idea of Resemblance as their main foundation, they necessarily take for granted in the course of their progress a further Idea of Natural Affinity. This appeared* by a general consideration of the nature of Science, by the recognition of natural species and genera, even in Artificial Systems of Classification†, and by the attempts of botanists to form a Natural System. It further appeared that among the processes by which endeavours have been made to frame a Natural System, some, as the method of *blind trial* and the method or *general comparison*, have been altogether unsuccessful; being founded only upon a collection of resemblances, casual in the one case and arbitrary in the other. In neither of these processes is there employed any general principle by which we may be definitely directed as to what resemblances we should employ, or by which the result at which we arrive may be verified and confirmed. Our object in the present chapter is to show that the Idea of Natural Affinity supplies us with a principle which may answer such purposes.

Families taken from the *typical* Genus. Thus the Family of *Carbonidiens* contains the following genera: *Calcispathum, Magnesispathum, Dolomispathum, Ferrispathum*, &c., *Malachita, Azuria, Gaylusacia.*

       * Art. 5.   † Art. 7.

I shall first consider the Idea of Affinity as exemplified in organized beings. In doing this, we may appear to take for granted Ideas which have not yet come under our discussion, as the Ideas of Organization, and Vital Function; but it will be found that the principle to which we are led is independent of these additional Ideas.

2. We have already seen that the attempts to discover the divisions which result from this Natural Affinity have led to the consideration of the *Subordination of Characters*. It is easy to see that some organs are more essential than others to the existence of an organized being; the organs of nutrition, for example, more essential than those of locomotion. But at the same time it is clear that any *arbitrary* assumption of a certain scale of relative values of different kinds of characters will lead only to an Artificial System. This will happen, if, for example, we begin by declaring the nutritive to be superior in importance to the reproductive functions. It is clear that this relation of importance of organs and functions must be collected by the study of the organized beings; and cannot be determined *à priori*, without depriving us of all right to expect a general accordance between our system and the arrangement of nature. We see, therefore, that our notion of Natural Affinity involves in it this consequence;—that it is not to be made out by an *arbitrary* subordination of characters.

3. The functions and actions of living things which we separate from each other in our consideration, cannot be severed in nature. Each function is essential; Life implies a collection of movements, and ceases when any of these movements is stopped. A change in the organization subservient to one set of functions may lead necessarily to a change in the organization belonging to others. We can often see this necessary connexion; and from a comparison of the forms of organized beings,—from the

way in which their structure changes in passing from one class to another, we are led to the conviction that there is some general principle which connects and graduates all such changes. When the circulatory system changes, the nervous system changes also: when the mode of locomotion changes, the respiration is also modified.

4. These corresponding changes may be considered as ways in which the living thing is fitted to its mode of life; as marks of *adaptation to a purpose*; or, as it has been otherwise expressed, as results of the *conditions of existence*. But at the present moment, we put forward these correspondencies in a different light. We adduce them as illustrations of what we mean by Affinity, and what we consider as the tendency of a Natural Classification. It has sometimes been asserted that if we were to classify any of the departments of organized nature by means of one function, and then by means of another, the two classifications, if each strictly consistent with itself, would be consistent with each other. Such an assertion is perhaps more than we are entitled to make with confidence; but it shows very well what is meant by Affinity. The disposition to believe such a general identity of all partial natural classifications, shows how readily we fix upon the notion of Affinity, as general result of the causes which determine the forms of living things. When these causes or principles, of whatever nature they are conceived to be, vary so as to modify one part of the organization of the being, they also modify another: and thus the groups which exhibit this variation of the fundamental principles of form, are the same, whether the manifestation of the change be sought in one part or in another of the organized structure. The groups thus formed are related by Affinity; and in proportion as we find the evidence of more functions and more organs to the propriety of our groups, we are more and more satis-

fied that they are Natural Classes. It appears, then, that our Idea of Affinity involves the conviction of the *coincidence of natural arrangements formed on different functions;* and this, rather than the principle of the subordination of some characters to others, is the true ground of the natural method of Classification.

5. For example, Cuvier, after speaking of the Subordination of Characters as the guide which he intends to follow in his arrangement of animals, interprets this principle in such a manner* as to make it agree nearly with the one just stated. "In pursuance of what has been said on methods in general, we now require to know what characters in animals are the most influential, and therefore those which must be made the grounds of the primary divisions." "These," he says, "it is clear must be those which are taken from the animal functions;—sensation and motion:"—But how does he confirm this? Not by showing that the animal functions are independent of, or predominant over, the vegetative, but by observing that they follow the same gradations. "Observation," he continues, "confirms this view, by showing that the degrees of developement and complication of the animal functions agree with those of the vegetative. The heart and the organs of the circulation are a sort of centre for the vegetative functions, as the brain and the trunk of the nervous system are for the animal functions. Now we see these two systems descend in the scale, and disappear the one with the other. In the lowest animals, when there are no longer any distinct nerves, there are also no longer distinct fibres, and the organs of digestion are simply hollowed out in the homogeneous mass of the body. The muscular system disappears even before the nervous, in insects; but in general the distribution of the medullary masses corre-

* *Règne Animal,* p. 55.

sponds to that of the muscular instruments; a spinal cord, on which knots or ganglions represent so many brains, corresponds to a body divided into numerous rings and supported on pairs of members placed at different points of the length, and so on.

"This *correspondence* of the general forms which result from the arrangement of the motive organs, from the distribution of the nervous masses, and from the energy of the circulatory system, must therefore form the ground of the first great sections by which we divide the animal kingdom."

6. Decandolle takes the same view. There must be, he says, *an equilibrium* of the different functions*. And he exemplifies this by the case of the distinction of monocotyledonous and dicotyledonous plants, which being at first established by means of the organs of reproduction, was afterwards found to coincide with the distinction of endogenous and exogenous, which depends on the process of nutrition. "Thus," he adds, "*the natural classes founded on one of the great functions of the vegetable are necessarily the same as those which are founded upon the other function;* and I find here a very useful *criterion* to ascertain whether a class is natural: namely, in order to announce that it is so, it must be arrived at by the two roads which vegetable organization presents. Thus I affirm," he says, "that the division of monocotyledons from dicotyledons, and the distinction of Gramineæ from Cyperaceæ, are real, because in these cases, I arrive at the same result by the reproductive and the nutritive organs; while the distinction of monopetalous and polypetalous, of Rhodoraceæ and Ericineæ appears to me artificial, because I can arrive at it only by the reproductive organs."

Thus the correspondence of the indications of different

* *Th. El.*, p. 79.

functions is the criterion of Natural Classes; and this correspondence may be considered as one of the best and most characteristic marks of the fundamental Idea of Affinity. And the Maxim by which all Systems professing to be natural must be tested is this:—that the *arrangement obtained from one set of characters coincides with the arrangement obtained from another set.*

This Idea of Affinity, as a natural connexion among various species, of which connexion all particular resemblances are indications, has principally influenced the attempts at classifying the animal kingdom. The reason why the classification in this branch of Natural History has been more easy and certain than that of the vegetable world is, as Decandolle says*, that besides the functions of nutrition and reproduction, which animals have in common with plants, they have also in addition the function of sensation; and thus have a new means of verification and concordance. But we may add, as a further reason, that the functions of animals are necessarily much more obvious and intelligible to us than those of vegetables, from their clear resemblance to the operations which take place in our own bodies, to which our attention has necessarily been strongly directed.

7. The question here offers itself, whether this Idea of Natural Affinity is applicable to inorganic as well as to organic bodies;—whether there be Natural Affinities among Minerals. And to this we are now enabled to reply by considering whether or not the principle just stated is applicable in such cases. And the conclusion to which our principle leads us is,—that there are such Natural Affinities among Minerals, since there are different sets of characters which may be taken, (and have by different writers been taken,) as the basis of classification. The hardness, specific gravity, colour, lustre,

* *Th. El.,* p. 80.

crystallization, and other *external* characters, as they are termed, form one body of properties according to which minerals may be classified; as has in fact been done by Mohs, Breithaupt, and others. The *chemical* constitution of the substances, on the other hand, may be made the principle of their arrangement, as was done by Haüy, and more recently, and on a different scheme, by Berzelius. Which of these is the true and natural classification? To this we answer, that *each* of these arrangements is true and natural, then, and then only, when it coincides with the other. An arrangement by external characters which gives us classes possessing a common chemical character;—a chemical order which brings together like and separates unlike minerals;—such classifications have the evidence of truth in their agreement with one another. Every classification of minerals which does not aim at and tend to such a result, is so far merely arbitrary; and cannot be subservient to the expression of general chemical and mineralogical truths, which is the proper purpose of such a classification.

8. In the History of Mineralogy I have related the advances which have been made among mineralogists and chemists in modern times towards a System possessing this character of truth. I have there described the mixed systems of Werner and Haüy;—the attempt made by Mohs to form a pure Natural History system;—the first and second attempt of Berzelius to form a pure chemical system; and the failure of both these attempts. But the distinct separation of the two elements of which science requires the coincidence threw a very useful light upon the subject; and the succeeding mixed systems, such as that of Naumann, approached much nearer to the true conditions of the problem than any of the preceding ones had done. Thus, as I have stated, several of Naumann's groups have both a common chemical character and great

external resemblances. Such are his *Anhydrous Unmetallic Haloids*—his *Anhydrous Metallic Haloids*—*Hydrous Metallic Haloids*—*Oxides* of metals—*Pyrites*—*Glances*—*Blendes*. The existence of such groups shows that we may hope ultimately to obtain a classification of minerals which shall be both chemically significant and agreeable to the methods of Natural History: although, when we consider how very imperfect as yet our knowledge of the chemical composition of minerals is, we can hardly flatter ourselves that we shall arrive at such a result very soon.

We have thus seen that in Mineralogy, as well as in the sciences which treat of organized bodies, we may apply the Idea of Natural Affinity; of which the fundamental maxim is, that *arrangements obtained from different sets of characters must coincide.*

Since the notion of Affinity is thus applicable to inorganic as well as to organic bodies, it is plain that it is not a mere modification of the Idea of Organization or Function, although it may in some of its aspects appear to approach near to these other Ideas. But these Ideas, or others which are the foundation of them, necessarily enter in a very prominent and fundamental manner into all the other parts of Natural History. To the consideration of these, therefore, we shall now proceed.

END OF THE FIRST VOLUME.

Milton Keynes UK
Ingram Content Group UK Ltd.
UKHW041520181024
449640UK00009B/99